FORMALDEHYDE ...

ADVANCES IN CHEMISTRY SERIES **210**

Formaldehyde
Analytical Chemistry and Toxicology

Victor Turoski, EDITOR
James River Corporation

Developed from a symposium sponsored by
the Division of Environmental Chemistry
at the 187th Meeting
of the American Chemical Society,
St. Louis, Missouri,
April 8–13, 1984

American Chemical Society, Washington, D.C. 1985

Library of Congress Cataloging in Publication Data

Formaldehyde: analytical chemistry and toxicology.
   (Advances in chemistry series, ISSN 0065-2393;
210)

   "Developed from a symposium sponsored by the
Division of Environmental Chemistry at the 187th
Meeting of the American Chemical Society, St. Louis,
Missouri, April 8-13, 1984."

   Includes bibliographies and indexes.

   1. Formaldehyde—Analysis—Congresses.
2. Formaldehyde—Toxicology—Congresses.

   I. Turoski, Victor, 1940-   . II. American
Chemical Society. Division of Environmental
Chemistry. III. American Chemical Society. Meeting
(187th: 1984: St. Louis, Mo.) IV. Series.

QD305.A6F67   1985   363.1'79   85-11205
ISBN 0-8412-0903-0

Advances in Chemistry Series

M. Joan Comstock, *Series Editor*

FOREWORD

The ADVANCES IN CHEMISTRY SERIES was founded in 1949 by the American Chemical Society as an outlet for symposia and collections of data in special areas of topical interest that could not be accommodated in the Society's journals. It provides a medium for symposia that would otherwise be fragmented because their papers would be distributed among several journals or not published at all. Papers are reviewed critically according to ACS editorial standards and receive the careful attention and processing characteristic of ACS publications. Volumes in the ADVANCES IN CHEMISTRY SERIES maintain the integrity of the symposia on which they are based; however, verbatim reproductions of previously published papers are not accepted. Papers may include reports of research as well as reviews, because symposia may embrace both types of presentation.

ABOUT THE EDITOR

VICTOR TUROSKI is a graduate of Wilkes College, Wilkes–Barre, Pennsylvania, and has pursued a career in analytical chemistry at both Carter Wallace and the American Can Company. Turoski presently heads the corporate Analytical Laboratory at the James River Technical Center in Neenah, Wisconsin, and is currently involved in both fundamental and applied research in papers and various coatings. He has also published several articles in the field of environmental chemistry, primarily in the area of gas chromatographic–mass spectrometric identifications and quantitation of resin acids and priority pollutants in complex matrices. Turoski is a frequent symposium chairman in the Environmental Division of the American Chemical Society at national meetings and performs article peer review for *Environmental Science and Technology*.

CONTENTS

EDITOR'S PREFACE

F ORMALDEHYDE IS A PRODUCT OF HUMAN METABOLISM and is present in every cell of our bodies. It is also ubiquitous in the environment as it is found in a great variety of consumer products such as cosmetics and permanent press fabrics. Furthermore, it is present in particle board, plywood, and urea–formaldehyde foam insulation. Formaldehyde occurs at many different concentrations; apples, grapes, and pears contain formaldehyde at 26, 80, and 275 ppb, respectively. Cigarette smoke contains formaldehyde at the 40-ppm level.

This book deals with the analysis of formaldehyde at trace levels and with its toxicity. It also deals with data obtained for determining formaldehyde levels in housing and data obtained with animal studies. Essentially, it is a reference of the current knowledge about this small organic chemical, CH_2O.

Early in 1984 the ACS decided to hold a symposium and thereby provide a public forum to discuss advanced techniques for the analysis of formaldehyde. At that time, as part of my work at James River, I was involved in some advanced trace-level formaldehyde quantitation and was asked by the ACS to chair the symposium. What occurred next was most interesting and is best stated by a direct quote from *Chemical & Engineering News*, April 30, 1984:

> When Victor Turoski first decided to organize a symposium on formaldehyde, he saw it as a half-day affair dealing only with analytical methods for measuring the compound. As word spread, chemists and nonchemists involved in the formaldehyde issue began clamoring to be included on the program. Before long, Turoski's modest idea had snowballed into a massive 4-day colloquium encompassing analytical methods, toxicology, and hazard assessment. In the end, the symposium dominated the Environmental Chemistry Division's program.
>
> Turoski, who heads the analytical department at James River Corp., thinks the response he got is due to the scientific community's concern about this controversial chemical.

The symposium proved to be quite successful. Papers were presented by many scientists in the private sector as well as government scientists from the Food and Drug Administration, the Occupational Safety and Health Administration, the U.S. Department of Agriculture,

Oak Ridge National Laboratory, and the National Cancer Institute. Unfortunately, representatives from the U.S. Consumer Product Safety Commission were unable to attend, present their data, and publicly defend their position.

All of the toxicology papers presented showed that when formaldehyde is used at "normal levels" by embalmers and gross anatomists, and in manufacturing processes, it is not a human carcinogen. During the Risk Assessment Panel portion of the open forum, a consensus on the relative safety of formaldehyde was also reached.

This book, dealing with the analytical complexities of formaldehyde and its suspected toxicity, is timely and will be useful to both the scientific and regulatory communities, as well as the public at large.

VICTOR TUROSKI
James River Corporation
Neenah, WI 54956

May 1985

PREFACE

By Bernard L. Oser

AFTER I HAD SUBMITTED A TITLE AND AN ABSTRACT of a paper for presentation at the American Chemical Society's Formaldehyde Symposium in St. Louis in April 1984, I became embroiled in a bout with cardiac surgery that left me hors de combat for many months. The paper was neither completed nor delivered. I regard it an honor, therefore, to have been invited to write a preface to this book of reports by the distinguished contributors to the symposium whom I regret having missed hearing in person.

Recent concern in the regulatory, industrial, and scientific arenas and in the news media over the potential risk of exposure to formaldehyde has given rise to numerous scientific conferences, symposia, and publications of which this book is one of the most comprehensive and authoritative. It was precipitated by chronic studies in which rodents, exposed to high levels of formaldehyde in air, developed squamous cell carcinomas in the nasal epithelium. Several of the papers presented at the symposium from which this book developed point out that no clear evidence links formaldehyde to nasal or any other focal or systemic type of cancer in humans. This finding may be attributable to the sensory and irritant properties of formaldehyde that serve to protect humans against excessive exposure, inasmuch as the effect is prodromal and self-limiting in humans. The odor of formaldehyde is detectable at levels less than 1 ppm, which is the threshold limit value currently adopted for workers exposed for 5 8-h days/week.

The formaldehyde issue is typical of many cases of alleged hazards of chemicals for which the multiplicity of options and judgments in the procedural steps leading to risk assessment in animals complicates the problem of arriving at an objective, scientific, and rational basis for the estimation of acceptable risk.

Retrospective epidemiology alone offers little prospect of evaluating the safety of low-level exposure to formaldehyde because of wide variations in duration, frequency, and dosage and the lack of adequate control data. A summary of experience in eight chemical plants was reported by the Epidemiological Panel of the Formaldehyde Workshop sponsored by the National Center for Toxicological Research in October

1983. It revealed that out of a total of 2059 male workers, 85% of the mortality was observed from all causes. However, nearly all deaths expected from all cancers [observed-to-expected ratio (O/E) = 512/527] or from lung cancer (O/E = 216/227) actually occurred. Of the not more than five that might have been predicted as due to nasal cancer, none were observed.

In animal experiments, inhalation of high levels is continuous, whereas human exposure, even among embalmers, pathologists, or workers in various industries and occupations, is generally intermittent and occurs at levels only a few orders of magnitude greater than that of ambient air.

Furthermore, in the interpretation of animal studies in terms of human safety, interspecies variations in respiratory metabolism must be taken into account. Significant differences are known to exist in minute volume and mucociliary function between rats and mice as well as between these species and large mammals, including humans. These differences in sensitivity among animal species raise the question of suitability of rodents as experimental surrogates for humans. Factors such as minute volume (respiratory rate × tidal volume), mucous flow rate, mucociliary activity, and covalent binding to glycoproteins and DNA have been shown to vary significantly among species. These findings emphasize the need for caution in extrapolating effects in rodents to humans.

In the interpretation of inhalation toxicity data in rats or mice it is essential to recognize that they are obligate nose breathers. When the toxicant in question is an extremely reactive chemical, the first target area is the anterior nasopharyngeal region. The pungency and irritant properties of formaldehyde at high levels in air tend to lower respiratory frequency and minute volume to a greater degree in the mouse than in the rat. At the 15-ppm exposure level the delivered dose has been estimated to be twice as great in rats as in mice. Mucous flow rate and ciliary activity are also diminished and thus serve to reduce further the defense against the focal toxicity of formaldehyde. These effects are a function of concentration and, at lower doses, are not cumulative.

Moreover, the fact that the inspired formaldehyde reacts immediately with the components of the nasal mucous layer and epithelium requires that account be taken of the dose delivered at the site of contact (which in practice is unmeasurable) rather than the administered (airborne) dose in the attempt to quantify the results in laboratory rodents in terms of occupational exposure in humans.

Research on mechanisms of HCHO toxicity in rats has revealed important findings among which is increased cell proliferation, considered to be a precursor to chemical carcinogenesis, peaking at the intermediate dose of 6 ppm. Other findings include the formation of DNA–protein cross-links in the nasal mucosa at concentrations greater than 2 ppm with no increase in blood formaldehyde and the nonlinear relationship of the administered dose to that delivered at the target area which is lower.

Toxicology is not an exact science. It has been developing as a regulatory discipline only with the past half of a century. Bioassays can, and do, vary in results because of differences in design, apart from choice of species and strains, namely in respect to number, size, and initial age of dosage groups; prenatal, postnatal, and test diets; route, frequency, duration, and range of dosages; choice of physical, behavioral, biochemical, hematological, functional, and pathological observations; teratologic findings; and metabolic pharmacokinetic, reproductive, and, when indicated, carcinogenicity studies. Some of these procedural conditions and observations are routine but nevertheless vary in extent or frequency. Some are optional. Some are continued through two or more generations. Some vary in methodology and in interpretation, particularly when findings are judgmental and reported in descriptive, but nonquantifiable terms. It is not surprising, therefore, that interlaboratory studies are seldom, if ever, exactly replicated.

It has been said that effective enforcement of safety regulations requires that risk assessment based on animal tests be expressed in numerical terms, such as exposure levels in dosage or time limits equivalent to a lifetime cancer incidence of one in a million. In the case of food additives, the traditional procedure has been to apply appropriate safety factors to "no observed adverse effect levels" in properly executed tests to arrive at safe doses (acceptable daily intakes [ADI]) for the human adult. This procedure has been used as the basis for regulations, including action levels, to avoid the Delaney clause proscribing the addition of cancer-inducing substances to foods. Whether these action levels can be sustained in the face of a recent challenge in the Court of Appeals remains to be seen.

The present practice recommended by EPA and OSHA is to extrapolate the dose–response data, including the response to the maximum tolerated dose, by using statistical equations based on assumed, but nonetheless hypothetical, mechanisms of carcinogenesis. The belief that this approach is more objective and can compensate for the fundamental uncertainties of the bioassay is a will-o'-the-wisp. Depending on the slope and assumed curvature of the dose response below the adverse effect level, calculated risk assessments via statistical models can vary by many orders of magnitude.

An eminent proponent of the use of mathematical models in risk assessment has pointed out that the "fundamental assumption is that the animals are surrogates for humans...*before methods of quantitative risk assessment are even considered*...The qualitative nature of the biological data...must be taken into account."

The design and execution of animal experiments are assumed to be valid and reproducible. Prolonged efforts have been directed over many years to establish guidelines for toxicological procedures and to set uniform standards to promote meaningful extrapolation of test results to the condition of human exposure. Experience has shown, however, that in many cases studies can vary sufficiently to become of critical importance when, in the interest of prudence, a regulatory agency considers a single adverse and possibly questionable study to outweigh many negative studies.

This prefatory outline of the steps involved in the toxicology and risk assessment of formaldehyde should conclude with some reference to the ultimate objective, namely the determination of the acceptability of the risk by those exposed to it and, of course, by the regulatory agencies.

Little concern should be felt for the inhalation of concentrations of formaldehyde less than the level of sensory detection (approximately 0.25 ppm). Formaldehyde is present at low concentrations in ambient air and in the smoke of frying foods and burning fuels where exposure is minimal and temporary. In laboratories, hospitals, and other occupational environments, levels may reach as high as 2–5 ppm; they are rarely higher. Although exposure is irritating at these levels, it is generally intermittent and the effects are transitory.

Acceptance of risks must take into account whether they are known or unknown, voluntary or involuntary, great or small, and occupational or nonoccupational and whether they have compensatory benefits or advantages. For example, the risks of drinking, smoking, transportation, or athletics are determined subjectively, whereas environmental and occupational risks may be beyond the control of the individual.

Risks may not be feared or consciously perceived. Technological benefits may not be understood or appreciated. Hence it falls upon society, through government, to determine a proper balance and whether benefit can be considered for regulatory purposes. A lesser risk may per se be a benefit. However, some agencies, claiming that they lack authority to weigh benefits except when concerned with health, maintain a firm, inflexible posture against risks no matter how theoretical or trivial they may be. Regulators must face realities and recognize that the beliefs that a substance must be carcinogenic unless it is "proved" otherwise, that zero tolerances can be validated, and that no benefit is worth any risk are fallacious, obsolete, and conceptually unattainable.

BERNARD L. OSER
2 Bay Club Drive
Bayside, NY 11360

June 3, 1985

ANALYTICAL CHEMISTRY

Industrial Hygiene Sampling and Analytical Methods for Formaldehyde

Past and Present

EUGENE R. KENNEDY, ALEXANDER W. TEASS, and
YVONNE T. GAGNON

National Institute for Occupational Safety and Health, Cincinnati, OH 45226

National Institute for Occupational Safety and Health (NIOSH) method P&CAM 125 (chromotropic acid), one of the first industrial hygiene sampling and analytical methods for formaldehyde, was adapted from Intersociety Method 116 and is still widely used today, although it is susceptible to many interferences. Several other adaptations of this method have been used in the development of passive monitors. Other methods for formaldehyde, including those using pararosaniline and 2,4-dinitrophenylhydrazine, have been developed and adapted to industrial hygiene sampling and do offer advantages over the chromotropic acid method. During field and laboratory evaluations of NIOSH method P&CAM 318 (oxidative charcoal), this method was found to have sample instability problems. Samples collected with NIOSH method P&CAM 354 [2-(ben-zylamino)ethanol] were found to be stable, but sensitivity was limited. Recent work on this method has allowed a sampling rate of 0.08 L/min for 8 h. On the basis of this work, measurement of formaldehyde at low levels (0.1 µg/L) should be possible with the 2-(benzylamino)ethanol method.

FORMALDEHYDE IS ONE OF THE MORE WIDELY PRODUCED and used chemical intermediates in the U.S. chemical industry. Production in 1983 was approximately 5.45 billion lb and has averaged around 5 billion lb over the past 3 years (1). Because of this large-volume production and usage of formaldehyde and its possible exposure-related health effects (2), there has been much concern over the sensitivity, accuracy, and applicability of sampling and analytical methodology for this compound. In this chapter a brief summary of some of the more widely used industrial hygiene sampling and analytical methods for formaldehyde is presented.

0065–2393/85/0210/0003$06.00/0

Chromotropic Acid Methods

One of the first industrial hygiene sampling and analytical methods for formaldehyde was adapted directly from air pollution monitoring. The impinger–chromotropic acid method [National Institute for Occupational Safety and Health (NIOSH) method P&CAM 125 (3)] for formaldehyde was taken from Intersociety method 116 (4). The method in its original form recommended the use of distilled water in an impinger for collection of formaldehyde. To an aliquot of this sample, chromotropic acid and concentrated sulfuric acid were added sequentially. The color of the reaction product of the chromotropic acid and the formaldehyde was purple. The absorbance of this purple chromophore was measured and related to the concentration of formaldehyde by the relationship of Beer's law. Further refinements of the method, such as use of a 1% sodium bisulfite solution as a collecting medium to increase collection efficiency from 80% to >95% and heating during the color development to ensure complete reaction, have been incorporated in the latest version of the method as published by NIOSH (5). With the original version of the method, a problem with long-term sample stability was encountered (6). Further work in our laboratory (7) indicated that when samples were stored in Nalgene cross-linked polyethylene (CPE) bottles with screw caps instead of glass scintillation vials, samples were stable for at least 2 weeks. The type of vial used for storage of samples was critical. If the vial cap did not seal well, then leakage of formaldehyde from the headspace was possible. With metal-lined caps, reaction of the bisulfite with metal cap liners and invalidation of the sample due to precipitate formation was possible (8). The measurement limit of the method was 1.5 μg/sample. A sample volume of 60 L collected at 1 or 0.25 L/min, in accordance with the sampling time desired, was recommended to keep samples in the linear range of the calibration curve without the need for dilution when measuring air concentrations around the Occupational Safety and Health Administration (OSHA) permissible exposure limit [3.7 mg/m^3 (9)]. This sample volume permitted a limit of quantitation of 0.025 mg/m^3. Although susceptible to many interferences, such as phenol, ethanol, higher molecular weight alcohols, and olefins, the method is still widely used today. These interferences create a negative bias in the method because they inhibit the color formation reaction.

A different sampling technique was combined with the chemistry of the chromotropic acid procedure in NIOSH method P&CAM 235 (10), which used an alumina sorbent tube instead of an impinger for collection of formaldehyde. The formaldehyde was desorbed from the tube with 1% aqueous methanol solution and determined by the chromotropic acid procedure described earlier. The reported measurement limit for this method was 1.0 μg/sample. The capacity of the sampler was limited to 6 L, collected at 0.2 L/min, to prevent sample breakthrough. On the basis of this volume, the limit of quantitation was 0.17 mg/m^3. The samples collected on alumina required immediate desorption at the field site to prevent loss of formaldehyde.

Several adaptations of the chromotropic acid procedure have been used in the development of diffusive monitors. The major advantage of these devices over more conventional sampling and analysis methods was that they did not require sampling pumps and impingers or sorbent tubes. The formaldehyde diffused into the monitor and was collected by adsorption onto a reactive sorbent or by absorption into a liquid medium separated from the air by a gas-permeable membrane. The sample was either desorbed with distilled water, or an aliquot of the absorbing liquid was taken and the formaldehyde was determined by modifications of the chromotropic acid procedure. Various modifications in the sample workup and measurement procedure, such as use of a more concentrated chromotropic acid solution and changes in aliquot volume, were incorporated to reduce the effect of interferences in these sampling devices (*11–12*). Also, because the diffusion coefficient for formaldehyde was larger than those of potential interferences, the ratios of sampling rates of formaldehyde to interferences were > 1, and the sampling devices exhibited a sampling selectivity for formaldehyde. The reported measurement limits for these types of devices were 0.25–0.8 μg/sample. The sampling rates for these devices were 0.0017–0.0569 L/min to give limits of quantitation of 0.3–0.03 mg/m^3 for 8-h sampling periods. Tests in our laboratory showed that the collection efficiency of one of these commercial badges was sensitive to relative humidity (*13*). Our findings indicated that the collection efficiency tended to decrease with decreasing relative humidity.

Pararosaniline Methods

The procedure for formaldehyde using pararosaniline was an adaptation of a method for sulfur dioxide (*14*). In this adapted method (*15*) formaldehyde is collected in sodium sulfite solution in an impinger and then reacted with sodium tetrachloromercurate and pararosaniline to yield a purple chromophore that is determined spectrophotometrically at 560 nm. Acetaldehyde and propionaldehyde were positive interferences at the 5-ppm levels, but not at low levels. A major drawback of this method was the use of the toxic tetrachloromercurate salt. In a subsequent modification of this method, the use of the salt was not required, but the modified method was sensitive to temperature and subject to interferences from hydrogen cyanide, sulfite ion, hydroxylamine, and sulfur dioxide (*16*). The reported measurement limits for the original method (*15*) and the modification (*16*) were 2.0 and 1.8 μg/sample, respectively. With sample volumes of 28 and 60 L, collected at 1 L/min, the limits of quantitation for the methods were 0.07 and 0.03 mg/m^3, respectively. This analytical procedure was used to analyze samples collected on molecular sieve sorbent tubes (*17*). The major drawback to this collection procedure was that the capacity of the sorbent varied inversely with the amount of water collected by the tube. In high-humidity environments, sampling time before breakthrough of formaldehyde was quite limited, with a maximum volume of 30 L collected over a 15-min period at 2 L/min recommended. This sample volume allowed a limit of quantitation of 0.03 mg/m^3. The pararosaniline method also was adapted for a commercial continuous air-monitoring device (18). Subsequently, this monitor was modified to improve sensitivity (19) and remove the need for the tetrachloromercurate salt (20).

2,4-Dinitrophenylhydrazine Methods

With the 2,4-dinitrophenylhydrazine (2,4-DNPH) procedure, formaldehyde reacted with 2,4-DNPH in either aqueous (21) or acetonitrile solution (22) to form the 2,4-dinitrophenylhydrazone, which was then determined by high-pressure liquid chromatography with UV detection. Other analytical techniques, such as spectrophotometry at 440 nm (23) and gas chromatography (24) have been used, but the liquid chromatographic analysis provided the sensitivity and selectivity required for formaldehyde. This method also was used for the simultaneous determination of certain other aldehydes and ketones. We experienced problems with low capture efficiency with an aqueous solution of 2,4-DNPH for low formaldehyde concentrations. This problem of low capture efficiency was overcome with the use of an acetonitrile solution of 2,4-DNPH (22); however, with the use of the acetonitrile solution, the inherent problems of solvent evaporation during sample collection and safe transport of samples back to the laboratory for analysis were encountered. The reported measurement limits for these methods with the aqueous and acetonitrile 2,4-DNPH solutions were 0.05 and 0.25 μg/sample, respectively. With a sample volume of 30 L collected at 0.5 and 1.5 L/min, the limits of quantitation were 0.002 and 0.008 mg/m^3, respectively.

The 2,4-DNPH sampling approach was also used in several sorbent tube methods for formaldehyde in which the 2,4-DNPH was coated on XAD–2 or silica gel (25–26). However, the stability of the 2,4-dinitrophenylhydrazone derivative on the 2,4-DNPH coated sorbents was questionable. In our laboratory, sample loss from the silica gel tubes has been observed after 7 days. Also, the tubes had a shelf life limited to 30 days. Experiments in our laboratory with the 2,4-DNPH coated silica gel tube for analysis of ketones indicated that the sampling tube did not trap methyl ethyl ketone. Although other ketones were not investigated, we concluded that other ketones would behave similarly. Reported measurement limits for formaldehyde with the XAD–2 and silica gel sorbents were 0.64 and 2.5 μg/sample, respectively. With sample volumes of 5 and 20 L collected at 0.2 L/min, the limit of quantitation was 0.13 mg/m^3 for each method.

3-Methyl-2-benzothiazolone Hydrazone Method

Another method for the measurement of formaldehyde adapted from air pollution monitoring was the 3-methyl-2-benzothiazolone hydrazone (MBTH) method (27). An air sample was collected in an impinger containing 0.05% aqueous MBTH solution. During analytical workup of the solution, addition of ferric chloride and acid caused the formaldehyde–MBTH reaction product to form a blue cationic dye. Formaldehyde concentration was then determined by the absorbance of this blue dye at either 635 or 670 nm by spectrophotometry. This blue color was not stable for periods longer than 4 h. Because other aldehydes reacted in a similar manner and, therefore, presented a positive interference in the method, this method was used for the determination of formaldehyde only when formaldehyde was the sole aldehyde present in the environment to be sampled. The method was used routinely for the measurement of total aldehydes. The reported measurement limit for formaldehyde was 5 μg/sample. With a sample volume of up to 250 L collected at 0.5 L/min, the limit of quantitation was 0.02 mg/m^3. A modification of this method in which sulfamic acid was added in the oxidation step to reduce sample turbidity has reduced the measurement limit to 2.0 μg/sample (28). With the same sampling conditions just described, the limit of quantitation for this modification was 0.008 mg/m^3.

Girard T Method

Under the joint NIOSH–OSHA Standards Completion Program a method for formaldehyde was developed and included in the NIOSH *Manual of Analytical Methods* as method S327 (29). Formaldehyde was collected with a bubbler filled with Girard T reagent (trimethylammoniohydrazide chloride) and determined by polarography of the resulting hydrazone. The major advantage of the method was that the sample required no preliminary workup and could be transferred directly to the polarographic cell for analysis. Because they form hydrazones, other volatile aldehydes, such as acrolein, crotonaldehyde, and benzaldehyde, were likely interferences in the analysis. The reported measurement limit for this method was 6 μg/sample. With a sampling rate of 0.1–0.2 L/min and a recommended sample volume of 18 L, the limit of quantitation was 0.3 mg/m^3.

Hydrazine Method

A method for formaldehyde developed recently by OSHA specifies sampling air with a bubbler containing aqueous methanol (30). The collected formaldehyde is reacted with hydrazine and determined by polarography. Because of the differences in half-wave potentials for the formaldehyde derivative and other low molecular weight aldehydes (e.g., acetaldehyde, propionaldehyde, and acrolein), the method is selective for formaldehyde.

The method has sufficient sensitivity for ceiling (15 min) measurements. The reported measurement limit is 1.8 μg/sample. When the maximum recommended sample volume of 160 L is collected at 0.7–1.15 L/min, the limit of quantitation is 0.01 mg/m^3. The major drawback of the method is that the collection media pose the problem of transportation of an organic solution back to the laboratory for analysis.

Oxidative Charcoal Tube Method

The NIOSH oxidative charcoal tube method [NIOSH P&CAM 318 (31)] used an oxidant-coated sorbent to react with the collected formaldehyde and partially oxidize it. The formaldehyde was recovered from the sorbent as formate by using aqueous hydrogen peroxide and determined by ion chromatography. Early commercial sample tubes had blank levels that were high and caused the analytical measurement limit to be high also. For later commercial lots of the tubes, the blank was greatly reduced and the reported measurement limit was 3 μg/sample. With a sampling rate of 0.2 L/min and a recommended sample size of 100 L, the limit of quantitation was 0.03 mg/m^3. Recent NIOSH field and laboratory evaluations of this method with commercial tubes revealed that the samples were not stable for more than 5 days (32). In the original work reported by NIOSH, samples were good for up to 30 days (33). Scientists believe that the original batch of sorbent with which the method was developed was never duplicated for the preparation of the commercial tubes. This method is no longer used by NIOSH.

2-(Benzylamino)ethanol-Coated Sorbent Tube Method

With the development of the 2-(benzylamino)ethanol (BAE)-coated sorbent tube method [NIOSH P&CAM 354 (34)], a different derivatization approach was taken for formaldehyde sampling. Formaldehyde was known to react with secondary aminoethanols to form cyclic derivatives called oxazolidines. This reaction was fast enough to be developed into a sampling method for formaldehyde. The major disadvantage of the method is the lack of sensitivity of the method, a result of the low sampling rate and volume and high blank values found in the unexposed tubes. The method is selective for formaldehyde because of the resolution of the capillary gas chromatographic analysis of the sample. Samples were stable for at least 4 weeks when stored at 25 $\pm$ 5 °C. Suspected interferences in the method are acid gases or mists, which will react with the BAE to convert it to the ammonium salts and thus prevent its reaction with formaldehyde. With tubes prepared in our laboratory, the measurement limit was approximately 5 μg/sample. With a sampling rate of 0.05 L/min and a recommended sample volume of 12 L, the limit of quantitation was 0.42 mg/m^3. Work by commercial vendors on the sampling tube has reduced the blank level to below 2 μg/sample with good precision. This reduction, in effect, has lowered the measurement limit to 2 μg/sample.

To attempt to improve the sensitivity of the method, capacity and sampling studies were conducted with the commercially available tubes. When an atmosphere of 9.4 mg/m^3 of formaldehyde was sampled at 0.08 L/min for up to 15 h, no breakthrough was observed. On the basis of this finding, the sampling volume can be increased from 12 to 38 L (0.08 L/min × 480 min) with the commercial tubes.

To study low sample loadings, an atmosphere of 0.36 mg/m^3 of formaldehyde, as determined by the chromotropic acid method (3), was generated. This atmosphere was sampled with three sets of six BAE-coated sorbent tubes for time intervals of 70, 286, and 482 min at flow rates around 0.08 L/min. The concentrations determined by these sets of samples with their 95% confidence limits were 0.33 ± 0.03, 0.31 ± 0.008, and 0.33 ± 0.012 mg/m^3, respectively. Levels of formaldehyde collected by the tubes were 1.6–10.8 μg/sample. The results of the 70-min set suggest that the limit of quantitation for a 38-L air sample is 0.05 mg/m^3.

Summary

In Table I several of the more commonly used sampling and analytical methods for formaldehyde are summarized. As can be seen in this table, the sampling devices most frequently used are the bubbler and impinger. These devices are selected because many of the industrial hygiene sampling and analysis methods have been adapted from methods of air pollution monitoring that were in use before sorbent tube sampling was common. Also, the impinger, instead of the bubbler, is specified in several of these methods. In the literature no rationale was ever given for this preference. The use of less than optimal flow rates and mismatched impinger stem–body assemblies may affect collection efficiency of the sampler adversely. For the collection of area samples, the impinger is adequate when the cautions stated earlier are observed, but when personal samples are required, this collection technique is much less appealing because of its cumbersome nature and potential for spillage. This device can also present major shipping problems when samples must be transported to the laboratory for analysis. If long-term samples are taken, then the volume of the absorbing solution should be measured both before and after sampling to correct for solution evaporation. If the volume loss is significant (e.g., >10%), then sampling efficiency may be reduced, and pollution of the environment with the solvent being used may occur. The method of choice for personal sampling is a solid sorbent-based device or a sealed liquid-containing device. This approach reduces the bulk of the sampling device and usually solves sample-handling and shipping problems.

Close study of Table I reveals that the most sensitive methods use an

Table I. Summary of Sampling and Analytical Methods for Formaldehyde

Method	Ref.	Sampler Type	Sampling Rate (L/min)	Sampling Volume (L)	Measurement Limit (μg/sample)	LOQ[a] (mg/m^3)
Chromotropic acid	3	impinger	1.0	60	1.5	0.025
	10	alumina sorbent tube				
Diffusive monitors	12	badge	0.2	6	1.0	0.17
	11	badge	0.0017	0.8	0.25	0.30
Pararosaniline	15	impinger	0.0569	27	0.8	0.03
	16	bubbler	5.6	28	2.0	0.07
	17	molecular sieve sorbent tube	1.0	60	1.8	0.03
2,4-DNPH	21	bubbler	2.0	30	1.0	0.03
	22	impinger	0.5–1.5	30	0.05	0.002
	26	coated silica gel sorbent tube	0.5–1.5	31	0.25	0.008
	25	coated XAD-2 sorbent tube	0.20	20	2.5	0.13
MBTH	27	impinger	0.20	5	0.64	0.13
	28	bubbler	0.5	250	5.0	0.02
Girard T	29	bubbler	0.5	250	2.0	0.008
Hydrazine	30	bubbler	0.1–0.2	18	6.0	0.3
Oxidative charcoal	31	coated charcoal sorbent tube	0.7–1.15	30	1.8	0.01
BAE-coated tube	34	coated XAD-2 sorbent tube	0.2	100	3.0	0.03
			0.08	38	2.0	0.05

[a]The limit of quantitation (LOQ) is based on the measurement limit divided by the recommended sample volume and is an estimate of the level that the method has the potential of measuring.

impinger or bubbler sampling device. The increased sensitivity is due mostly to the larger sample volume allowed by the impinger or bubbler as compared to the sorbent tube. The analytical measurement limit of a method (e.g., micrograms per sample detectable) is not as important in determination of the limit of quantitation of the method as the allowable sample volume. In those methods having a blank, the analytical measurement limit of the method is restricted by the blank levels found. A recent article reported that the measurement limit of the BAE-coated sorbent tube method for 3-benzyloxazolidine had been lowered to 6 ng/sample (35). Unfortunately in this instance, the blank level found on the tube still remained at 1 μg/sample and thus precluded measurement below this level.

The methods discussed in this chapter are probably a majority of the total number of methods that have been used for formaldehyde sampling. No one optimum method for the determination of formaldehyde exists. Several factors, such as potential interferences, type of sample (personal vs. area) and sampling device, length of sampling time (8-h time-weighted average vs. 15-min short-term exposure), atmospheric parameters, sample storage, and analytical capability of the laboratory, must be considered when planning to sample for formaldehyde for an industrial hygiene survey. In the selection of a method for formaldehyde, some trade-offs of advantages and disadvantages will be necessary. We hope the information presented in this chapter will assist industrial hygienists in their selections of methods for formaldehyde.

Literature Cited

1. Kiefer, D. M. *Chem. Eng. News* 1983, *61* (*51*), 26.
2. Committee on Aldehydes "Formaldehyde and Other Aldehydes"; National Academy: Washington, D.C., 1981; pp. 175–220.
3. "National Institute for Occupational Safety and Health Manual of Analytical Methods"; Taylor, D. G., Ed.; NIOSH: Cincinnati, Ohio, 1977; DHHS(NIOSH) Publication No. 77–157–A, P&CAM 125.
4. "Methods of Air Sampling and Analysis," 2d ed.; Katz, Morris, Ed.; American Public Health Association: Washington, D.C., 1977; pp. 303–7, Method 116.
5. "National Institute for Occupational Safety and Health Manual of Analytical Methods," 3d ed.; Eller, P. M., Ed.; NIOSH: Cincinnati, 1984; DHHS(NIOSH) Publication No. 84–100, Method 3500.
6. National Institute for Occupational Safety and Health–Occupational Safety and Health Administration Standards Completion Program Failure Report for Formaldehyde, S327, NIOSH contract report, 1976.
7. Kennedy, E. R.; Smith, D. L.; Geraci, C. L., Jr., Chapter 11 in this book.
8. Skisak, C. M. *Am. Ind. Hyg. Assoc. J.* 1983, *44*, 948–50.
9. "Labor Code of Federal Regulations" Title 29, Pt. 1910.1000, p. 594, Government Printing Office: Washington, 1980.
10. "National Institute for Occupational Safety and Health Manual of Analytical Methods"; Taylor, D. G., Ed.; NIOSH: Cincinnati, Ohio, 1977; DHHS(NIOSH) Publication No. 77–157–A, P&CAM 235.
11. Rodriguez, S. T.; Olson, P. B.; Lund, V. R., presented at the Am. Ind. Hyg. Conf., Portland, Oreg., May 1981.

12. Kring, E. V.; Thornley, G.; Dessenberger, C.; Lautenberger, J. *Am. Ind. Hyg. Assoc. J.* **1982**, *43*, 786–95.
13. Kennedy, E. R.; Hull, R. D., presented at the Am. Ind. Hyg. Conf., Philadelphia, Penna., May 1983.
14. Nauman, R. V.; West, P. W.; Tron, F. *Anal. Chem.* **1960**, *32*, 1307–9.
15. Lyles, G. R.; Dowling, F. B.; Blanchard, V. J. *J. Air Pollut. Control Assoc.* **1965**, *15*, 106–8.
16. Miksch, R. K.; Anton, D. W.; Fanning, L. Z.; Hollowell, C. D.; Revzan, K.; Glanville, J. *Anal. Chem.* **1981**, *53*, 2118–23.
17. Matthews, T. *Anal. Chem.* **1982**, *54*, 1495–98.
18. CEA 555 TGM Formaldehyde Monitor, CEA Instruments, Inc. Westbrook, New Jersey.
19. Matthews, T. *Am. Ind. Hyg. Assoc. J.* **1983**, *43*, 547–52.
20. Walters, Robert B. *Am. Ind. Hyg. Assoc. J.* **1983**, *44*, 659–61.
21. Kuwata, K.; Uebori, M.; Yamaski, Y. *J. Chromatogr. Sci.* **1979**, *17*, 264–68.
22. Lipari, F.; Swarin, S. J. *J. Chromatogr.* **1982**, *247*, 297–306.
23. Papa, L. J. *Environ. Sci. Technol.* **1969**, *3*, 397–98.
24. Fracchia, M. F.; Schuette, F. J.; Mueller, P. K. *Environ. Sci. Technol.* **1967**, *1*, 915–22.
25. Andersson, G.; Andersson, K. *Chemosphere* **1979**, *10*, 823–27.
26. Beasley, R. K.; Hoffmann, C. E.; Rueppel, M. L.; Worley, J. W. *Anal. Chem.* **1980**, *52*, 1110–14.
27. Sawicki, E.; Hauser, T. R.; Stanley, T. W.; Elbert, W. *Anal. Chem.* **1961**, *33*, 93–96.
28. Hauser, T. R.; Cummins, R. L. *Anal. Chem.* **1964**, *36*, 679–81.
29. "National Institute for Occupational Safety and Health Manual of Analytical Methods"; Taylor, D. G., Ed.; NIOSH: Cincinnati, Ohio, 1978; DHHS(NIOSH) Publication No. 78–175, P&CAM S327.
30. Septon, J. C.; Ku, J. C. *Am. Ind. Hyg. Assoc. J.* **1982**, *43*, 845–52.
31. "National Institute for Occupational Safety and Health Manual of Analytical Methods"; Taylor, D. G., Ed.; NIOSH: Cincinnati, Ohio, 1980; DHHS(NIOSH) Publication No. 80–125, P&CAM 318.
32. Smith, D. L.; Bolyard, M.; Kennedy, E. R. *Am. Ind. Hyg. Assoc. J.* **1983**, *44*, 97–99.
33. Kim, W. S.; Geraci, C. L.; Kupel, R. E. *Am. Ind. Hyg. Assoc. J.* **1980**, *41*, 340–44.
34. "National Institute for Occupational Safety and Health Manual of Analytical Methods"; Taylor, D. G., Ed.; NIOSH: Cincinnati, Ohio, 1981; DHHS(NIOSH) Publication No. 82–100, P&CAM 354.
35. Matthews, P.; Penton, Z. *Ind. Hyg. News* **1984**, *January*, 32–34.

RECEIVED for review September 28, 1984. ACCEPTED January 7, 1985.

Ion Chromatography with Pulsed Amperometric Detection

Simultaneous Determination of Formic Acid, Formaldehyde, Acetaldehyde, Propionaldehyde, and Butyraldehyde

ROY D. ROCKLIN

Dionex Corporation, Sunnyvale, CA 94088–3603

Formaldehyde, acetaldehyde, propionaldehyde, butyraldehyde, and formic acid can be determined in a single analysis by ion chromatography. The aldehydes and formic acid are separated on a fully functionalized cation-exchange resin in the potassium form. They are detected electrochemically by oxidation using pulsed amperometric detection at a platinum electrode. Detection limits range from 1 to 3 ppm. Methanol and ethanol interfere with the analysis.

$\mathbf{M}$ANY METHODS OF DETERMINING FORMALDEHYDE IN AIR involve its sorption on a solid sorbent followed by desorption by a liquid (*1*). The formaldehyde in the liquid is then assayed. For this method to be reliable, the formaldehyde concentration on the sorbent must be directly proportional to its concentration in the air. The desorption step must be quantitative, and finally, the assay of the solution must be accurate. The methods used today to determine formaldehyde in air suffer from two problems. First, during the time between the sampling and the final assay, formaldehyde will slowly be oxidizing to formic acid. Second, all known assay procedures are subject in some degree to interferences. For formaldehyde, these are usually other aldehydes, alcohols, and phenol (*1*).

The major advantage of a chromatographic assay is the improved selectivity attained when a physical separation is present between the analyte species and the potential interferences. For this reason, two ion chromatographic methods were developed and have been reported previously. In each, the formaldehyde is reacted during the desorption process to produce a stable product.

The first method is NIOSH P&CAM 318 (*1, 2*). Formaldehyde is ad-

0065–2393/85/0210/0013$06.00/0

sorbed onto activated charcoal and desorbed with a solution containing hydrogen peroxide. The hydrogen peroxide is added to quantitatively convert the formaldehyde to formic acid to solve the oxidation problem. The formic acid is then assayed by ion chromatography. Although this method has worked well, difficulties were reported once. During NIOSH evaluations (3), samples were found to be stable for only a few days, and the quality of the sorbent affected sample stability.

In the second method, formaldehyde is adsorbed onto molecular sieves and desorbed with a solution containing bisulfite ion (4). Formaldehyde and bisulfite react to form hydroxymethanesulfonate ($HOCH_2SO_3^-$, often called formaldehyde bisulfite), which is assayed by ion chromatography at standard anion conditions. Because the addition product coelutes with sulfite ion, the remaining bisulfite in the solution is oxidized to sulfate by hydrogen peroxide. One disadvantage of this method is that any formaldehyde that oxidizes before the desorption step will be lost.

For this reason, a chromatographic method in which both formaldehyde and formic acid can be determined in a single injection has been developed and is reported here. This method requires the formaldehyde and formic acid to be in aqueous solution. For example, they could have been desorbed from a solid sorbent. The species are then determined by ion chromatography by using pulsed amperometric detection. Other aldehydes such as acetaldehyde, propionaldehyde, and butyraldehyde can also be determined. Methanol and other alcohols can interfere with the analysis.

Experimental

All chromatography was performed on a Dionex system 2011i ion chromatograph, which consists of a pump, a chromatography module, and a pulsed amperometric detector. A platinum working electrode, a glassy carbon counterelectrode, and a silver–silver chloride (1 M NaCl) reference electrode were used in the amperometric flow-through detector cell. The applied potentials (volts) and pulse durations (milliseconds) were the following: $E1(t1)$, 0.2(60); $E2(t2)$, 1.3(60); and $E3(t3)$, $-0.3(240)$. The sample loop volume was 50 μL. The column was a Dionex HPICE–AS1 with the cation-exchange resin converted to the potassium form. The eluant was 0.10 N H_2SO_4 and 0.05 M K_2SO_4 at a flow rate of 1.0 mL/min. All chemicals were reagent grade except the sodium formaldehyde bisulfite, which was technical grade and was obtained from Eastman. Formaldehyde was 37% in water.

Results and Discussion

Pulsed Amperometric Detection. In 1981, Hughes et al. introduced pulsed amperometric detection. By using a strongly acidic solution, alcohols, glycols, and formic acid could be detected (5). The same conditions used to detect alcohols and formic acid can also be used to detect aldehydes, and are described next.

The most commonly used form of electrochemical detection is single potential amperometry, sometimes called DC amperometry. In this method, a single potential is applied to the working electrode of a flow-through cell, and the resulting current is continuously monitored. Pulsed amperometric detection is a new form of electrochemical detection used to detect those species that cause the electrode to become poisoned when they are oxidized. Pulsed amperometric detection uses a repeating sequence of three potentials. The analyte molecules are oxidized at the first potential ($E1$), and the current is measured. The potential is then stepped to a more positive value ($E2$), and an oxide layer is formed on the electrode surface. The third potential ($E3$) is a negative potential at which the oxide layer is reduced to produce the bare metal. This sequence is repeated several times per second, and one point on the chromatogram is acquired during each cycle. The advantage of pulsed amperometric detection is the electrode cleaning provided by the alternating positive and negative pulsing. When only a single potential is used, peak heights from a series of injections will quickly decrease as the electrode becomes coated by the products of the oxidation reaction.

Information regarding the choice of the three applied potentials is obtained from electrochemical experiments such as cyclic voltammetry and rotated-disk voltammetry. Steady state cyclic voltammetry for formaldehyde in acid on a platinum electrode is shown in Figure 1. The curve is very similar to that of formic acid, shown in Figure 2 and in Reference 5. Hughes et al. have developed a theory for the mechanism of alcohol and formic acid oxidation that probably applies to aldehydes. They conclude that analyte molecules are adsorbed onto the platinum electrode surface at negative potentials and oxidized (probably to carbon dioxide and water for formic acid, formaldehyde, and methanol) on the positive going scan. When platinum oxide is formed on the electrode surface at approximately 0.7 V, the analyte oxidation reaction stops. Rotated-disk voltammetry with formic acid and alcohols shows that the decrease in current is caused by platinum oxide blocking the electrode, and is not just the decreasing current normally seen with diffusion control in potential-scan voltammetry. This situation probably occurs with aldehydes. At potentials beyond 1.0 V, oxidation continues. After the reversal of the potential-scan direction, the current will actually reverse from cathodic to anodic, reaching a peak at approximately 0.2–0.3 V. This result is caused by the resumption of the analyte oxidation reaction as platinum oxide is reduced ($E_p = 0.44$) to produce a bare and, therefore, more active surface.

The cyclic voltammetry of propionaldehyde (Figure 3) is quite different. Little deviation from the background current occurs except when the potential is positive of 0.5 V, where oxidation (probably to propionic acid) takes place. Because propionaldehyde can be oxidized when the electrode is coated with platinum oxide, the reaction mechanism is probably different from that of the single carbon molecules. The lack of current between 0 and 0.5 V implies that propionaldehyde cannot be detected if $E1$ is set be-

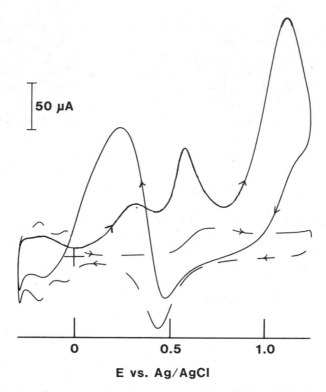

50 µA

0   0.5   1.0

E vs. Ag/AgCl

Figure 1. Steady state cyclic voltammetry of 20 mM formaldehyde in 0.1 N
H_2SO_4 and 0.05 M K_2SO_4 at a platinum working electrode. Sweep rate is 200
mV/s. Analyte is indicated by the solid line; background current is indicated
by the dashed line.

tween 0 and 0.5 V. However, this situation is not the case. The difference
between the pulsed amperometric detection and the cyclic voltammetry
could be caused by the large difference in time scale for the two experiments. The cyclic voltammetry of acetaldehyde and butyraldehyde are
similar to that of propionaldehyde.

Information from cyclic voltammetry is now used to choose the three
applied potentials. The potentials and pulse durations used are illustrated
in Figure 4. Of the three potentials, $E1$ is the most important because the
current is measured at this potential. $E1$ is chosen to be not only on a large
oxidation peak for the analyte, but also at a potential where little, if any,
faradaic background current occurs. On a platinum electrode in an acidic
solution, this region is between the hydrogen adsorption and reduction reactions that are negative of 0 V and the platinum oxide formation that begins at approximately 0.5 V. Then, $E2$ is set to 1.3 V, a value chosen to be
within 0.1 V of the positive potential limit, and $E3$ is set to -0.3 V, within
0.1 V of the negative limit. The optimum value for $E1$ is determined by
making a series of injections while varying $E1$. A plot is then made of peak

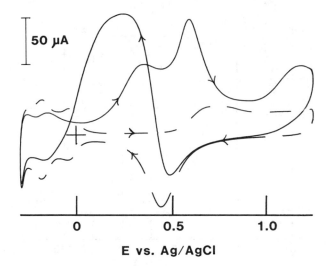

Figure 2. Steady state cyclic voltammetry of 20 mM formic acid. Conditions are the same as in Figure 1.

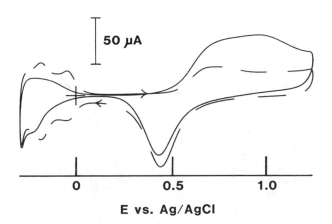

Figure 3. Steady state cyclic voltammetry of 40 mM propionaldehyde. Conditions are the same as in Figure 1.

heights versus $E1$. This plot for formaldehyde, acetaldehyde, propionaldehyde, and formic acid is shown in Figure 5. The background current is also shown with a minimum at 0.1 V. For formaldehyde and formic acid, one would expect from the cyclic voltammetry that maximum peak heights would result if $E1$ was set near 0.35 V, which is the peak current for the first oxidation reaction. Actual optimum potentials are lower: 0.2 V for formaldehyde and 0.1 V for formic acid. A possible explanation is that adsorbed molecules are known to decrease the rate of the oxide formation

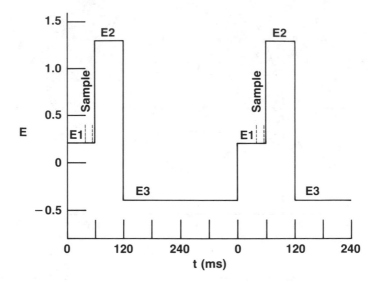

Figure 4. Applied potentials and pulse durations for aldehyde and formic acid determination. Following a 40-ms delay to allow charging current to decay, the current is sampled for 16.7 ms during E1.

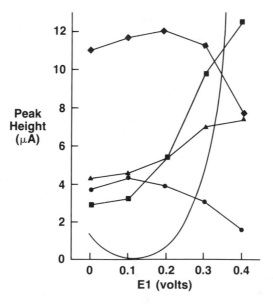

Figure 5. Peak height vs. E1 for 100 ppm of formic acid (●), 100 ppm of formaldehyde (◆), 300 ppm of acetaldehyde (■), and 300 ppm of pro-pionaldehyde (▲). The solid line is background current.

reaction (6). At potentials negative of 0.5 V, some platinum oxide forms, even though the reaction rate increases considerably as the potential is made more positive. Because only the sum of analyte oxidation current and background current can be measured, the decrease in peak height as $E1$ is increased toward 0.5 V could be caused by adsorbed analyte lowering the background current.

For both acetaldehyde and propionaldehyde, the current increases as $E1$ is made more positive. Because of the lack of any oxidation current shown on the cyclic voltammetry of propionaldehyde, it is surprising that current can be detected at all with $E1$ set negative of 0.4 V. Although greater currents can be obtained by further increases of $E1$, the large increase in background current caused by the formation of platinum oxide increases the noise more than the signal.

Separation. Pulsed amperometric detection at a platinum electrode in an acidic solution is only semispecific; that is, only small oxidizable organic molecules are detected. These include aldehydes, alcohols, glycols, phenols, and formic acid. (Other aliphatic acids such as acetic acid are not detected.) The selectivity for quantitative analysis must be provided by the chromatography. Separating species with similar properties such as aldehydes and alcohols by liquid chromatography can be a difficult task. Because these molecules are nonionic, initial attempts at separation were performed on a neutral macroporous resin designed to maximize differences in adsorptive interactions. (The column used was the Dionex MPIC–NS1.) Because of the hydrophilic nature of the analytes, they were largely unretained. Better retention was achieved on a high-capacity, fully functionalized, strong-acid cation-exchange resin in the potassium form (Dionex HPICE–AS1, potassium form). With this resin, the retention mechanism is a combination of adsorption and the weak interactions between the oxygen-containing functional groups and the potassium on the resin. The separation between formaldehyde and formic acid was poorer on sodium- or hydrogen-form resins.

A chromatogram of a standard solution of formaldehyde, acetaldehyde, formic acid, propionaldehyde, and butyraldehyde in water is shown in Figure 6. Complete baseline separation occurs between formaldehyde and formic acid; however, alcohols (not shown) interfere. Methanol elutes between formaldehyde and acetaldehyde, and ethanol coelutes with formic acid. If the methanol concentration is more than twice that of the formaldehyde, the formaldehyde peak will be obscured. Phenol is very strongly retained on the styrene–divinylbenzene resin and does not interfere.

One characteristic of this chromatographic method is the presence of a small dip between the formaldehyde and acetaldehyde peaks (Figure 7). This dip only appears on the higher sensitivity current scales, and it effectively limits the minimum detection limit by obscuring the formaldehyde

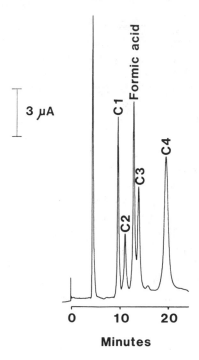

Figure 6. Chromatography of formaldehyde (C₁) through butyraldehyde (C₄) and formic acid. Concentrations are 30 ppm for formaldehyde and 100 ppm for others. See experimental section for conditions.

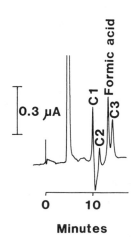

Figure 7. Chromatography of low concentrations of formaldehyde (3 ppm) and formic acid and other aldehydes (10 ppm). See experimental section for conditions.

and acetaldehyde peaks when low concentrations are determined. In spite of this situation, detection limits are approximately 1 ppm for formaldehyde, and approximately 2–3 ppm for formic acid and the other aldehydes.

Formaldehyde can be selectively trapped without trapping alcohols in a solution of bisulfite in which it will form hydroxymethanesulfonic acid. This product can be separated into its original components in strong base. Although a thorough study of the optimum conditions for this reaction was not performed, several solutions were prepared as a test. A twice molar excess of bisulfite was added to standard solutions of formaldehyde to form the addition product. Also, standard solutions of sodium hydroxymethanesulfonate (obtained as the solid) were prepared. After adjusting the pH to 13.5 to release the formaldehyde, the solution was injected into the ion chromatograph. The percent recovery of formaldehyde from hydroxymethanesulfonate solid was 85% at 10 ppm and 86% at 100 ppm. The percent recovery of formaldehyde from bisulfite added to formaldehyde was 92% at 10 ppm and 91% at 100 ppm. (These values are the average of two samples, ±2%.) Thus, the percent recovery is consistent, although it is less than 100%. The solution was stable with time; no change in the formaldehyde peak height occurred after 30 min at pH 13.5. Raising the solution pH did not result in a more quantitative recovery, although less formaldehyde was recovered at pH 12. One method of increasing the accuracy of formaldehyde determination is to match the matrix of the standards to the sample by preparing the standards in bisulfite solution and adjusting to the same high pH as the sample. If this procedure cannot be done, the standard addition method of analysis should be used.

Conclusion

Ion chromatography with pulsed amperometric detection is a sensitive and selective technique for the simultaneous determination of formic acid, formaldehyde, and other aldehydes. The usefulness of this technique will increase as new chromatographic methods that further increase the separation between alcohols and aldehydes are developed.

Literature Cited

1. "National Institute for Occupational Safety and Health Manual of Analytical Methods"; Taylor, D. G., Ed.; NIOSH: Cincinnati, 1977; Vol. 1, DHHS(NIOSH) Publication No. 77–157–A.
2. Kim, W. S.; Geraci, C. L.; Kupel, R. E. In "Ion Chromatographic Analysis of Environmental Pollutants"; Mulik, J. D.; Sawicki, E., Eds.; Ann Arbor Science: Ann Arbor, 1979; Vol. 2.
3. Smith, D. L.; Bolyard, M.; Kennedy, E. R. *Am. Ind. Hyg. Assoc. J.* 1983, *44*, 97.
4. Dolzine, T. W.; Williams, K. E.; Gaffney, R., unpublished data.
5. Hughes, S.; Meschi, P. L.; Johnson, D. C. *Anal. Chim. Acta* 1981, *132*, 1.
6. Austin, D. S.; Polta, J. A.; Polta, T. Z.; Tang, A. P.-C.; Cabelka, T. D.; Johnson, D. C. *J. Electroanal. Chem.* 1984, *168*, 227.

RECEIVED for review September 28, 1984. ACCEPTED February 14, 1985.

Analysis of Formaldehyde in an Industrial Laboratory

S. A. SCHMIDT, M. F. ANTLOGA, and M. MARKELOV

The Standard Oil Company (Ohio), Research Center, Cleveland, OH 44128

This chapter describes two analytical methods that we use to determine formaldehyde (HCHO). The first is a polarographic method that measures the reduction of HCHO at a dropping mercury electrode (DME) in a 0.1 N LiOH–0.01 N LiCl electrolyte. Use of hydrazine derivatization to confirm the presence of HCHO is shown. The second method, based on the reaction of HCHO with pararosaniline, has been automated by use of a Zymate robot and Brinkman colorimeter with a fiber-optic probe. The polarographic method is used when small batches of samples are submitted or if the concentration range is unknown. The automated colorimetric method is used when large numbers of samples are submitted. Comparison of the results obtained on a variety of industrial samples with these techniques is made. Initial results obtained with headspace gas chromatography and a photoionization detector are discussed.

Requests for formaldehyde analyses come from many varied groups in our company. Researchers want to know formaldehyde concentration in reactor effluents and automobile exhaust condensates from test engines; engineers request its analysis in various chemical plant streams and raw materials; and environmentalists and toxicologists request its determination in the workplace atmosphere.

To handle these varied requests, we have found it useful to have a variety of analytical methods available. The choice of the analytical procedure depends upon many factors: sample matrix, expected concentration range, number of samples, availability of analysts and equipment, etc. Analysis of any new or unusual samples by at least two independent methods is also desirable.

This chapter will discuss two of the procedures we use routinely: a differential pulse polarographic technique based on work by Vesely and Brdicka quoted by Kolthoff and Lingane (1) and an automated colorimetric method based on the reaction with pararosaniline as modified by

Miksch et al. (2). We will also describe method development currently underway in our laboratory using headspace gas chromatography with a photoionization detector.

Experimental

Polarographic Method. INSTRUMENTATION. PAR 174A polarographic analyzer with PAR 303 DME operated in the differential pulse mode is used at a scan rate of 5 mV/s, a modulation amplitude of 25 mV/p-p, a drop rate of 1/s, and a scan range of -1.5 to -2.25 V.

PROCEDURE: DIRECT MEASUREMENT OF FORMALDEHYDE. An aliquot of the sample is added to a polarographic cell. Electrolyte is added so that the final concentration of the electrolyte in the cell is 0.1 N LiOH–0.01 N LiCl. The solution is purged with 99.99% Ar for 4–8 min. The reduction of formaldehyde at a dropping mercury electrode is measured with differential pulse polarography. The voltammogram is recorded from -1.5 to -2.25 V versus a Ag–AgCl electrode.

PROCEDURE: INDIRECT MEASUREMENT OF FORMALDEHYDE AS THE HYDRAZINE DERIVATIVE. The hydrazine derivative is prepared in the polarographic cell by adding 1 mL of 2% hydrazine sulfate in distilled water and 4 mL of acetate buffer (0.1 M NaOAc–0.1 M HOAc) to 5 mL of the sample. The voltammogram is recorded from -0.5 to -1.5 V versus a Ag–AgCl electrode.

Pararosaniline Colorimetric Method. MANUAL PROCEDURE. One milliliter of acidified pararosaniline solution (0.16 g of pararosaniline/100 mL of 0.24 N HCl) and 1 mL of freshly prepared sodium sulfite solution (0.1 g of Na_2SO_3/100 mL of distilled water) are added to 10 mL of the sample. The resulting solution is mixed and allowed to stand for 1 h, and the absorbance at 570 nm is read from a Beckman Acta III spectrophotometer with a 1-cm cell. A blank determination is made with the same water used for sample dilutions.

AUTOMATED PROCEDURE. The manual method is automated by use of a Zymate robot (Zymark) with the equipment listed in the box. The layout is shown in Figure 1. The analyst places the samples (which are in 23-mL headspace vials) in the sample rack. The robot transfers an aliquot of the sample from the sample vials to a 16 × 100-mm culture tube in the dilution rack with a syringe hand with a disposable pipette tip (Figure 2A). Ten milliliters of distilled water is added from an automatic dispenser station (Figure 2B). The sample can be diluted again at this point, if necessary, by repeating this procedure. Once the sample concentration is within the calibrated range of the test, the test tube is moved to the dispensing station (Figure 2C) and 1 mL of pararosaniline and 1 mL of sodium sulfite solution are added. The test tube is placed in a vortex mixer (Figure 2D) for 30 s and returned to the dilution rack where it is allowed to stand for 1 h. The robot places the color probe from the Brinkman probe colorimeter in the sample test tube (Figure 3), and the percent transmittance (% T) is read by sending an analog signal directly to the power- and event-control module, which has a built-in analog-to-digital (A/D) converter. The digitized signal is then sent to the robot controller which has an Intel 8088 microprocessor board that can handle simple calculations. The % T is converted to concentration by using a calibration curve that has been previously entered. If the absorbance reading does not fall in the calibrated range, the robot will rerun the sample at a different dilution. The color probe is rinsed between samples.

HEADSPACE GAS CHROMATOGRAPHIC PROCEDURE. A 10-mL aliquot of sample is placed in a standard 23-mL headspace vial containing approximately 6.5 g of sodium sulfate. The vial is sealed with a Teflon-lined septum and placed into a

Robot Pararosaniline Method
Equipment
A. Zymark Robot
   1. Controller
   2. Robot Hands
      a. Syringe hand with 1-mL syringe mounted
      b. General-purpose hand with fingers to handle 16–20-mm test tubes
      c. Blank hand to mount color probe
   3. Master Laboratory Station
      a. 10-mL syringe in Dispenser 1
      b. 1-mL syringe in Dispenser 2
      c. 1-mL syringe in Dispenser 3
   4. Power- and Event-Control Station
   5. Vortex Station
   6. Printer
B. Brinkman PC–1000 Colorimeter with Fiber-Optic Probe

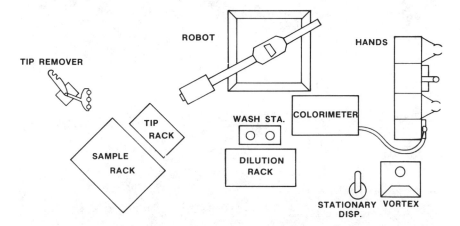

Figure 1. Configuration of the Zymate robot work station.

temperature bath at 50 °C for 1 h. A 0.5-mL aliquot of the headspace over the sample is introduced into a Photovac gas chromatograph (GC) equipped with a 4-ft. $\times \frac{1}{8}$-in. i.d. Teflon column packed with Carbopak B–HT and a special photovac photoionization detector. Sample introduction is either manual via a gas-tight syringe or automatic via a Perkin–Elmer F42 headspace analyzer.

Discussion

Polarographic Method. In aqueous solution, most of the formaldehyde is present in the form of the monohydrate, methylene glycol, and a series of low molecular weight polymeric hydrates or poly(oxymethylene) glycols. Only a small fraction, less than 0.1%, is present as the free aldehyde. Because only the aldehyde form of formaldehyde is reduced to meth-

A

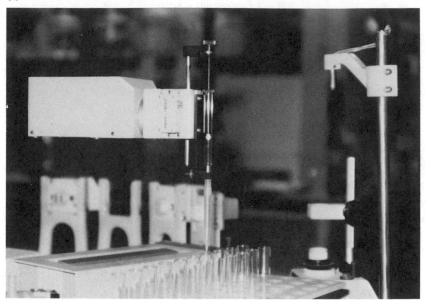

B

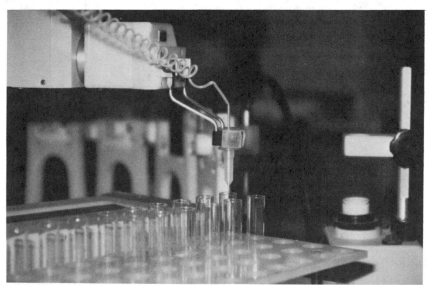

Figures 2A and 2B. Automation of the color development steps of the pararosaniline method for formaldehyde by use of the Zymate robot: A, sample transfer; and B, sample dilution.

C

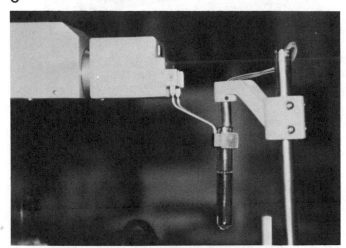

D

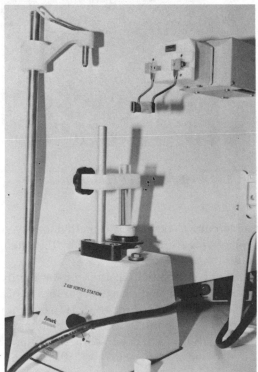

Figures 2C and 2D. Automation of the color development steps of the pararosaniline method for formaldehyde by use of the Zymate robot: C, addition of reagent; and D, sample mixing.

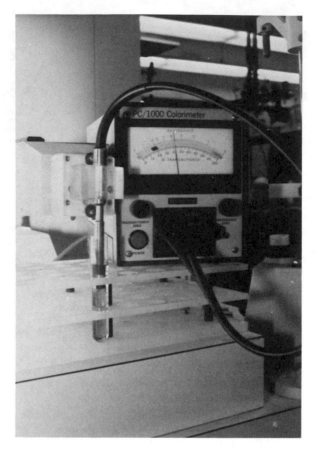

Figure 3. Automatic measurement of the color by using the Zymate robot and Brinkman probe colorimeter.

anol at a dropping mercury electrode, and methylene glycol is not reducible, the limiting current measured is governed by the rate of dehydration of the methylene glycol (Scheme I).

Hydroxide ions catalyze the dehydration of the methylene glycol to the aldehyde and thus make the analysis very sensitive to pH. The limiting current increases as the pH increases (Figure 4). However, if the concentration of hydroxide is increased too much, the reaction of hydroxide with the methylene glycol to form the anion begins to become important, and a decrease in the limiting current is seen. At pH < 7, the formation of the nonelectroactive species, methylene glycol, is favored. At room temperature, a maximum in the limiting current is observed at pH 13. The pH also affects the peak potential (Figure 5). As the pH increases, the peak potential becomes more negative.

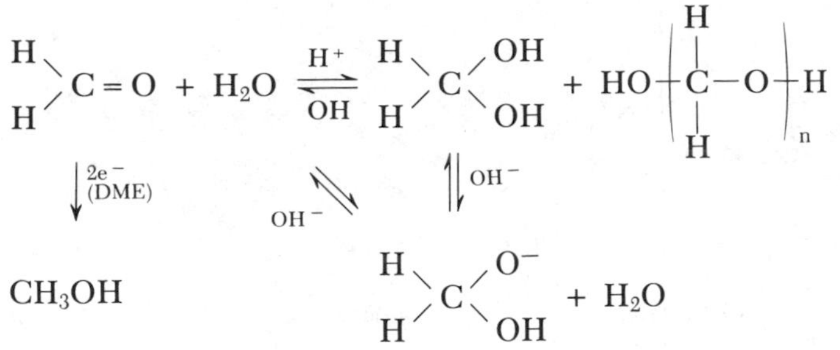

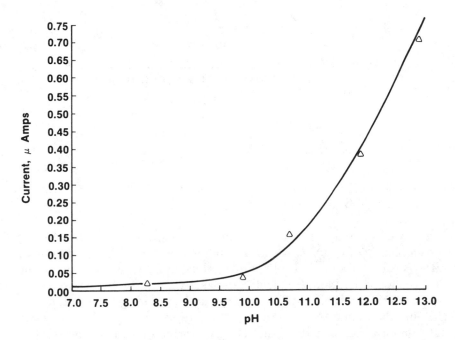

Scheme I.

Figure 4. Dependence of peak height (current) of the formaldehyde peak on pH.

The limiting current is also dependent on anything else that affects the equilibrium of the dehydration reaction such as the choice of solvent, buffer, and analysis temperature. For instance, variations as large as 6% in the response for formaldehyde for every degree change in temperature have been reported. We have found the 0.1 M LiOH–0.01 M LiCl buffer suggested by Kolthoff and Lingane (1) to give satisfactory results. However, because of the enhanced sensitivity of pulse polarography over the

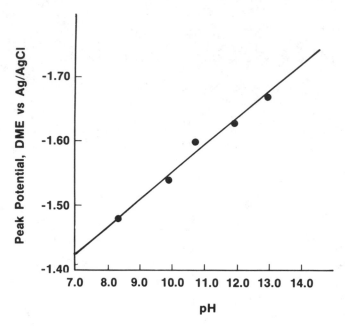

Figure 5. Dependence of the limiting potential of the formaldehyde peak on pH.

classical direct current techniques and improved instrumentation, we work at lower concentrations and do not need to add a maximum suppressor for quantitation.

If we have only a few samples or expect matrix effects or variations in the ambient temperature, we use a standard addition technique to quantitate. We observe a response of approximately 0.02 $\mu A/\mu g$ or 0.056 $\mu A/\mu mol$ of formaldehyde. More frequently, especially if we have several samples, we use an external calibration procedure and prepare the calibration curve the same day. Figure 6 shows a typical calibration curve obtained for 0.07–1.1 ppm of formaldehyde in the cell. The correlation coefficient is 0.995. The voltammogram shown in Figure 7 was obtained for 0.073 ppm of formaldehyde in the cell, which is our quantification limit. Table I shows typical results obtained with the polarographic method. Relative standard deviations (RSD) of 3–5% are usually obtained. However, the RSD increases as the concentration decreases.

Scientists were concerned that formaldehyde was being introduced as an impurity in the polymerization inhibitor added in the industrial process for making HCN. We were requested to determine it in the liquid HCN product. Because working with liquid HCN is extremely hazardous, we attempted to avoid this problem by extracting the formaldehyde into an aqueous phase and then removing the HCN. We decided that we would try

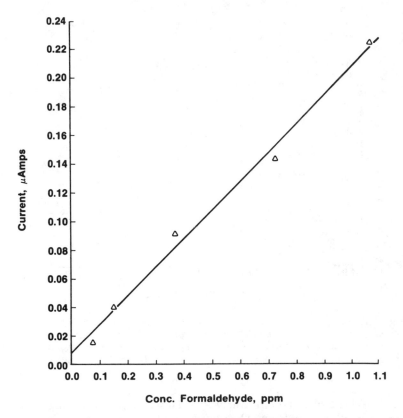

*Figure 6. Typical calibration curve for formaldehyde via polarographic
analysis at DME vs. Ag–AgCl in 0.1 N LiOH–0.01 N LiCl.*

mixing an equal volume of water with the HCN in a 16 × 150-mm test tube
to hold the formaldehyde and applying a gentle stream of nitrogen to the
surface to remove the HCN. We assumed that the adduct of formaldehyde
and HCN would decompose under these conditions. To test the feasibility
of this approach, we spiked three samples of HCN with formaldehyde,
added an equal volume of water, and removed the HCN by evaporation
with a stream of nitrogen. The resulting aqueous solution of formaldehyde
was analyzed with the polarographic method outlined earlier. The results
are shown in Table II. Adequate recoveries were observed at low formaldehyde levels. The fact that recoveries are lower at higher concentrations is
an unusual phenomenon for most analytical procedures and is usually attributed to limited solubilities of the components in question or to increased
rates of side reactions at higher concentrations. However, in the case of
formaldehyde and formaldehyde cyanohydrin, solubility in either the water or the HCN is not a problem. The low recoveries at higher formalde-

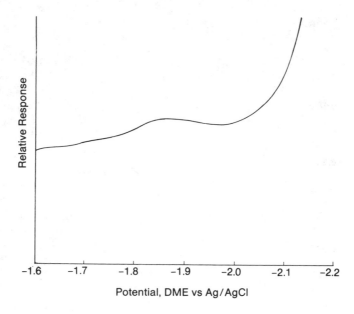

Figure 7. Voltammogram at the lower detection limit, 0.073 ppm of formaldehyde in the cell. Differential pulse polarographic conditions: scan rate, 5 mV/s; modulation amplitude, 25 mV/p-p; drop time, 1/s; display direction, positive; 0.1 N LiOH–0.01 N LiCl; DME vs. Ag–AgCl.

Table I. Results of the Analysis for Formaldehyde in Various Aqueous Samples as Determined with the Polarographic Method

Sample	Conc. HCHO[a] (ppm)	Standard Deviation	Rel. Std. Dev. (%)
A	990	46	4.7
B	1190	30	2.5
C	1100	28	2.5
D	27	2	7.5
E	57	2	3.5
F	31	3	10

[a] Average of three determinations.

hyde concentrations are probably attributable to increased effects of side reactions with impurities present in HCN.

Although any compound that can be reduced in the 0.1 M LiCl–0.01 M LiOH (pH 12.7) buffer at a dropping mercury electrode between − 1.5 and − 1.80 V versus a Ag–AgCl electrode will interfere, the method is still very selective for our types of samples. Saturated aldehydes and α- and β-

Table II. Recovery of Formaldehyde Spike from Liquid HCN as Determined by Polarographic Analysis

Formaldehyde Added (ppm)	Formaldehyde Recovered (ppm)	% Recovery
74	71	96
148	126	85
740	540	73

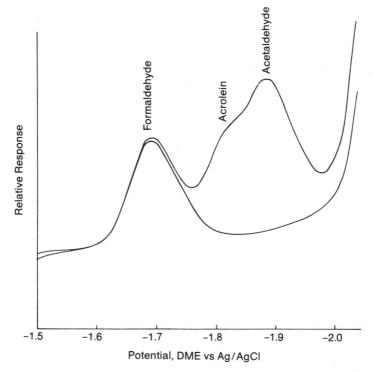

Figure 8. Voltammogram obtained for 3.7 ppm of formaldehyde in the presence of 1.0 ppm of acetaldehyde and 1.0 ppm of acrolein. Polarographic conditions: mode, differential pulse; scan rate, 5 mV/s; modulation amplitude, 25 mV/p-p; drop rate, 1/s; display direction, positive; 0.1 N LiOH–0.01 N LiCl; DME vs. Ag–AgCl.

unsaturated aldehydes are reduced at more negative potentials than formaldehyde and do not interfere if their concentration is comparable to the concentration of formaldehyde. Figure 8 is the voltammogram obtained for 3.7 ppm of formaldehyde in the presence of 1.0 ppm of acetaldehyde and acrolein. Aromatic aldehydes are reduced at more negative potentials than formaldehyde and do not normally interfere.

Confirmation of the presence of formaldehyde can be made by forming the hydrazone and rerunning the sample in an acetate buffer at pH 4.
Figure 9 shows the voltammogram obtained for the hydrazone of a 3.7
ppm solution of formaldehyde. The derivatization shifts the peak at 0.7 V
for less negative value. The peak potential of the other aldehydes will also
be shifted. Figure 10 shows the voltammogram for 3.7 ppm of formaldehyde in the presence of 1 ppm of acrolein and acetaldehyde.

Pararosaniline Colorimetric Method.   This method is based on the
reaction of formaldehyde with pararosaniline in the presence of sulfite. Although the mechanism of the reaction is not completely understood,
Miksch suggested the formation of a Schiff base intermediate between
acidified pararosaniline and formaldehyde (2). This colorless Schiff base
then combines with sulfur dioxide under acidic conditions to form the al-
kylsulfonic acid chromophore. The dehydration of the pararosaniline–
formaldehyde adduct to the carbinolamine adduct is acid catalyzed
(Scheme II).

We use a Zymate robot to automate this procedure. The analyst places
the samples in the sample rack and starts the robot. Everything else, including reporting results and deciding if the sample needs to be rerun because the concentration fell outside the expected range, is done by the robot
automatically.

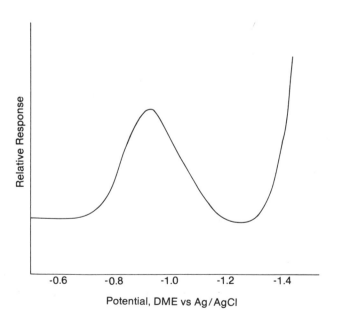

*Figure 9.   Voltammogram of the hydrazone of 3.7 ppm of formaldehyde. Po-
larographic conditions: scan rate, 10 mV/s; modulation amplitude, 25 mV/
p-p; drop rate, 1/s; display direction, positive; 0.1 M NaOAc–0.01 M HOAc
(4 mL) at pH 4.16; 2 % hydrazine sulfate (1 mL).*

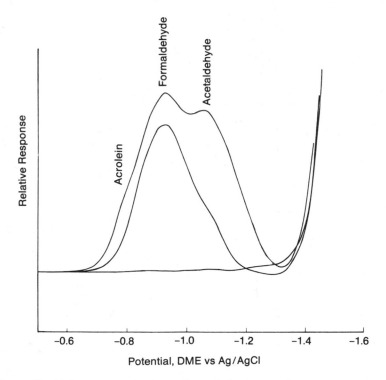

Figure 10. *Voltammogram obtained for the hydrazine derivatives of formaldehyde (3.7 ppm) in the presence of acrolein (0.1 ppm) and acetaldehyde (1.0 ppm). Polarographic conditions: scan rate, 10 mV/s; modulation amplitude, 25 mV/p-p; drop rate, 1/s; display direction, positive; 0.1 M NaOAc–0.1 M HOAc (4 mL); 2% sodium hydrazine (1 mL); pH 4.16; DME vs. Ag–AgCl.*

For robotic applications, we felt that using a Brinkman probe colorimeter equipped with a fiber-optic color probe would be more convenient than using a conventional spectrophotometer such as our Beckman Acta III equipped with a flow-through cell. Smaller sample aliquots could be used, and the Brinkman probe colorimeter is significantly less expensive and takes up less room in the robot's work area.

The colorimeter has six built-in interference filters in the following wavelengths: 450, 470, 520, 570, 620, and 670 nm. We did this work at 570 nm. It was designed as an end-point sensor for photometric titrations and quality control work. It uses a fiber-optic light probe to measure the transmittance of light in a solution. The robotic blank hand was modified to accommodate the color probe which the robotic arm could then dip in a sample solution or water rinse solution. The analog signal from the colorimeter is sent to the microprocessor of the robot via the power- and event-

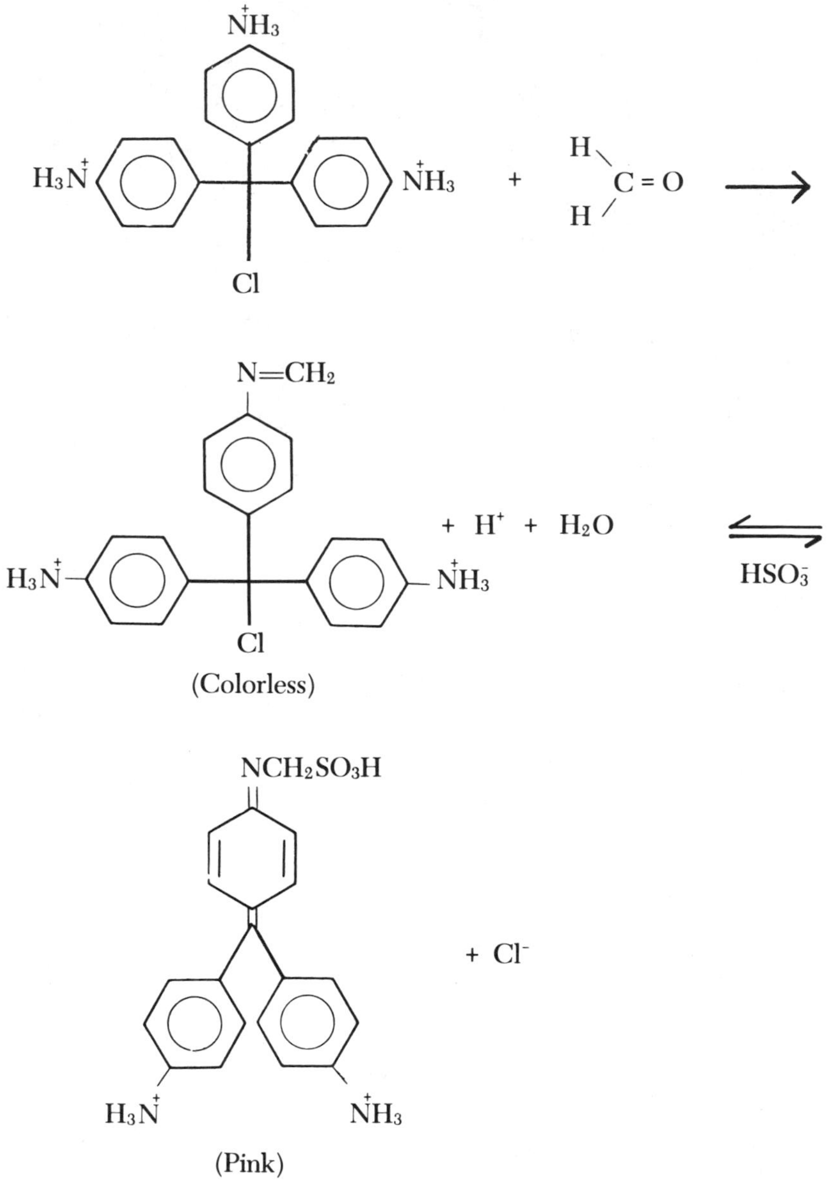

$\overset{+}{N}H_3$
$H_3\overset{+}{N}$
$\overset{+}{N}H_3$
+
H
C = O
H
Cl
N=CH2
$H_3\overset{+}{N}$
$\overset{+}{N}H_3$
Cl
(Colorless)
+ H+ + H2O
HSO_3^-
NCH2SO3H
$H_3\overset{+}{N}$
$\overset{+}{N}H_3$
(Pink)
+ Cl-

Scheme II.

control module. A calibration curve was prepared from 0.1 to 1.6 ppm of formaldehyde (Figure 11, curve A).

Because the analog signal is in % T, the resulting curve for % T versus formaldehyde concentration is exponential. This curve can be converted to the more common absorbance versus concentration form from the relationship $A = \log(1/T)$ where A is absorbance and T is absolute temperature (Figure 11, curve B). Our curve gives a correlation coefficient of 0.999. We have linearity from 0.1 to 1.6 ppm of formaldehyde. The data used to generate our curve have a percent relative standard deviation for five calibration curve determinations ranging from 0.9 to 3.1.

We measured the repeatability of the method by determining the formaldehyde levels in known solutions. In 10 determinations 0.53 ± 0.02 ppm and 1.31 ± 0.03 ppm were measured for 0.5- and 1.3-ppm solutions, respectively. These are recoveries of 106% and 100%, respectively.

We compared the results obtained by the robot to those obtained manually with a Beckman Acta III UV–vis spectrophotometer. The results for 17 samples are shown in Figure 12. These data have a correlation coefficient of 0.979.

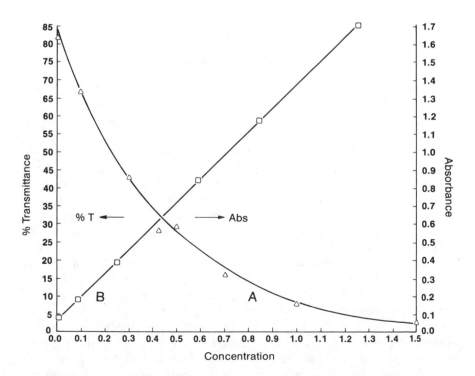

Figure 11. Calibration curve obtained for formaldehyde via the automated pararosaniline procedure: A, percent transmittance vs. formaldehyde concentration; and B, absorbance vs. formaldehyde concentration.

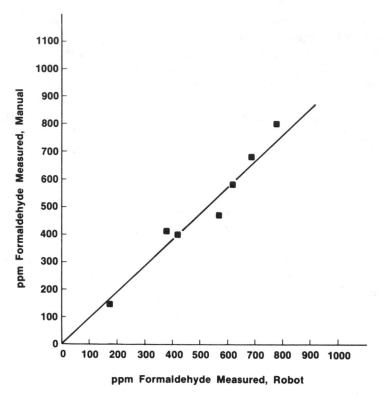

Figure 12. Correlation of the results obtained for formaldehyde by the manual pararosaniline method vs. the automated pararosaniline method.

The mixing step following the sample dilution is not needed if the disposable tip is placed under the surface of the water during the addition of the concentrated sample, and the diluted sample is drawn into the tip at least three times. This rinsing action creates sufficient swirling to mix the sample thoroughly. The fiber-optic probe could be rinsed sufficiently to prevent cross-contamination by dipping it in the rinse water in a test tube and then placing it under a stream of nitrogen. The rinse water in the test tube needs to be replaced after approximately 25 samples. The sample containers should be capped during storage in the sample rack. Capping was easily accomplished by using 23-mL headspace vials with crimped septum caps. A hole was drilled in each septum large enough to allow the disposable pipette tip to pass through it. This procedure effectively forms a washer. A piece of plastic wrap is placed over the mouth of the vial, the washer is placed on top of the plastic wrap, and the metal cap is crimped onto the vial. The crimping action pulls the plastic wrap tightly across the vial, and the washer provides support. When the syringe hand with the disposable tip is brought down into the sample vial it is able to puncture the plastic wrap to allow the sample to be withdrawn.

Comparison of the Data Obtained by the Automated Pararosaniline Colorimetric and Polarographic Methods. Figure 13 shows a comparison of the results obtained on a series of samples by the polarographic method and by the automated colorimetric method. The results obtained for the determination of formaldehyde in glycolic acid were interesting because the color of the resulting pararosaniline solution was not the expected pink. We assumed that interference with the colorimetric method occurred. However, the results obtained on this sample by the colorimetric method agreed very well with those obtained by the polarographic method.

Headspace Gas Chromatographic Method. We decided to test the applicability of a Photovac GC for the determination of formaldehyde directly in air. Gas chromatographic techniques are not applicable to the di-

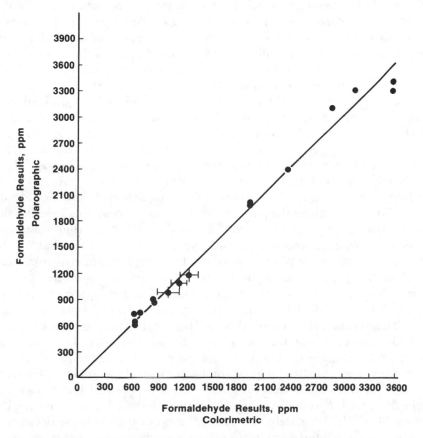

Figure 13. Correlation of formaldehyde determined by the polarographic method vs. the automated pararosaniline technique.

rect determination of trace levels of formaldehyde because of the inherent insensitivity of the thermal conductivity detector and poor response of the flame ionization detector to formaldehyde. The Photovac has a photoionization detector that our previous experience has shown is very sensitive to hydrocarbons in the atmosphere.

A gaseous standard of formaldehyde was prepared by placing a known concentration of formalin in a headspace vial that was then sealed and equilibrated at 50 °C. The headspace vapors over the formaldehyde solution were injected into the Photovac by connecting the inlet of the Photovac to the outlet of the injector on our Perkin–Elmer–42 headspace GC. When the headspace over the formalin solutions was injected, we obtained a peak at 11.7 min that was proportional to the formalin concentration. This peak was tentatively identified as formaldehyde and gave us a detection limit of 0.5 ppm of formaldehyde in water.

This method, although developed for direct injection of gaseous samples, could also be used to analyze aqueous samples by using conventional headspace gas chromatographic techniques. We decided to apply it to samples that had already been analyzed by the polarographic and pararosaniline methods. The results obtained by the headspace method were lower than those obtained by the other two methods by a factor of 40. We thought that this difference might be due to a matrix effect, and that the partitioning of the formaldehyde from the sample matrix might be different than from the aqueous standard used to calibrate the procedure. Any matrix effects should be eliminated by the use of a standard addition technique. However, results obtained by standard addition still did not agree with the other results. Addition of pararosaniline to the headspace vial removed the peak tentatively identified as formaldehyde. However, addition of hydrogen peroxide, which reacts with formaldehyde to produce formic acid, did not eliminate the peak. An increase in the peak that matches the retention time of formic acid was observed. Similarly, sodium bisulfite, which forms a complex with formaldehyde, did not affect the peak identified as formaldehyde. We also tested a solution of formaldehyde prepared by dissolving paraformaldehyde in water at elevated temperature. A sharp formaldehyde odor was noted, but injection of the headspace over this solution did not produce the peak at the expected retention time.

These observations have led us to conclude that the peak is not formaldehyde, but is a species that can react with pararosaniline. It is present in formalin solutions but not in paraformaldehyde. When sodium sulfate was added to the solution, the response of this peak was increased by a factor of 6. This result indicates that the compound is very polar. Using retention time data, we have eliminated acetaldehyde, methanol, and formic acid as possibilities. Unfortunately, the configuration of the Photovac detector is not amenable to interfacing to our mass spectrometer, and no further identification attempts were made. However, these results raise questions concerning the validity of the pararosaniline calibration curve obtained with formalin solutions.

Conclusions

We would choose to use the polarographic method when the number of samples is small, or when little is known about the expected concentration range or sample matrix, because the analysis is rapid (less than 10 min per sample) and analysis conditions are easy to change. If we have a large number of samples or if we know what concentration range to expect, we generally use our automated pararosaniline method. Using the robot, we can analyze a batch of samples in minutes.

These bulk methods are very useful for the analysis of routine samples for which the matrix is well understood. However, in unknown samples with complex matrices, interferences that might not be recognized may be encountered, and incorrect results could be obtained. Analysis by more than one analytical technique is suggested for any new sample types.

Literature Cited

1. Kolthoff, I. M.; Lingane, J. J. "Polarography"; Interscience: New York, 1952; Vol. 2, pp. 652–61.
2. Miksch, R. R.; Anthon, D. W.; Fanning, L. Z.; Hollowell, C. D.; Revzan, K.; Glanville, J. *Anal. Chem.* **1981**, *53*, 2118.

RECEIVED for review September 28, 1984. ACCEPTED January 2, 1985.

Method Development for the Determination of Trace Levels of Formaldehyde in Polluted Waters

JOHANNES T. GRAVEN, MAURIZIO F. GIABBAI, and
FREDERICK G. POHLAND
Environmental Engineering Program, School of Civil Engineering, Georgia
Institute of Technology, Atlanta, GA 30332

An analytical methodology was developed for the determination of trace amounts of formaldehyde in complex aqueous solutions. By exploiting the reaction specificity of 2,4-dinitrophenylhydrazine for carbonyl functional groups, formaldehyde was converted to a hydrophobic derivative, 2,4-dinitrophenylhydrazone, to enhance its isolation. Moreover, the use of glass capillary columns proved effective in separating the C_1–C_2 aldehyde homolog derivatives. With this method, formaldehyde concentrations in leachate samples generated from simulated landfills containing municipal solid wastes and urea–formaldehyde foam as an intermediate cover material were monitored. The detection limit for formaldehyde in these samples was estimated to be 100 µg/L, and the average recovery for the derivative was 55%. The formaldehyde detected in the leachate samples (3–40 mg/L) was determined not to adversely affect the microbially mediated processes of stabilization within the landfills, and as stabilization progressed, formaldehyde was converted and reduced in concentration.

IN THE OVERALL PLANNING, DESIGN, AND OPERATION of sanitary landfills, requirements for daily, intermediate, and final cover often impose a costly and problematic element. Although soil is the most frequently used material for intermediate and final cover, the use of alternative cover materials may be more advantageous and economical in many applications. Urea–formaldehyde foam has gained recent popularity as an alternative cover material because of its ease and homogeneity of application, its ability to increase otherwise unavailable landfill capacity and extend service life, and its frequent economic advantage over alternative cover methods. Although efforts to determine the general applicability of urea–formaldehyde foam to landfilling operations appear promising (1, 2), they have not ade-

0065–2393/85/0210/0043$06.00/0

quately addressed the relatively long-term issues of the impacts of the foam and its leachable constituents (e.g., formaldehyde) on the landfill environment and vice versa.

Because of some evidence of carcinogenicity of formaldehyde in experimental animals, as well as numerous health-related complaints concerning the release of formaldehyde from consumer products (3, 4), public and regulatory concern over releases of formaldehyde to the environment have increased (5). Although most of the concerns are related to indoor atmospheric releases, an increased focus on potential releases to the aqueous environment, including the leaching of formaldehyde from landfills, has resulted. Furthermore, because of its toxicity to lower animals and microorganisms (6), formaldehyde leached from landfills may have an adverse impact on the microbial processes within the landfill. Hence, the leaching of formaldehyde during and after landfilling operations has been an unresolved issue that, in the case of the associated acceptability of foam applications, was in need of resolution within a process-assessment as well as environmental-health and regulatory perspective.

In response to the need for reliable measurement of formaldehyde in complex aqueous solutions such as landfill leachate, several candidate analytical methods were evaluated. These included a chromotropic acid method (7); a direct aqueous injection, packed column, gas chromatographic (GC) method; and a direct aqueous injection, glass capillary column, gas chromatographic–mass spectroscopic (GC–MS) method. All of these methods were found to have limitations that ultimately restricted their use, particularly for the determination of trace quantities of formaldehyde in leachate samples originating from the landfill disposal of solid wastes. However, further investigation revealed that derivatization techniques prior to GC or high-pressure liquid chromatography (HPLC) analysis could be used successfully for the determination of trace amounts of aldehydes and other carbonyl compounds. Mansfield et al. (8) used the formation of 2,4-dinitrophenylhydrazone (2,4-DNPH) from the reaction of formaldehyde with 2,4-dinitrophenylhydrazine (2,4-DNP) to detect formaldehyde in tobacco smoke. Fung and Grosjean (9), as well as Selim (10), have similarly used this derivatization to separate and quantify nanogram amounts of carbonyl compounds.

Because none of the investigators applied the 2,4-DNP derivatization technique to complex aqueous solutions, this study included an emphasis on the feasibility of applying this approach for the specific determination and quantification of formaldehyde to polluted waters such as sanitary landfill leachates. Therefore, the analytical method thus established was used for monitoring formaldehyde in leachate samples from simulated landfill cells and to assess the relative suitability and applicability of urea–formaldehyde foam as an alternative cover material.

Experimental

Analytical Reagent Preparation. Concentrated formaldehyde solutions were purchased from Fisher Scientific, and the reagent solution was prepared by adding 0.25 g of 2,4-DNP (Eastman Kodak) to 100 mL of 6 N hydrochloric acid solution (10). The "organic-free" water used in reagent preparation and throughout these experiments was prepared by passing tap water through a series of treatment steps composed of a Millipore 360 activated carbon cartridge (Continental Water Systems); a Millipore 300 deionizer cartridge; a glass column (60 × 3 cm i.d.) packed with 50 g of 16 × 30 mesh activated carbon (Calgon Filtrasorb F–400); and two modified UV disinfection modules (Model H–40, Ultraviolet Technology) supplemented with hydrogen peroxide. The formaldehyde–2,4-DNPH reference was prepared by standard procedures (11) and purified by recrystallization from ethanol.

Derivatization and Extraction Procedure for Leachate Samples. To 100 mL of leachate, 5 mL of 2,4-DNP reagent solution was added; 1 mL of reagent was used for the formaldehyde-spiked reference samples prepared in organic-free water. The mixtures were stirred for 5 min, then 50 mL of methylene chloride (Burdick and Jackson) was added to the mixture, and stirring was continued for 1.5 h. The two-phase mixture was then separated in a separatory funnel. Two additional extractions with 50-mL aliquots of methylene chloride were performed, and the organic solvent was finally concentrated to approximately 4 mL in a Kuderna–Danish apparatus at 70 °C and then concentrated to 1.0 mL by evaporation with nitrogen.

Instrumentation. A Hewlett–Packard 5830A GC equipped with a split–splitless capillary injection system and a flame ionization detector (FID) was used for derivative quantification. The separation was performed on a glass capillary column (30 m × 0.25 mm i.d.) which was deactivated according to the procedure proposed by Grob (12) and coated by the static method (13) with SE–54 silicone gum phase (ca. 0.2 μm film thickness). The GC operating conditions were the following: injector temperature, 250 °C; transfer line temperature, 285 °C; splitless injection; and volume injected, 1 μL.

Confirmation of derivative formation was accomplished with a Finnigan 4023 mass spectrometer (MS) interfaced with a Hewlett–Packard 5830A GC, as described in Reference 14. The MS operating conditions were the following: ionization mode, electron impact; electron multiplier, 1500 V; electron energy, 70 eV; emission current, 0.5 mA; mass range, 40–450 amu; and scan rate, 1.0 s/decade. The GC conditions were identical to those employed in the GC analysis. The MS was calibrated with perfluorotributylamine, and a solution of decafluorotriphenylphosphine (1 μL) was subsequently injected onto the chromatograph to verify the calibration thus obtained.

Simulated Landfill Cell Construction and Operation. To provide an opportunity to assess the potential impact of formaldehyde that could be leached from urea–formaldehyde foam during its use as a cover material during landfilling operations, two simulated landfill cells with the necessary appurtenances to permit leachate and gas collection for a single-pass cell, as well as leachate recycle for a recycle cell, were constructed as shown in Figure 1. The single-pass cell was intended to simulate the impact of rainfall-induced leaching events during conventional landfill operations; the recycle cell was intended to simulate conditions under which leachate is formed, contained, collected, and recycled. Thus, the landfill was used as an in situ leachate treatment system as well as to accelerate microbially mediated stabilization processes. In both cells, a 5-cm intermediate layer of urea–formaldehyde foam was placed between two layers of shredded residential-type

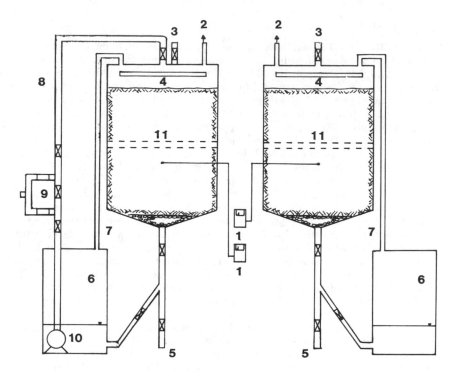

Figure 1. Schematic of simulated landfill cells: left, recycle cell; right, single-pass cell. Key: 1, thermocouple–temperature recorder; 2, to gas meter; 3,
water addition port; 4, flow distribution plate; 5, leachate drain pipe–sampling port; 6, leachate reservoir; 7, leachate reservoir gas line; 8, leachate
recycle line; 9, pH–ORP measuring loop; 10, leachate recycle pump; and 11,
urea–formaldehyde foam layer.

solid waste. A total of 60 kg (39 kg dry weight) of solid waste and approximately
2.89 g of foam/dry kg of solid waste were placed in each cell. In an actual landfill,
the amount of foam could typically range from 1–3 g of foam/dry kg of solid waste,
depending on the operational and management techniques employed (i.e., cell
depth and thickness of foam).

After the cells were sealed from the atmosphere, moisture was added to initially bring them to field capacity and to subsequently simulate rainfall events.
Accumulated leachate was collected and recycled in the recycle cell and collected
and removed to storage from the single-pass cell. Leachate and gas samples were
collected at periodic intervals and analyzed to determine the progress of landfill
stabilization processes within the cells. Leachate analyses included pH, oxidation-
reduction potential (ORP), conductivity, alkalinity, volatile organic acids, chemical oxygen demand (COD), 5-day biochemical oxygen demand (BOD_5), total organic carbon (TOC), sulfides, and selected metals. In addition, the volume and
characteristics of gases generated (CH_4, CO_2) and temperatures were monitored on
a daily basis for both cells.

To determine if the formaldehyde detected in landfill leachate was attributable only to the foam or also to other constituents in the solid waste, two additional

bench-scale simulated landfill cells were constructed. In one cell, a 2.5-cm layer of foam was placed between two layers of shredded solid waste; only shredded solid waste was placed in the other cell. A total of 5.5 kg (3.9 kg dry weight) of solid waste was placed and compacted in each cell. Subsequent to sealing of the cells to the atmosphere and initial moisture addition to bring the cells to field capacity, moisture addition was provided to ensure that sufficient volumes of leachate would be generated for formaldehyde determinations at periodic intervals. To document that the environmental conditions within each of these cells remained relatively similar and representative of actual leaching conditions, additional analyses indicative of landfill stabilization processes were performed on selected samples.

Results and Discussion

Analytical Methodology. The formaldehyde–2,4-DNPH prepared and purified by recrystallization from ethanol was used for the development of the GC–MS method. A preliminary verification of the formation, extraction, and chromatographic analysis of trace amounts of the formaldehyde–2,4-DNPH was performed with 100-mL organic-free water solutions and leachate samples spiked with formaldehyde to approximately 500 μg/L. The concentrated organic extracts from these samples were analyzed by GC and GC–MS.

Typical reconstructed ion chromatograms (RIC) of the extracts are illustrated in Figures 2 and 3, and the mass spectrum of the derivative is presented in Figure 4.

Although potential temperature or catalytic degradation of this derivative during instrumental analysis, particularly by GC (*15*), has led investigators to develop and propose HPLC methods (*8–10*), the literature is essentially devoid of reports advancing the use of capillary columns for GC analysis of this derivative. Therefore, the need for high-resolution and on-line MS confirmation of trace amounts of this derivative in the complex samples of this study led to the development of suitable capillary columns for GC–MS analysis. As shown by Figures 2 and 3, nonpolar glass capillary columns deactivated by the persilylation method (*12*) provided the required inertness, temperature stability, and resolution for GC analyses of the formaldehyde–2,4-DNPH derivative. From GC–FID analysis, a lower detection limit of 5 ng was estimated, and good linearity was observed over a concentration range of 5–200 ng/μL. Furthermore, the use of a fused-silica, direct-transfer line between the GC column and ionization source of the MS proved to be satisfactory for preserving chemical integrity and thus allowed structural identification of the derivative.

Little information is available concerning the electron-impact mass spectra of the carbonyl 2,4-DNPHs of aldehydes possessing a straight-chain alkyl group of three or fewer carbon atoms. Actually, these derivatives were reported to have no intense fragment peaks that are readily associated with a particular structural feature (*16*). However, the presence of an intense molecular ion (i.e., m/e = 210) and minor fragments corresponding

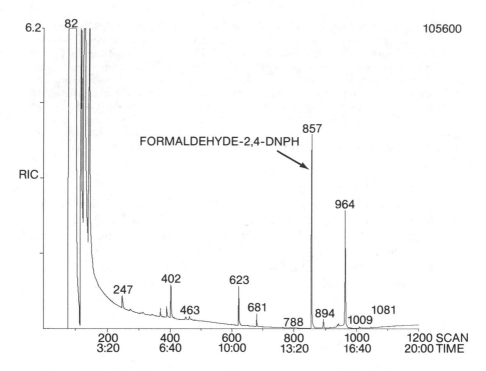

Figure 2. RIC of organic-free water extract of formaldehyde–2,4-DNPH. The GC and MS operating conditions are described under "Instrumentation," and the temperature program was 40 °C (2.0 min) at 15 °C/min to 290 °C (5 min).

to the elimination of NO (i.e., m/e = 180) and NO_2 (i.e., m/e = 164), which are typical of 2,4-DNPH (*16*), was considered supportive evidence for the identification of the formaldehyde–2,4-DNPH derivative (*see* Figure 4). No structural explanation was determined for the intense peaks in the lower mass range (i.e., m/e = 122, 79, 63, and 51).

In conjunction with the analyses of formaldehyde-spiked samples, blank determinations with organic-free water as well as with the distilled water added to the simulated landfill cells were conducted. No trace of the formaldehyde–2,4-DNPH derivative was detected in any of these samples. Moreover, solutions spiked with C_1–C_2 aliphatic aldehydes confirmed that adequate separation of these derivatives could be achieved as shown by the RIC in Figure 5 and the acetaldehyde–2,4-DNPH derivative mass spectrum in Figure 6. As with the formaldehyde derivative, this mass spectrum presented an intense molecular ion (i.e., m/e = 224) and the same intense peaks in the lower mass range (i.e., m/e = 122, 79, 63, and 51). The major chromatographic peak that is present in the total ion trace, particularly in

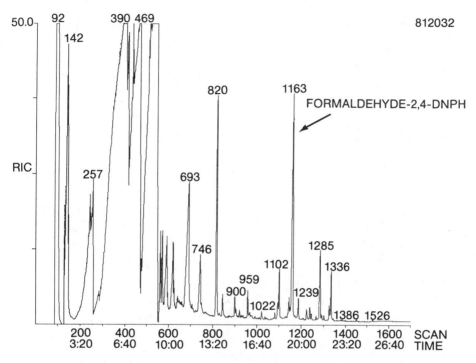

Figure 3. RIC of leachate extract containing formaldehyde–2,4-DNPH. The GC and MS operating conditions are described under "Instrumentation," and the temperature program was 40 °C (2.0 min), 15 °C/min (6 min), and 8 °C/min to 290 °C (5 min).

Figure 5, is the excess 2,4-DNP reagent that is partially coextracted from the water solution.

The derivatization was carried out in essentially the same manner as that described by Selim (*10*) in which a two-phased reaction system was used to improve reaction yields at trace levels of carbonyl compounds. Even though Selim had demonstrated that total conversion of propionaldehyde occurred within 30 min, the reaction time in these studies was extended to 1.5 h to ensure that all the formaldehyde present in solution could be converted to the 2,4-DNPH derivative. The reproducibility of the derivatization was verified for both organic-free water solutions spiked with known amounts for formaldehyde as well as leachate samples that already contained formaldehyde. These results are presented in Table I. Initial method reproducibility tests for leachate samples revealed a large variation in results. This variation was primarily attributed to the presence of high concentrations of short-chain fatty acids (ca. 10 g/L) and to other organics leached from the solid waste and causing formation of thick emul-

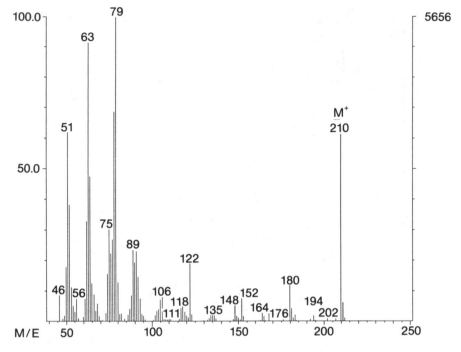

Figure 4. Mass spectrum of formaldehyde–2,4-DNPH.

sions during separatory funnel extraction. Attempts to overcome this problem included the addition of saturated calcium chloride solutions and the use of a vapor–vapor solvent extraction apparatus (*17*). Both alternatives failed to improve the results because the electrolyte solution did not "break" the emulsions, and steam distillation extraction gave very poor recoveries of the derivative. This emulsion formation attributed to the presence of short-chain fatty acids was only of concern for leachate samples collected during the early stages of landfill stabilization (i.e., acid fermentation phase) when these acids were generated in high concentrations as a result of the conversion of readily biodegradable organics present in the landfill (*18*).

To resolve this problem, smaller aliquots of leachate (i.e., 25 and 50 mL) were used and diluted with distilled water prior to analysis, to minimize emulsion interferences. Leachate samples collected during the subsequent phase of landfill stabilization (i.e., methane fermentation phase) contained considerably lower concentrations of short-chain fatty acids, and emulsion formation was minimal. The results presented in Table I were obtained with leachate samples collected during the early stages of landfill stabilization when high volatile acid concentrations prevailed.

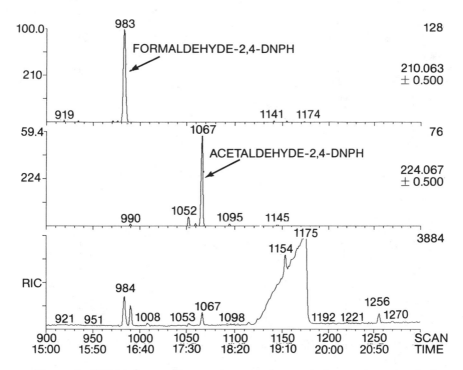

Figure 5. RIC and mass fragmentogram of organic-free water extract of formaldehyde–2,4-DNPH and acetaldehyde–2,4-DNPH (same conditions as Figure 2).

A recovery study was also conducted to determine the effectiveness of the solvent extraction and concentration procedure once the formaldehyde–2,4-DNPH was formed. Both organic-free water solutions and leachate samples were spiked with known amounts of the derivative (490 and 450 μg/L, respectively) and solvent extracted with methylene chloride. The leachate samples, which already contained formaldehyde, were first subjected to derivatization with 2,4-DNP followed by the addition of a known amount of derivative prior to solvent extraction. These results are presented in Table II.

The level for recovery attained for the formaldehyde–2,4-DNPH derivative from leachate samples was considered to be satisfactory because recoveries of micro amounts of organic priority pollutants in environmental samples of similar matrix complexity (i.e., raw sewage and sludge) were reported to range between 30 and 100% (*17*). Therefore, the detection limit of the analytical method for formaldehyde in leachate samples was estimated to be 100 μg/L or better. The unavailability of formaldehyde solution of sufficiently accurate concentration prevented the evaluation of

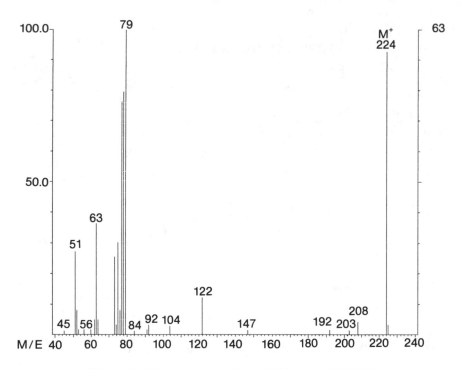

Figure 6. Mass spectrum of acetaldehyde–2,4-DNPH.

Table I. Reproducibility of Results for the Derivatization of Formaldehyde–2,4-DNPH in Organic-Free Water Solutions and Leachate Samples

Sample	Amount of Formaldehyde–2,4-DNPH (ng)	Mean (ng)	Standard Deviation (ng)	Coefficient of Variation (%)
Organic-free water[a]				
W–1	86.3			
W–2	99.4	93.5	6.0	6.4
W–3	94.7			
Leachate[b]				
L–1	125.4			
L–2	128.4	132.5	7.5	5.7
L–3	142.7			

[a] Samples consisted of organic-free water spiked with formaldehyde and were subsequently divided into three aliquots.
[b] Samples consisted of a leachate sample diluted 50 % with distilled water and were subsequently divided into three aliquots.

Table II.   Recovery Studies for Formaldehyde–2,4-DNPH in Organic-Free Water
Solutions and Leachate Samples

Sample	Initial Amount (ng)	Amount Spiked (ng)	Amount Recovered (ng)[a]	Recovery (%)
Organic-free water	0.0	49.0	42.6 ± 4.4	86.9
Leachate	131.1 ± 6.1[a]	45.0	156.0 ± 4.4	55.3

[a] Mean ± standard deviation.

derivatization efficiencies in leachate samples. Hence, in this study a 100%
derivatization efficiency was assumed because a two-phase system that favored the formation of the reaction product by continuous removal of the
2,4-DNPH from the aqueous phase and a longer reaction time than had
been reported as being necessary for the complete conversion of propionaldehyde to its 2,4-DNPH (*10*) were used.

**Presence and Fate of Formaldehyde Within the Simulated Landfill
Cells.**   After the development of the DNPH derivatization method, a
monitoring program for formaldehyde in leachate samples from the simulated landfill cells was initiated and implemented during the duration of
the study. This program permitted some definitive conclusions regarding
the presence and fate of formaldehyde over the period of investigation. Under the conditions of the experiments simulating conventional landfill management both with and without recycle, the total mass of formaldehyde
removed from each of the simulated landfill cells containing 2.8 g of foam/
dry kg of solid waste was determined to be 16.5 and 18.4 mg/g of foam,
respectively. Some of the formaldehyde (3 mg/dry kg of solid waste) detected in the leachate from the simulated landfill cells originated from the
solid waste itself. This result was not considered to be particularly unusual
or unexpected because of the widespread use of products containing form-
aldehyde-based resins (e.g., pressed wood products, insulation, paper, fabric, and carpet) (*4, 6, 19*), many of which may eventually be disposed of in
sanitary landfills. Because formaldehyde is extremely soluble in water, it
could be readily extracted from these materials as field capacity was
reached and leachate was generated within the landfill. With similar re-
movals of formaldehyde from solid waste within actual landfills, approximately 25% of the total mass of formaldehyde leached from conventional
landfill cells covered with an equivalent amount of foam could originate
from the solid waste.

Formaldehyde and other possible constituents leached from the foam
in the simulated landfill cells did not preclude anaerobic microbial production of methane during landfill stabilization. This result also was anticipated because the maximum concentrations of formaldehyde of 28.6 and
41.8 mg/L measured during this study in the leachates of the single-pass

and recycle cells, respectively, were much less than those that have been reported as being inhibitory to anaerobic degradation processes (20).

The formaldehyde concentrations in the leachate from the recycle simulated landfill decreased from a high of 41.8 mg/L to less than 3 mg/L. This decrease was primarily attributed to microbially mediated conversion of the formaldehyde within the landfill cell because the leachate was contained within the cell during this period and no "wash-out" of formaldehyde was permitted. In addition, volatilization during recycle was not considered to be significant because of the high solubility of formaldehyde in water (40 g/100 mL of water at 20 °C) (7). Furthermore, the decrease in formaldehyde concentrations of the leachate coincided with the period of accelerated methanogenic stabilization of the landfill where the formaldehyde was present as one of many available substrates (21).

On the basis of these findings, urea–formaldehyde foam used as an alternative to daily landfill cover did not adversely impact landfill stabilization, and associated concerns over adverse health and environmental impacts from the possible release of formaldehyde containing leachate to the environment were concomitantly reduced. Because of the vastness of the hydrogeological setting, such in situ treatment protocol could not only be used to reduce the potential for migration, but also to render formaldehyde concentrations to levels well below those reported for many food products (22).

Conclusions

On the basis of the evaluations conducted during this study, it was concluded that derivatization of formaldehyde in aqueous solution to formaldehyde–2,4-DNPH, followed by extraction and concentration of the organic extract and subsequent analysis by glass capillary column, GC–FID, and GC–MS, is a viable method for the determination and quantification of formaldehyde in leachates and other similarly contaminated and polluted waters. In addition, urea–formaldehyde foam may be used as an alternative to daily landfill cover without posing adverse environmental impacts attributable to the release of formaldehyde from the foam during accelerated landfill stabilization processes.

Acknowledgments

This research project was sponsored in part by research grant E–20–H01 from SaniFoam, Inc., Costa Mesa, Calif.

Literature Cited

1. Allan, G. G. "Environmental Impact of the Use of Plastic Foam as a Daily and Interim Landfill Cover"; Univ. of Washington Press: Seattle, 1980.
2. Firstman, S. I.; Pohland, F. G. "Operational Test of Sani-Blanket Daily and Intermediate Plastic Foam Landfill Cover"; Georgia Institute of Technology: Atlanta, 1981.

3. National Academy of Sciences, "Formaldehyde and Its Health Effects"; Government Printing Office: Washington, 1980.
4. National Research Council, "Formaldehyde and Aldehydes"; Government Printing Office: Washington, 1981.
5. Hileman, B. *Environ. Sci. Technol.* **1982**, *16*, 543A–47A.
6. Environmental Protection Agency, "Investigation of Selected Potential Environmental Contaminants: Formaldehyde"; Government Printing Office: Washington, 1976; EPA Report 560/2–76–009.
7. Walker, J. F. "Formaldehyde"; Reinhold: New York, 1964; p. 368.
8. Mansfield, C. T.; Hodge, B. T.; Hege, R. B.; Hamlin, W. C. *J. Chromatogr. Sci.* **1977**, *15*, 301–2.
9. Fung, K.; Grosjean, D. *Anal. Chem.* **1981**, *53*, 168–71.
10. Selim, S. *J. Chromatogr.* **1977**, *136*, 271–77.
11. Shriner, R. L. "The Systematic Identification of Organic Compounds"; John Wiley: New York, 1980; p. 263.
12. Grob, K. *J. High Resolut. Chromatogr. Chromatogr. Commun.* **1978**, *3*, 493–96.
13. Giabbai, M. F.; Shoults, M.; Bertsch, W. *J. High Resolut. Chromatogr. Chromatogr. Commun.* **1978**, *1*, 277.
14. Giabbai, M. F.; Roland, L.; Chian, E. S. K. In "Chromatography in Biochemistry Medicine and Environmental Research"; Figerio, A., Ed.; Elsevier Science: Amsterdam, 1983; pp. 41–52.
15. Papa, L. J.; Turner, L. P. *J. Chromatogr. Sci.* **1972**, *10*, 744–47.
16. Budwikiewicz, H.; Djerassi, C.; Dudley, H. W. "Mass Spectrometry of Organic Compounds"; Holden Day: San Francisco, 1967; pp. 399–405.
17. Environmental Protection Agency, "Presence of Priority Pollutants in Sewage and Their Removal in Sewage Treatment Plants," by DeWalle, F. G.; Chian, E. S. K.; Kalman, D. A.; Giabbai, M. F.; Cross, W. H.; Government Printing Office: Washington, 1981; Final Report, EPA Grant R806102.
18. Pohland, F. G. "The Use of Plastic Foam as a Cover Material During Landfilling of Solid Wastes"; Georgia Institute of Technology: Atlanta, 1983; Final Report E20–H01, pp. 42–43.
19. Pickrell, A.; Mockler, R. V.; Griffis, L. C.; Hobbs, C. H. *Environ. Sci. Technol.* **1983**, *17*, 753–57.
20. Parkin, G. F.; Speece, R. E.; Yang, C. H. J.; Kocher, W. M. *J. Water Pollut. Control Fed.* **1983**, *55*, 44–53.
21. Wolfe, R. S. *Adv. Microb. Physiol.* **1971**, *6*, 106–46.
22. Lawrence, J. F.; Iyengar, J. R. *Int. J. Environ. Anal. Chem.* **1983**, *15*, 47–52.

RECEIVED for review September 28, 1984. ACCEPTED December 26, 1984.

Quantitative Analysis of Formaldehyde Condensates in the Vapor State

DAVID F. UTTERBACK[1], AVRAM GOLD[2], and DAVID S. MILLINGTON[3]

[1]Department of Health Science, California State University—Fresno, Fresno, CA 93740
[2]Department of Environmental Sciences and Engineering, University of North Carolina, Chapel Hill, NC 27514
[3]Division of Genetics and Metabolism, Duke University Medical Center, Durham, NC 27710

Quantitative analysis of vapor phase formaldehyde condensates with water and methanol was performed by reaction with N,O-bis-(trimethylsilyl)trifluoroacetamide to form trimethylsilyl derivatives of the condensates, which were then analyzed by capillary gas chromatography with flame ionization detection. Methylal, methylene glycol, and a series of poly(oxymethylene) glycol monomethyl ethers accounted for the total formaldehyde content in vapors at equilibrium with a formalin solution. Structures of the formaldehyde species were confirmed by capillary gas chromatography-chemical ionization mass spectrometry with ammonia as the reactant gas.

FORMALDEHYDE IS A HIGHLY REACTIVE COMPOUND that equilibrates between many molecular forms in aqueous solutions. The addition of methanol to formaldehyde solutions as a stabilizer further increases the number of molecular species in which formaldehyde may exist (*1–3*), although molecular or monomeric formaldehyde is present only at very low concentrations (*1*). Many of the formaldehyde condensates with water and methanol have significant vapor pressure that results in their evolution into the gas phase, yet the molecular state of formaldehyde as it exists in the gas phase has not been verified.

Formation of formaldehyde condensates, that is, poly(oxymethylene) glycols and the glycol monomethyl ethers in formalin solution, are represented by the following equilibria:

0065–2393/85/0210/0057$06.00/0

$$CH_2O + H_2O \rightleftharpoons CH_2(OH)_2$$

$$CH_2(OH)_2 + CH_2O \rightleftharpoons HO(CH_2O)_2H$$

$$HO(CH_2O)_nH + CH_2O \rightleftharpoons HO(CH_2O)_{n+1}H$$

Scheme I

$$HO(CH_2O)_nH + CH_3OH \rightleftharpoons HO(CH_2O)_nOCH_3$$

$$CH_2O + 2 CH_3OH \rightleftharpoons CH_3OCH_2OCH_3$$

Scheme II

Both oligomeric series, the poly(oxymethylene) glycols (Scheme I) and the poly(oxymethylene) glycol monomethyl ethers (Scheme II), are thermally unstable and decompose readily upon separation by gas chromatography (3). Trimethylsilylation of the hydroxyl groups of the oligomers in solution by N,O-bis(trimethylsilyl)trifluoroacetamide (BSTFA) allows separation, identification, and quantitation of the respective trimethylsilyl (TMS) ethers through combined analysis by capillary gas chromatography–chemical ionization mass spectrometry (GC–CIMS) and gas chromatography (GC) with flame ionization detection (FID) (2).

Indications of the presence of methylene glycol and low molecular weight oligomeric forms of formaldehyde in the gas phase appear in the literature. Hall and Piret (4) determined an expression for the equilibrium constant, K_p, over the temperature range of 40 to 160 °C for a ternary gas phase system of water, formaldehyde, and methylene glycol: $\log(K_p) = (3200/T) + 9.8$, where T is temperature. On the basis of vapor pressure data, Hall and Piret predicted that at 10 °C above its boiling point, methylene glycol would be 95% dissociated into water and formaldehyde. Hence, formaldehyde may exist in the hydrated form even at 106 °C. Their calculations assumed that poly(oxymethylene) glycols were not present in significant concentrations, although no experimental evidence supported this assumption.

By indirect methods, Iliceto (5) investigated the formation of the dimeric species, oxydimethylene glycol, from formaldehyde and methylene glycol and calculated the enthalpy of formation to be -11.6 kcal/mol in the gas phase.

This equilibrium has been supported by the work of Bryant and Thompson (6). Using techniques similar to those of Iliceto, they predicted an exothermic reaction for the formation of both methylene glycol and oxydimethylene glycol in the gas phase and also calculated the Gibbs free energies and the equilibrium constants for the reaction. The thermodynamic data are consistent with those reported by Iliceto and by Hall and Piret.

Sawicki and Sawicki (7) assert that the 2–4% water present in para-formaldehyde causes repolymerization of formaldehyde generated in a gas stream that has passed over a paraformaldehyde permeation tube. Giesling et al. (8) also found that the decomposition of paraformaldehyde involves emission of water along with formic acid and methyl formate, and Walker (9), citing similar results, adds that the presence of these products in the gas phase of formaldehyde promotes rapid polymerization at temperatures below 100 °C. Schnizer et al. (10) reported that the yield of monomeric form-aldehyde from acid-catalyzed depolymerization of trioxane was limited to 89% because of deposition of solid paraformaldehyde on the surfaces of the generating apparatus. This deposition was said to be catalyzed by the presence of water. In all these investigations, solid paraformaldehyde, the polymeric form of formaldehyde, was observed in the gas phase emissions from a formaldehyde source. Short- and intermediate-length oligomers of formaldehyde must be involved in the gas phase equilibrium formation of the long-chain polymers from the monomer. A method was developed for the detection and quantitative analysis of these gas phase formaldehyde condensates.

Experimental

Analysis was performed on a Varian model 3700 capillary gas chromatograph equipped with an FID. Separation was accomplished with a 30-m DB-5 (methyl phenyl silicone) fused silica capillary column. Helium carrier gas flow rate was 1.5 mL/min, and a Grob type split–splitless injector was operated at a 10:1 split ratio. The injector was heated to 250 °C, and the detector was heated to 300 °C. After a 6-min isothermal period at 30 °C, column temperature was programmed to rise 6 °C/min to a limit of 230 °C. Detector output was recorded by a Hewlett Packard model HP3380 integrator. Previously reported retention times (2) for oligomers on the same chromatographic system were used for identification of the silylated oligomers.

Headspace derivatization was performed by injecting 5:1::BSTFA:DMF through a serum stopper into an evacuated reaction vial (70 mL). Headspace above a reagent grade formalin solution (Fisher Scientific) at 24 °C in a 2-L flask equipped with a hypodermic needle (Figure 1) was then bled into the reaction vial, and the pressure was equalized with atmospheric pressure. Quantitative analysis was performed on both the liquid and gas phases in the reaction vial by GC–FID. A derivatized gas phase aliquot (100 μL) was taken for analysis directly from the reaction vial with a gas-tight syringe. The liquid phase from a separate derivatization reaction was transferred by pipette to a vial equipped with a screw-top septum cap and a Teflon-lined septum from which an aliquot (1 μL) was taken for analysis.

Structures were confirmed by analysis of a liquid phase aliquot with capillary GC–CIMS with ammonia as the reactant gas (2). The GC–CIMS system consisted of a Hewlett Packard model HP5710A gas chromatograph with a 30-m DB-5 fused silica capillary column interfaced with a VG micromass model 7070F mass spectrometer and VG 2035 F–B data system. Ionizing voltage was 70 eV (tungsten filament, 200-μA trap current), and the scan cycle time over the mass range 20 to 300 was 3.0 s. Source temperature was 200 °C, the injector was 250 °C, and the temperature program was the same as described earlier.

Quantitative analysis of the total formaldehyde equivalents in the formalin

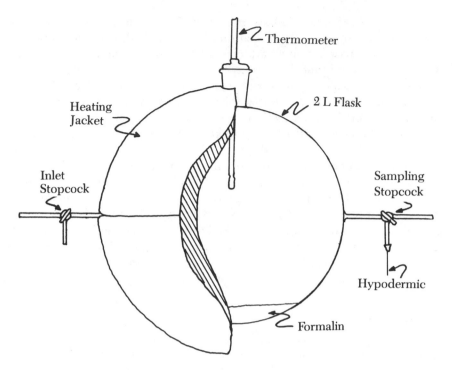

Figure 1. Schematic representation of headspace generator used in gas-phase derivatization technique.

headspace was performed by bubbling vapor (70 mL) from the headspace generator through an impinger containing distilled water (20 mL) (*12*). An aliquot was transferred to a volumetric flask (25 mL) and diluted with distilled water to obtain a concentration in the linear range of the calibration curve. Standard solutions were generated by dilution of 37% formalin solution (1 mL) containing 10 to 15% methanol (Fisher Scientific) in a volumetric flask (500 mL) with distilled water. Aliquots (1 mL) of this stock solution were then diluted to generate solutions of 3.2, 1.6, and 0.4 µg/mL formaldehyde equivalents. Aliquots (4 mL) of the sample, standards, and a blank of distilled water were then concurrently derivatized by addition of 0.1% chromotropic acid (1 mL) and concentrated sulfuric acid (6 mL). Absorbance of the solutions was measured at 580 nm by a Coleman 124 UV–vis spectrophotometer.

Results and Discussion

Gas phase trimethylsilylation of vapors in equilibrium with a formalin solution yields derivatives having sufficient vapor pressure to appear in the headspace as well as in the excess liquid BSTFA contained in the reaction vial. Quantitative measure, therefore, requires analysis of both gas and liquid phases in the reaction vial.

Distribution of the reaction vial contents between the liquid and gas phases after headspace derivatization was approximated by injection of serial amounts of the 5:1::BSTFA:DMF solution into evacuated reaction vials followed by introduction of room air. The volume of the reaction solution that totally vaporized (20 μL) was assumed to equal the volume of derivatization mixture in the gas phase of the reaction vial after derivatization. Hence, the volume of the liquid phase in the reaction vial after derivatization and equilibration could be estimated, and the total amount of derivatized formaldehyde oligomers in the headspace could be determined.

Because oligomers of formaldehyde in the equilibrium mixture may not be isolated and purified, the FID response factors for TMS derivatives of the formaldehyde condensates were estimated by BSTFA derivatization of chemically similar compounds (2). These derivatives indicate FID response to TMS derivatives of poly(oxymethylene) glycols and the glycol monomethyl ethers is a function of the TMS groups present (2, 3, 11). The response factor of the FID was determined to be 8.16 $\times$ 10^{12}/mol TMS.

Quantitative analysis of methylal, a thermally stable compound present in formalin solution, and headspace was readily accomplished by determination of the molar response factor of reagent grade methylal (Fisher Scientific) dissolved in methanol. Detector response for methylal was determined to be 1.77 $\times$ 10^{13}/mol.

Gas chromatograms from the derivatization of formalin headspace by BSTFA are shown in Figure 2. Amounts of the formaldehyde condensation products as TMS ethers and methylal calculated with FID molar response factors are shown in Table I. Formaldehyde equivalents are calculated and compared with the total formaldehyde concentration in the headspace determined by the chromotropic acid method (12). Within the accuracy of measurement, all the formaldehyde equivalents in the vapor phase in equilibrium with formalin solution were in the form of methylal, methylene glycol, and the oligomers of poly(oxymethylene) glycol monomethyl ethers containing one, two, and three formaldehyde units.

Mass spectra for the TMS derivatives of the formaldehyde condensates and methylal from the formalin headspace derivatization liquid phase appear in Figure 3 and are identical to the fragmentation patterns of the TMS derivatives obtained for formaldehyde condensates in formalin solution (2).

Conclusions

Total formaldehyde content of the formalin headspace measured by a gas phase trimethylsilylation procedure closely approximated total formaldehyde content of the formalin solution headspace determined by chromotropic acid analysis. The vapor phase species in equilibrium with a formalin solution were shown to be methylal, methylene glycol, and three oligomeric poly(oxymethylene) glycol monomethyl ethers.

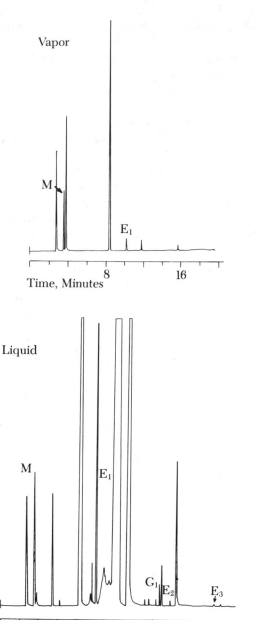

Figure 2. Chromatograms of the liquid phase and the vapor phase in the reaction vial after gas-phase derivatization of formalin headspace with BSTFA indicating the presence of methylene glycol TMS derivatives (G₁), methylal (M), and the first three oligomers of polyoxymethylene glycol monomethyl ether TMS derivatives (E₁₋₃).

Table I. Formaldehyde Condensation Products with Water and Methanol Found in Vapors (70 mL) at Equilibrium with a Formalin Solution (mol)

| | Analysis I | | | Analysis II | | |
| | Derivatization Reaction Phase | | Total Formaldehyde Content | Derivatization Reaction Phase | | Total Formaldehyde Content |
Compound	Liquid	Gas		Liquid	Gas	
$CH_3OCH_2OCH_3$	2.30×10^{-7}	5.43×10^{-6}	5.66×10^{-6}	4.70×10^{-7}	6.26×10^{-6}	6.73×10^{-6}
CH_3OCH_2OTMS	1.37×10^{-6}	2.17×10^{-6}	3.54×10^{-6}	2.22×10^{-6}	1.73×10^{-6}	3.95×10^{-6}
$TMSOCH_2OTMS$	4.56×10^{-8}	n.d.	4.56×10^{-8}	1.08×10^{-7}	n.d.	1.08×10^{-7}
$CH_3(OCH_2)_2OTMS$	1.83×10^{-7}	n.d.	3.65×10^{-7}	2.46×10^{-7}	n.d.	4.91×10^{-7}
$CH_3(OCH_2)_3OTMS$	8.44×10^{-9}	n.d.	2.53×10^{-8}	n.d.	n.d.	n.d.

NOTE: n.d. means not determined. The total formaldehyde by the gas phase derivatization method is 9.64×10^{-6} mol for Analysis I, and 1.13×10^{-5} mol for Analysis II. The total formaldehyde equivalents by the chromotropic acid method are 1.08×10^{-5} mol.

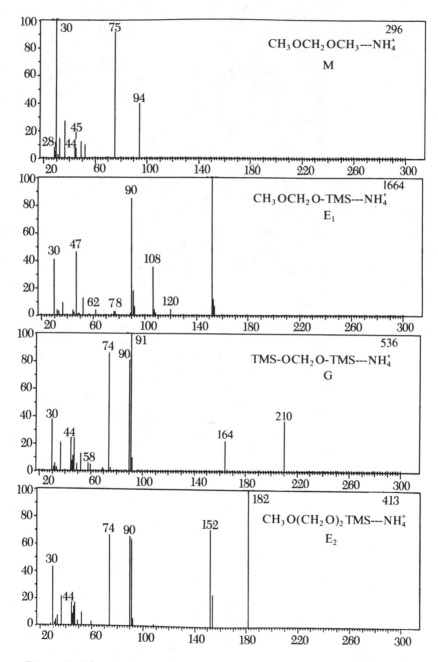

Figure 3. Chemical ionization mass spectra with ammonia as the reacting gas obtained from the liquid phase of the gas-phase derivatization of forma-lin headspace by BSTFA.

Acknowledgments

This work was supported by University of North Carolina Biomedical Research Support Grant 2S07 RR05450–21.

Literature Cited

1. Walker, J. F. "Formaldehyde," 3d ed.; Reinhold: New York, 1964; p. 96.
2. Utterback, D. F.; Millington, D. S.; Gold, A. *Anal. Chem.* **1984**, *56*, 470–73.
3. Dankelman, W.; Daeman, J. M. H. *Anal. Chem.* **1976**, *48*, 401–4.
4. Hall, M. N.; Piret, E. L. *Ind. Eng. Chem.* **1949**, *41*, 1277–86.
5. Iliceto, A. *Gazz. Chim. Ital.* **1954**, *84*, 536–52; *Chem. Abstr.* **1954**, *48*, 9318.
6. Bryant, W. M. D.; Thompson, J. B. *J. Polm. Sci. Part A–1* **1979**, *9*, 2523–40.
7. Sawicki, E.; Sawicki, C. R. "Aldehyde—Photometric Analysis"; Academic: New York, 1978; p. 194.
8. Geisling, K. L.; Miksch, R. R.; Rappaport, S. M. *Anal. Chem.* **1982**, *54*, 140–42.
9. Walker, J. F. "Formaldehyde," 3d ed.; Reinhold: New York, 1964; p. 44.
10. Schnizer, A. W.; Fisher, G. J.; McLean, F. *J. Am. Chem. Soc.* **1953**, *75*, 4347–48.
11. Sevcik, J. "Detectors in Gas Chromatography," translated by K. Stulik; Elsevier Scientific: New York, 1976; p. 94.
12. "National Institute of Occupational Safety and Health Manual of Analytical Methods," 2d ed.; NIOSH: Cincinnati, 1977; Vol. 1, p. 127.

RECEIVED for review September 28, 1984. ACCEPTED April 1, 1985.

Evolution of Testing Methodology for Atmospheric Formaldehyde in the Home Environment

EDWARD N. LIGHT

Public Health Sanitation Division, West Virginia Department of Health, Charleston, WV 25312

Unique requirements of residential formaldehyde sampling programs include a low limit of detection, an averaging time comparable to suspected dose–response relationships, a consideration of environmental factors that determine indoor air quality, and the need for simple, low-cost screening techniques. NIOSH P&CAM 125 and passive dosimeters have been the most commonly used home-sampling methods. Accuracy of these tests is subject to a variety of limitations. Meteorological and occupant-related factors cause significant fluctuations in home formaldehyde levels. Standard test conditions involving temperature, ventilation, and transient sources have been developed. Models for the adjustment or interpretation of home test results are also in use. Health assessments of measured residential formaldehyde levels have included a 0.1-ppm guideline, a generalized statement of risk, and site-specific medical diagnoses.

T HE EVALUATION OF INDOOR AIR QUALITY is a rapidly evolving science still in its infancy. Although some aspects are common to the monitoring of air in the workplace and the management of outdoor air quality, the study of indoor air pollution is unique in many respects. As one of the potentially most significant indoor air pollutants, formaldehyde has been the subject of considerable investigation. However, even the most comprehensive home-testing programs appear to provide, at best, only a rough estimate of true occupant exposure and health risks. This chapter will trace the development of residential formaldehyde testing procedures with an evaluation of common sampling methods, the consideration given to environmentally induced fluctuations, and the assessment of health risks.

0065–2393/85/0210/0067$06.00/0

Background

Sources.   Formaldehyde is a common component of many materials found in the home, but generally only urea–formaldehyde foam insulation and pressed-wood products containing urea–formaldehyde resin are potentially significant sources of indoor air pollution (*1*). Homes containing formaldehyde sources generally have air concentrations in the range of 0.01–1.0 ppm (*2*). The highest levels tend to be in manufactured housing that combines a high loading of pressed-wood products with low infiltration of outside air (*3*). The amount of formaldehyde offgassing from new building materials varies in accordance with the type of formulation and the quality of production. Offgassing generally decreases with age (*4*). An exception to this relationship may occur when the resin is affected by an ongoing source of moisture in the home that may accelerate the hydrolysis of urea–formaldehyde resin.

Health Effects.   Acute symptoms of exposure to low concentrations of formaldehyde include irritation of the eyes, throat, and nose; headaches; nausea; congestion; asthma; and skin rashes (*5, 6*). These symptoms are generally building related and tend to clear up outside of the home (*7, 8*). In one controlled exposure study, 19% of healthy adult volunteers had irritation symptoms at 0.25 ppm (*9*). Formaldehyde symptoms may occur at levels as low as 0.05 ppm in very sensitive individuals (*9, 10*). These individuals may include infants, children, the elderly, those with preexisting allergies or respiratory diseases, and persons who become sensitized (*11, 12*). Studies used to develop this information on the acute health effects of formaldehyde were based on either short-term controlled exposure tests (5 min to 5 h) or short-term measurements of residences in conjunction with epidemiological studies (samples ranging from instantaneous to 90-min averaging time). Homes tested in conjunction with epidemiological studies were generally tested under conditions that approached peak exposure levels (windows closed, etc.) (*12, 13*).

Because formaldehyde is a suspect carcinogen, long-term average exposure to formaldehyde has also been of interest. Cancer risk is hypothesized to be negligible at low levels of formaldehyde exposure, and to increase in proportion to long-term average exposure (*9*).

Public Concern.   Investigation of airborne formaldehyde in the home environment has been reported as far back as 1961 (*6, 7*). In 1969, the U.S. Department of Agriculture issued a report on the possible health hazards associated with formaldehyde release from paneling (*14*). In the early 1970s, investigations of problems caused by formaldehyde exposure in European residences were published (*15, 16*).

A sharp increase in the number of residential formaldehyde complaints in the United States was apparent during the mid-1970s (*7*). Contributing factors included increased use of high-formaldehyde-emitting products (paneling, particle board, urea–formaldehyde foam insulation, etc.); energy conservation practices that decreased the infiltration of outside air into homes; and increased awareness of environmental factors that may affect health.

Since 1980, a declining trend in the average level of formaldehyde measured in homes is apparent. This decrease may be due to the ban on urea–formaldehyde foam insulation and the introduction of lower emitting pressed-wood products (*17*). However, health problems associated with residential formaldehyde exposure continue to be reported.

Testing Programs

General Discussion.   A variety of home-testing programs have been initiated in recent years in response to concern with formaldehyde. These programs generally fall into one of four categories:

1. Screening. Low-cost methods are used to estimate the formaldehyde level. This testing is generally done by either the home owner with a mail-order test kit or the local health department.
2. Health hazard assessment. Several air samples are taken to determine formaldehyde exposure with more detailed testing methodology. This testing may be accompanied by a medical examination. Such tests are generally conducted by a commercial laboratory or consulting firm.
3. Product compliance. This testing is conducted by the manufacturer to ensure compliance with product safety standards. Accurate test methods are required.
4. Research. Advanced monitoring techniques have been used by various institutions to study the behavior of formaldehyde in the indoor environment.

Each type of sampling program has a unique set of requirements in regard to cost, technical expertise, accuracy, detail, and ultimate use for the sampling data. These requirements have prompted the development of a number of diverse approaches to residential formaldehyde testing.

State Health Departments.   Formaldehyde screening programs offered by public health agencies have been responsible for a large number of home tests. In 1982, the Center for Disease Control surveyed state health departments to determine activity in this area (*18*). At that time, 39 states provided some coverage of indoor air pollution, and 29 offered formaldehyde sampling under some circumstances (*3*). One of the earliest sampling programs began in Minnesota in 1979 (*19*). Of these programs, 68% offered unrestricted sampling (*18*), and others required a request from the family physician (*12, 19, 20*). Sampling is generally, but not always, provided at no cost (*20, 21*).

Sampling methods used by these programs include NIOSH P&CAM 125 (72%), gas detector tubes (17%), dosimeters (7%), and direct-reading meters (4%) (18, 22). The cost and technical expertise required to conduct pump–impinger sampling have discouraged some agencies from initiating formaldehyde programs and caused others to discontinue or limit their sampling efforts. In 1982, the Canadian urea–formaldehyde foam insulation program switched from a pump–impinger method to passive dosimeters for most home testing because of their simplicity and low cost (23).

One-half of the state formaldehyde testing programs collected no other environmental data at the site, whereas the remainder generally recorded such parameters as relative humidity, temperature, and source specifications. Sixty-four percent had the capability to sample for other indoor air pollutants (18). Some programs specify that when building-related symptoms are occurring and formaldehyde is apparently not the cause, the investigator is to look for evidence of combustion gases, pesticides, etc. in the home (24).

Approximately one-half of the programs measured exposure levels only, whereas the remainder provided some type of evaluation of health status (18). Some health departments advise occupants to consult their family physician if symptoms are suspected of being formaldehyde related (9, 21, 25). The following criteria were used by the programs in evaluating test results: 38% recommended the lowest feasible level; 34% had no specified criteria; 24% specified criteria ranging from 0.05 and 0.50 ppm; and 3% used the OSHA standard (3 ppm). One-half of the programs went on to provide formaldehyde-control information to occupants (18). In individual assessments of home formaldehyde levels, some programs have urged sensitive individuals to move out if formaldehyde is above a certain level (25), whereas others have presented a generalized risk statement (i.e., formaldehyde ≤0.10 ppm: symptoms unlikely; 0.11–0.39: symptoms may occur in sensitive individuals; ≥0.40: symptoms common) (26).

Regulatory Programs.   Formaldehyde sampling under governmental regulatory programs has primarily involved manufactured housing. In 1980, Wisconsin required new mobile homes to meet a standard of 0.40 ppm (1-h sample) prior to sale (27, 28). In 1982, Minnesota adopted a 0.50-ppm formaldehyde standard for new housing (invalidated in 1984) (19). Both standards specify that sampling is to be done with NIOSH P&CAM 125 under standardized test conditions. The Department of Housing and Urban Development recently promulgated federal regulations governing formaldehyde in manufactured housing; these regulations supersede state standards. Product standards are set with a goal of achieving 0.40 ppm. This level was designated to reduce, not prevent, health problems (29).

Sampling Methods

Ideally, residential formaldehyde sampling methods should meet the following minimum criteria:

1. The detection limit should not exceed 0.05 ppm.
2. Accuracy should be at least $\pm 25\%$ in an intermediate range.
3. Sample averaging time should be of the proper duration to determine health risks.
4. When shipping samples from remote locations is necessary, the samples must be stable under the most extreme conditions.
5. Air samples of formaldehyde at known concentrations should be available for quality control purposes.
6. Other pollutants present in the residential environment should not cause a significant interference.
7. When used in a screening program, the method should be low cost and demand no special expertise.

Many home formaldehyde test methods currently in use fail to meet one or more of these criteria. The following section provides a brief critique of the more common testing methods.

P&CAM 125. The most popular sampling method has been NIOSH P&CAM 125. When this industrial hygiene procedure was first applied to the residential environment, air samples were generally collected in two impingers with distilled water at room temperature (7). Around 1980, accuracy problems in the low range became apparent with this method. Modifications during this period included placing the impingers in an ice bath (21) and changing to a 1% sodium bisulfite collection medium (30, 31). Another recommendation at this time was to eliminate the second impinger because it was only needed as a backup for high-range occupational sampling (32).

Detection limits for NIOSH P&CAM 125 ranging from 0.03 ppm (33) to 0.10 ppm (21) have been reported. Accuracy is generally stated to be $\pm 5\%$ (21, 34), although considerably more error can occur. Loss of formaldehyde has resulted for samples collected and shipped in distilled water at room temperature (35, 36). The sodium bisulfite modification has been found to increase collection efficiency to 98% and sample stability to at least 4 weeks (37). One problem observed with the use of sodium bisulfite has been the occurrence of high blanks (21, 38). This result has been traced to an interference from certain bottle caps and can be avoided by using Nalgene sample bottles (34, 37). One suspected interference with the P&CAM 125 method is trioxane, another compound released into the home by urea–formaldehyde resin. One investigator reported that trioxane appeared to increase the actual formaldehyde level by 30% (39). The typical quality control procedure associated with P&CAM 125 uses only aqueous-phase standard formaldehyde solutions for calibration. Such an approach is insensitive to sampling or shipping error and interferences. Another criticism of this method has been corrosion of pumps by sodium bisulfite (21), although such corrosion can be prevented. A general drawback is the relatively high cost and expertise required (*40*). Despite these limitations, NIOSH P&CAM 125 is currently the most widely accepted formaldehyde air-sampling method and should be considered reasonably accurate when following procedures appropriate for the testing situation.

Alternative Active Sampling Techniques. Several alternative impinger and solid sorbent techniques have been introduced with advantages over NIOSH P&CAM 125. These include Chromosorb 102 coated with *N*–benzylethanolamine (*41*), pararosaniline (*42*), Purpald reagent (*30, 43*), acetylacetone (*44*), 3-methyl-2-benzothiazolone hydrazone (*2*), and 13× molecular sieve (*2*). The primary application of these methods in home testing has been in conjunction with research projects. Cost, expertise required, and lack of general acceptance have discouraged their use in other home-testing programs.

Several direct-reading instruments and detector tubes have also been used for residential formaldehyde sampling. These include the CEA toxic gas monitor, the Lyon formaldemeter, Draeger gas detector tubes, and vacuum tubes. A modified CEA model 555 combines good accuracy and reliability with a 0.01-ppm detection limit. It has not seen widespread use because of cost and expertise required (*45*). The relatively low-cost Lyon formaldemeter has only a 0.30-ppm detection limit and is subject to interferences (*28*). Gas detector tubes provide a low-cost screening tool for some indoor air pollutants, but having a detection limit of 0.50 ppm and potential moisture interference, they are not generally suitable for measuring formaldehyde in the home (*3*). At least one commercial laboratory has marketed vacuum tubes in which a client collects a sample of air in his or her home and then mails it to a commercial laboratory for gas chromatographic analysis (*46*). One general limitation of these four techniques is that when they are used to obtain a few instantaneous formaldehyde measurements, the level may not be a good representation of occupant exposure.

Passive Dosimeters. Dosimeters provide a low-cost, easy-to-use tool for home sampling. Dosimeters currently used to measure formaldehyde in the home can be divided into long-term [3–M formaldehyde monitor, Du Pont Pro-Tek, air quality research (AQR) PF–1] and short-term (Envirotech) dosimeters. The short-term dosimeter collects formaldehyde on the basis of a principle of hydrogen bonding to a cellulose surface in an acidic solution with analysis by the Purpald method (*47*). Long-term dosimeters are based on collection by diffusion followed by chromotropic acid analysis. All four dosimeters have been found to have accuracies within ±25% in the intermediate range. However, they can be distinguished by differing sensitivities, in addition to varied responses to humidity and face velocity.

Long-term dosimeters have a relatively low sensitivity and thus require long sampling times when used in the residential environment (detection limits: 3–M, 0.03 ppm/24 h (48); Du Pont, 0.08 ppm/24 h (33); and AQR, 0.01 ppm/7 days (49). Short-term dosimeters have a higher sensitivity (Envirotech detection limit: 0.03 ppm/1 h). The 3–M dosimeter was found to give low readings in low humidity because of moisture loss (50). Conversely, the AQR dosimeter may not be accurate in very high humidity (49, 51). Humidity is reported not to significantly affect short-term dosimeters (32). Face velocity may affect long-term dosimeters in that a minimum air flow is needed in the room to achieve a dependable collection efficiency (52). The normal range of air velocities found in the indoor environment is reported to not significantly affect short-term dosimeters (32).

Environmental Factors

Although the variability inherent in the sampling methods discussed can be significant, this variability is generally overshadowed by the impact of environmental factors. Changes in temperature, ventilation, humidity, transient formaldehyde sources, and formaldehyde sinks cause wide diurnal and seasonal variations in home formaldehyde levels. Unless these fluctuations are taken into account, a formaldehyde test made with even the most accurate sampling method may not provide a meaningful measure of occupant exposure.

Meteorological Factors.   Changes in temperature and humidity play important roles in determining both the emission rate of formaldehyde sources and the migration of formaldehyde into and out of the home. Temperature fluctuations inside the home normally range from 20 to 40 °C, whereas outside temperatures range from −35 to 45 °C. In controlled chamber experiments, a rise in temperature generally increases the rate of formaldehyde offgassing until an equilibrium is reached (53). Short-term fluctuations in home formaldehyde levels have often been observed to parallel changes in the indoor temperature (54). Relative humidity generally varies from 20% to 90% indoors and from 20% to 100% outdoors. Raising humidity has also been observed to increase formaldehyde offgassing. This increase is generally not as significant as that caused by temperature. The time period required to achieve an equilibrium level following an increase in humidity is longer than for temperature (53).

Infiltration of outside air into the home tends to dilute airborne formaldehyde. Infiltration is regulated by the type of structure, meteorological factors, and occupant activity. The effect of infiltration on formaldehyde levels is generally less than would be predicted with a simple dilution model. This result occurs because dilution also increases the rate of offgassing in a tendency to maintain equilibrium. This phenomenon was illustrated when a tripling of the air-exchange rate was found to reduce the formaldehyde level by only 20% (39, 55). An important factor regulating the infiltration of outside air into a home is the difference between indoor and outdoor temperature (ΔT). Air exchange tends to be lowest when ΔT is zero, and increases with higher temperature differentials (39, 54). High wind speeds can override ΔT in determining air exchange (56).

Several models have been developed to explain the role of meteorological factors in determining residential formaldehyde levels. One use for such models has been the interpretation of one-time formaldehyde tests. Models have also been used to adjust a measured formaldehyde level to that expected to occur under standardized conditions. These models are based on widely varying assumptions as to which meteorological factors play the dominant role. Those currently receiving attention include the following:

INDOOR TEMPERATURE AND HUMIDITY PREDOMINATE ("T,RH MODEL"). This model quantifies the relationship between increasing temperature or humidity and increasing formaldehyde. It was originally presented as the "Berge equation" on the basis of the response of boards under controlled chamber conditions and was recently updated following analysis of additional data (53). This model predicts, for example, that where a formaldehyde concentration of 0.60 ppm was observed at 30 °C, the level at 25 °C would be 0.37 ppm $\pm$ 50%. Effects of humidity change are similarly quantified but are less reliable than temperature (53). One investigator reported that some recently produced lower emitting pressed-wood products may have a much flatter response to temperature change than predicted by these earlier efforts (32). Several home testing programs have assumed that the T,RH model controls the level of formaldehyde when other conditions (i.e., ventilation) are held constant (27).

INDOOR–OUTDOOR TEMPERATURE DIFFERENCE PREDOMINATES ("ΔT MODEL"). This model predicts that the amount of outside air infiltration as influenced by the difference in indoor and outdoor temperature is often the critical factor regulating the level of formaldehyde (54). A modification divides ΔT by indoor relative humidity. This modified model predicts, for example, that when values of $\Delta T/RH$ are less than 0.10, the formaldehyde in the home is at a peak level (57).

INDOOR–OUTDOOR VAPOR PRESSURE DIFFERENCE PREDOMINATES ("ΔVP MODEL"). This model places primary emphasis on the movement of atmospheric formaldehyde associated with water vapor to explain seasonal changes in formaldehyde levels. Vapor pressure (VP) is determined from temperature and relative humidity. A difference between indoor and outdoor VP (ΔVP) will force a migration of water vapor from an area of high VP to an area of low VP. Formaldehyde is highly soluble and will tend to migrate with moisture. According to this model, water vapor migrating into a home through the wall and floor cavities will transport formaldehyde into the living space. Conversely, water vapor migrating out of the home will tend to remove formaldehyde. Factors that might affect the rate of this exchange include the use of vapor barriers and the opening of doors and windows (32).

The three models make different predictions regarding seasonal changes in residential formaldehyde levels. Insufficient data have been collected to validate any models. More research in this area would not only provide for more meaningful interpretation of formaldehyde test results, but might also help develop more effective strategies for the control of formaldehyde.

Occupant-Related Factors. Occupant-related factors can also affect residential formaldehyde exposure. These include transient sources, formaldehyde sinks, and ventilation.

Sources of residential formaldehyde not related to building materials are generally transitory. These include cigarette smoke, natural gas combustion, deodorants, cosmetics, and the heating of certain cooking oils (*32, 58, 59*). Other transitory sources can also act to scavenge formaldehyde from the air (i.e., ammonia-containing products react with formaldehyde to form hexamethylenetetramine) (*60*). When more potent formaldehyde sources associated with building materials are present, the impact of transitory sources is generally low (*59*).

Porous surfaces in the home play an important role in determining the formaldehyde equilibrium. Materials that originally contained little or no formaldehyde will sorb the pollutant from the air under some conditions. Acting as a sink, they may later release this formaldehyde when environmental conditions change (*61*). In this manner, changes in home furnishings that serve as sinks can affect formaldehyde levels.

Homes constructed or retrofitted to minimize energy use have low air-exchange rates. For example, manufactured housing has been found to average 0.30 air changes/h, and values as low as 0.10 air changes/h have been reported (*39*). Opening windows and doors increases the infiltration of outside air into the home (*24, 54*). Conversely, the longer the time doors and windows in a home remain closed, the higher the formaldehyde level (*62*). The operation of a forced air heating–cooling system in the continuous fan mode will increase the air-exchange rate (*63*). Adding make-up air from the outside to a forced air system will increase the air-exchange rate even further. Use of exhaust fans or a fireplace also increases the infiltration of outside air into a home (*24*).

The combined impact of all these environmental factors is major diurnal and seasonal variations in residential formaldehyde levels. High and low formaldehyde levels over a 24-h period have been found to differ up to a factor of 6 (*47, 54, 55*). Peak levels were often, but not always, observed in the late afternoon (*55, 59, 64*). Over a period of 1 year, formaldehyde levels have been reported to vary by factors of 4 (*56*) and 10 (*55*). Lowest levels have generally been found in the winter (*59*). Peaks have occurred in March, July, and October (*55*).

Sampling Protocol

Ideally, residential formaldehyde sampling protocol should reflect periods of occupant exposure that are critical to the determination of health effects in addition to taking into account both diurnal and seasonal variation.

Measures of occupant exposure that are critical to the determination of health risks have not been clearly defined. Short-term peaks have generally been used in the study of acute symptoms, whereas the long-term average is thought to determine carcinogenic risk. In this regard, when the primary goal of a formaldehyde test is to assess or prevent the occurrence of acute symptoms, a short-term test (30 min to 3 h) near peak conditions is most consistent with the limited research in this area. In other situations, a long-term test (at least 24 h) under average conditions would be appropriate for the estimation of carcinogenic risk.

In practice, sampling strategy has often been based purely on convenience and the limitations of the selected test method. The NIOSH P&CAM 125 method has been used for residential testing with sampling times ranging from 30 min to 90 min (7, 21, 23). Long-term dosimeters generally collect samples over a 1–7 day period (51, 65). Short-term dosimeters can be used over a 1–4 h period (47). Direct-reading instruments and detector tubes measure a single instantaneous formaldehyde level, although multiple samples have been used to estimate longer averaging times.

Testing protocol should also ensure that sampling sites are representative of occupant exposure. In general, one to three rooms per home have been tested; the testing of two rooms per home is the most common (5, 19, 21, 27). In manufactured housing very close agreement generally exists between formaldehyde readings taken in different rooms (5, 39) because of the uniform distribution of sources and lack of outside air infiltration. This result suggests that one sampling site per mobile home could be considered representative of overall exposure. Sampling devices are generally situated in the breathing zone of each room (3–5 ft) (19), although one dosimeter must be hung 20 in. from the ceiling (66). The sample site is also specified to be either away from walls (67) or in the middle of the room (27). Protocols can also require the sample to be located away from heat registers, air conditioners, fans, and materials containing formaldehyde (67). Tests located near an outside door used during the sampling period may not be representative of closed-home conditions (39).

Some protocols require an outdoor measurement to be taken in conjunction with indoor sampling (2). Outdoor formaldehyde levels seldom exceed 0.02 ppm except in areas of very high traffic density (23, 24, 26). Because this level is generally a small fraction of the indoor formaldehyde level and is also within the range of error for most testing methods, outside air is often not sampled in home-testing programs.

Some of the more detailed testing protocols suggest additional sampling in wall cavities, floor cavities, and cabinets (5, 24). These measurements help to delineate the critical sources of formaldehyde within a home and to prescribe corrective measures. Such diagnostic sampling has limited value in manufactured housing for which formaldehyde sources are generally well defined.

Standard Test Conditions

Early residential formaldehyde sampling efforts generally had no standard protocol for taking into account environmental factors. Such tests simply obtained a short-term sample under the prevailing conditions. Two strategies to promote the collection of more representative samples have evolved since that time. The first uses long-term sampling to obtain a time-weighted average under prevailing meteorological conditions and occupant activities. This sample should generally represent long-term average exposure unless extreme meteorological or occupant-related conditions occur during the test. Short-term formaldehyde peaks that may be responsible for acute health effects will not be determined by this approach. The second strategy is to control environmental factors in order to simulate a peak formaldehyde level that the occupants will experience on a repetitive basis. This control is generally accomplished by conducting the test under a narrow range of environmental parameters. The test results are sometimes also corrected to the formaldehyde level that would be expected to occur at a standard temperature (24).

The variables generally selected for control during residential formaldehyde testing are ventilation, temperature, and transient sources. The home is often conditioned in some manner prior to testing. Such conditioning may begin with a period of maximum ventilation (all doors and windows open). Reasons for this conditioning include control of formaldehyde from transient sources and reduction of excessive accumulations of formaldehyde in homes that were previously closed up for a long period of time. After ventilation, the home is closed up (all doors and windows shut) for a standard period of time before and during the sample collection. Two-hour close-up time is specified in some procedures. This time is an approximation of the time required for the formaldehyde level to reach equilibrium after a change in temperature (58). This time also represents a typical period that homes are maintained in a completely closed condition during the daytime. Some procedures close up the home for 24 h prior to testing. This close-up period may create a higher formaldehyde condition than the occupants will normally experience for any substantial period of time.

Indoor temperature is often restricted during testing. At lower temperatures formaldehyde offgassing can be negligible, whereas at higher temperatures an excessive amount of formaldehyde may be released. Neither of these extremes would be experienced by occupants under most circumstances. Transient sources such as cigarette smoking are sometimes prohibited, both during the preparation of the home and the sampling. The contribution of such sources is generally only brief or minor. Testing in the absence of transient sources reflects exposure from the more stable and significant sources of formaldehyde. Testing for regulatory and litigation purposes must often be restricted to the contributions of building-related materials.

In 1980, Wisconsin was one of the first states to systematically develop standardized conditions for formaldehyde testing. The original purpose of the protocol was to ensure a reasonable probability of compliance under most environmental conditions with the state's 0.40-ppm formaldehyde standard for manufactured housing (28). The Wisconsin protocol required that homes be aired out for 2 h, closed up for 2 h, and maintained at 21–29 °C; that gas appliances be turned off and smoking be prohibited; and that the observed formaldehyde level be adjusted to 25.6 °C with the Berge equation (27).

The Minnesota regulatory program basically adopted this same procedure (25). Testing conditions that have been required in other testing programs include the following:

1. Ventilation. Test when winds are less than 25 km/h (24). Maintain air conditioner in recirculation mode (24). Avoid use of fireplaces (66) and exhaust fans (24). Employ close-up times prior to testing of 18 h (24) and 3 h (57). Air out homes less than 60 days old for 24–72 h (68).
2. Temperature. Maintain home at 24–27 °C (46, 65). Maintain home at 20 °C during conditioning period and 14 °C during test (24).
3. Transient Sources. Avoid use of aerosols (24), paints, or solvents (66). Do not generate cooking smoke or operate vehicle in attached garage (24).
4. Time. Sample in mid (24) or late afternoon, if possible (67).

Exposure Assessment

Three basic approaches have been used for relating the measured level of formaldehyde in a home to occupant safety. The first approach involves comparison with specific standards or guidelines. Such levels are generally set with the assumption that exposure below that level is safe or relatively safe. The second approach involves an estimate of health risks at the measured formaldehyde level (i.e., probability of cancer, 1 in 10,000; irritation effects, 1 in 3). The third approach involves a site-specific assessment of the occupants' health status. This assessment generally includes a determination of the relationship between observed symptoms and formaldehyde exposure. Substantial uncertainty is involved in all three types of exposure assessments.

Most residential formaldehyde standards designed to protect the general population from health problems have been set at the level of 0.10 ppm. Standards or guidelines in this category include Canada (69), West Germany, the Netherlands (6), the American Society of Heating, Refrigerating, and Air Conditioning Engineers (ASHRAE) (45), and NASA (spacecrafts) (64). As discussed earlier, this level may not protect certain sensitive individuals.

A Comprehensive Home Testing Program

The West Virginia Department of Health initiated a residential screening program for formaldehyde in 1982. In response to requests from the general public, investigations of a relatively large number of homes and buildings are conducted. The program is constrained by a limited budget and by field personnel with little background in air testing.

Air samples are collected with the Envirotech dosimeter, and a 2-h averaging time is used. Quality control includes the analysis of air samples spiked in a room with a stabilized, known formaldehyde concentration.

Prior to sampling, homes are aired out for at least 10 min and then closed up for at least 2 h. If the home was previously unoccupied, it is warmed up the day before the test and aired out for at least 2 h. Temperature must be in the range of 21–29 °C, and the observed formaldehyde level must be adjusted to 25 °C with the equation developed by Myers (53). Cigarette smoking, the unvented combustion of natural gas, and the use of ammonia cleaners are prohibited.

A questionnaire is administered to obtain the following data: periods of time during which apparent building-related symptoms have occurred; type of home, date moved in, and information on remodeling; and other sources of indoor air pollution.

Exposure histories and symptom histories are compared to determine any potential correlations. Sampling of other indoor air pollutants such as combustion products or pesticides is conducted if they are a likely contributor to any of the observed health problems.

Measured formaldehyde levels are categorized into four ranges for reporting purposes: background <0.05 ppm; low, 0.05–0.19 ppm; moderate, 0.20–0.40 ppm; and high, >0.40 ppm. These classifications are based on the apparent incidence of formaldehyde-related symptoms in the general population.

If the symptom history appears to track an occupant's exposure to formaldehyde, and other indoor air pollutants appear not to be significant, the Health Department classifies the symptoms as suspect formaldehyde related. Verification requires examination by a physician followed by observation of the patient after sources of formaldehyde are controlled.

The selection of remedial measures is left to the occupant, and the Health Department provides the following guidelines:

1. No action is needed at levels below 0.20 ppm if no symptoms are occurring.
2. Basic measures are needed only when minor symptoms are occurring (i.e., keep windows cracked, seal particle board).
3. Additional controls may help reduce the potential risks of long-term exposure.
4. Advanced measures are needed when symptoms are major or

formaldehyde level is high (i.e., ammonia fumigation of the rooms and wall cavities).

5. Moving out of the home may be needed in cases of extreme sensitization (on the basis of the advice of a physician).

Literature Cited

1. U.S. Consumer Product Safety Commission, "Formaldehyde in Indoor Air: Sources and Toxicity" by Gupta, K. C.; Ulsamer, A. G.; Preuss, P. W.; Government Printing Office: Washington, 1981.
2. Matthews, T. G.; Hawthorne, A. R.; Howell, T. C.; Metcalfe, C. E.; Gammage, R. B. *Environ. Int.* 1982, 8, 143.
3. Godish, T. *Natural Resources Notes* Ball State Univ., Muncie, Ind., 1983, 7.
4. Godish, T. *Natural Resources Notes* Ball State Univ., Muncie, Ind., 1982, 4.
5. Breysse, P. A., presented at the Am. Med. Assoc. Congr. Occup. Health, Chapel Hill, N.C., 1979.
6. Breysse, P. A. "Health Hazard Implications of Proposed and Adopted Indoor Air Standards for Formaldehyde"; Univ. of Washington Press: Seattle, 1983.
7. Breysse, P. A. *Environmental Health and Safety News*, Univ. of Washington, Seattle, 1977, 26, 1.
8. Godish, T. *Natural Resources Notes*, Ball State Univ., Muncie, Ind., 1983, 3.
9. U.S. Consumer Produce Safety Commission, "Health Sciences Input for the First Quarter Commission Briefing on Pressed Wood Products," memorandum from Cohn, M. S., to Medford, R., November 1982.
10. "Formaldehyde—An Assessment of Its Health Effects"; National Academy of Sciences: Washington, D.C., 1980.
11. Grunby, P. *J. Am. Med. Assoc.* 1980, 243, 1697.
12. U.S. Consumer Product Safety Commission, "Health Sciences Input for the Regulatory Options for Urea–Formaldehyde Foam Insulation," memorandum from Ulsamer, A. G., to Cohen, H., July 1980.
13. Hanrahan, L. P.; Anderson, H. A.; Dally, K. A.; Eckmann, A. D.; Kanarek, M. S. "A Multivariate Analysis of Health Effects in a Cohort of Mobile Home Residents Exposed to Formaldehyde"; Wisconsin Division of Health: Madison, 1981.
14. Gillespie, R. H. "That Panel Odor—What You Can Do About It"; USDA Forest Products Laboratory: Madison, 1969.
15. Burdach, S. T.; Wechselbery, K. *Fortshr. Med.* 1980, 98, 377.
16. Anderson, I.; Lundquist, G. R.; Molhave, L. *Atmos. Environ.* 1975, 9, 1121.
17. Archibald, E. *Adhes. Age*, 1982, *July*, 27–30.
18. Houk, O. V. N. "A Survey of State Health Department Programs for Hazard Evaluations of Nonoccupational Indoor Air Pollution"; U.S. Public Health Service Center for Disease Control: Atlanta, 1983.
19. "Statement of Need and Reasonableness in the Matter of a Proposed New Rule Relating to Formaldehyde"; Minnesota Dept. of Health: Minneapolis, October 1981.
20. "Public Health Advisory"; Michigan Dept. of Public Health: Lansing, 1982.
21. Anderson, H. A.; Dally, K. A. "Formaldehyde Sampling," memorandum to Local Health Agency Officials; Wisconsin Division of Health: Madison, March 1982.
22. "National Institute for Occupational Safety and Health Manual of Analytical Methods," Taylor, D. G., Ed.; NIOSH: Cincinnati, 1974; Vol. 1.
23. Consumer and Corporate Affairs, UFFI Centre, letter from Hall, S. F., to Van Zelst, T. W., Hull, Quebec, July 1983.
24. Bowen, R. P.; Shirtliffe, C. J.; Chown, G. A. "Urea–Formaldehyde Foam Insulation: Problem Identification and Remedial Measures for Wood-Frame

Construction"; National Research Council of Canada: 1981; Building Practice Note No. 23.
25. "Formaldehyde Facts"; Minnesota Department of Health: Minneapolis, November, 1979.
26. Wisconsin Division of Health, letter from Anderson, H. A., to Wisconsin residents, Madison, 1982.
27. "Wisconsin Administrative Code," Chapter Ind. 14; Wisconsin Dept. of Industry, Labor and Human Relations: Madison, 1982.
28. "Formaldehyde Proposed Standard: Report Summarizing Public Hearings and Explaining Department Position"; Wisconsin Dept. of Industry, Labor and Human Relations: Madison, October 1980.
29. Dept. of Housing and Urban Development, "Manufactured Home Construction and Safety Standards"; Government Printing Office: Washington, 1984; 24CF Part 3280.
30. Seymour, J. W., presented at the Wood Adh. Symp., Madison, Wis., September 1980.
31. Envirotech Services, Inc., letter to Meyer, R. from Seymour, J. W., Madison, March 1981.
32. Seymour, J. W., personal communication.
33. Godish, T. "Monitoring Indoor Formaldehyde Levels Using the Chromotropic Acid Method," Ball State Univ.: Muncie, Ind., 1982.
34. Defendant's brief, MHI v. Wisconsin (Dane Cy. Cir. Ct. 81–CV–4933); Wisconsin Dept. of Industry, Labor and Human Relations: Madison, 1982.
35. "Report of the Hearing Examiner on Proposed Rules Governing Formaldehyde in Housing Units"; Minnesota Office of Administrative Hearings: Minneapolis, February 1982.
36. Ludlam, P. R.; King, J. G. *Analyst* 1981, *April*, 488.
37. Meadows, G. W.; Rusch, G. M., *Am. Ind. Hyg. Assoc. J.* 1983, *44*, 71.
38. Eckmann, A. D.; Dally, K. A.; Hanrahan, L. P.; Anderson, H. A. "Comparison of the Chromotropic Acid and Pararosaniline Methods for HCHO Determination Using Various Collection Techniques"; Wisconsin Division of Health: Madison, 1982.
39. Singh, J.; Walcott, R.; St. Pierre, C.; Ferrel, T.; Garrison, S.; Gramp, G.; Groah, W. "Evaluation of the Relationship Between Formaldehyde Emissions from Particle Board Mobile Home Decking and Hardwood Plywood Wall Paneling"; prepared by Clayton Environmental Consultants for U.S. Dept. of HUD, Washington, 1982.
40. Geisling, K. L.; Miksch, R. R.; Rappaport, S. R.; Tashima, M. K., "A New Passive Monitor for Determining Formaldehyde in Indoor Air"; Lawrence Berkley Laboratory: Berkley, 1982.
41. "A Review of Workplace Measurement Methods for Formaldehyde"; National Council of the Paper Industry for Air and Stream Improvement: New York, 1982; Technical Bulletin No. 380.
42. Miksch, R. R.; Anthon, D. W.; Fanning, L. Z.; Hollowell, C. D.; Revzah, K.; Glanville, J. *Anal. Chem.*, 1981, *53*, 2118.
43. Myers, G. E.; Nagaoka, M., *Wood Sci.* 1981, *13*, 140.
44. Myers, G. E., *For. Prod. J.* 1982, *32*, 20.
45. Matthews, T. G. *Am. Ind. Hyg. Assoc. J.* 1982, *43*, 47.
46. Jacoby, M., personal communication.
47. "Air Quality Testing Equipment"; Envirotech Services: Prarie de Sac, Wis., 1983.
48. "3M Brand Formaldehyde Monitor #3750 Product Information and Usage Guide"; Occupational Health and Safety Products Division, 3M Company: St. Paul, 1981.
49. Rappaport, S. M.; Miksch, R. R. "The Measurement of Airborne Formaldehyde in Homes and Office Buildings with PF–1 Passive Monitors"; Air Quality Research: Berkley, 1982.

50. *Morbidity and Mortality Weekly Report*, 1983, *32*, 615.
51. Hodgson, A. T.; Geisling, K. L.; Remijn, B.; Girman, J. R. "Validation of a Passive Sampler for Determining Formaldehyde in Residential Indoor Air"; Lawrence Berkeley Laboratory: Berkeley, 1982; Report No. LBL-1462 prepared for U.S. Dept. of Energy.
52. Rose, V. E.; Perkins, J. L. *Am. Ind. Hyg. Assoc. J.* 1982, *43*, 605.
53. Myers, G. E. "The Effects of Temperature and Humidity on Formaldehyde Emission from U-F Bonded Boards: A Literature Critique," USDA Forest Products Laboratory: Madison, 1984.
54. Myers, G. E.; Seymour, J. W.; Khan, T. "Formaldehyde Air Contamination in Mobile Homes: Variation with Interior Location and Time of Day"; USDA Forest Products Laboratory: Madison, 1982.
55. Gammage, R. B.; Hingerty, B. E.; Matthews, T. G.; Hawthorne, A. R.; Womack, D. R.; Westley, R. R.; Gupta, K. C., Proceedings: Measurement and Monitoring of Non-Criteria Contaminants in Air, 1983, 453-62.
56. Godish, T. *Natural Resource Notes* Ball State Univ., Muncie, Ind., 1983, 7.
57. Godish, T., Proceedings: Measurement and Monitoring of Non-Criteria Contaminants in Air, 1983, 463-70.
58. Plaintiff's Trial Memorandum, MHI v. Wisconsin (Dane Cy. Cir. Ct. 81-CV-4933), Manufactured Housing Institute: Madison, 1982.
59. Godish, T. *Natural Resources Notes*, Ball State Univ., Muncie, Ind., 1983, 3.
60. Sardinas, A. V.; Most, R. S.; Giuletti, M. A.; Honchar, P. *J. Environ. Health* 1979, *41*, 270.
61. Meyer, B. "Urea-Formaldehyde Resins," Addison-Wesley: Reading, Mass., 1979; p. 260.
62. U.S. Environmental Protection Agency, "Indoor Air Pollution in the Residential Environment" by Moschandreas, D. J.; Stark, J. W. C.; McFadden, J. E.; Morse, S. S.; Government Printing Office: Washington, 1978.
63. "Analysis of the Proposed Formaldehyde Standard for Wisconsin"; Manufactured Housing Institute: Washington, D.C., 1980.
64. National Research Council "Formaldehyde and Other Aldehydes"; Government Printing Office: Washington, 1981.
65. "Formaldehyde Sampling Instructions"; Environmental Research Group: Ann Arbor, 1983.
66. "Instructions for Home Use of Dosimeters"; Consumer and Corporate Affairs, UFFI Centre: Hull, Quebec, 1982.
67. "Formaldehyde Dosimeter Instructions"; SERCO Laboratories: Roseville, Minn., 1983.
68. "CP-3600 Formaldehyde Dosimeter Directions"; Chemical Products Development: Oklahoma City, 1982.
69. Hanson, D. *Chem. Eng. News*, 1982, *60*, 34.

RECEIVED for review September 28, 1984. ACCEPTED February 8, 1985.

Predicting Release of Formaldehyde from Cellulosic Textiles

B. A. KOTTES ANDREWS and ROBERT M. REINHARDT

Southern Regional Research Center, Agricultural Research Service, U.S. Department of Agriculture, New Orleans, LA 70179

Because of their low cost, availability, and effectiveness, the reaction products of formaldehyde and cyclic amides are generally used to impart durable press properties to textiles. Therefore, test methods for estimation of formaldehyde release from fabrics have received widespread attention. Sources of measurable formaldehyde release are residual formaldehyde from finishing, free formaldehyde from the established cellulose–ether equilibria, and formaldehyde produced by hydrolysis of the cellulose cross-links and non-cross-linking substituents during end use or testing. Test methods for release vary in their assessment of free and hydrolysis-induced formaldehyde on the basis of the relative severity of the incubation or extraction step. Determinations of formaldehyde are very precise, but interferences with accuracy of the analyses can occur in several of the methods. Selectivity, versatility, and scope of the various tests are detailed.

CELLULOSE-CONTAINING TEXTILES require chemical treatment to impart smooth-drying and durable-press properties, and also certain functional properties such as dimensional stability. These chemical treatments comprise cross-linking adjacent microstructural units of cellulose to improve the resiliency of the cotton fiber, and thus the textile. The most common class of cross-linking agents used in these treatments is the methylolamide class, both because of economic considerations and because these compounds have consistently surpassed other types of cellulose cross-linkers in efficiency of reaction (*1, 2*). The methylolamide class of compounds is important to the textile industry also for use as pigment binders, a use not requiring cross-linking.

Methylolamides are the products of reaction between formaldehyde and an amide. Because this reaction and the reaction between the methylolamide and cellulose are reversible and subject to hydrolysis during end-use conditions, chemically finished cellulosic-containing textiles can serve as sources of releasable formaldehyde.

A survey of the advances in testing for formaldehyde release over the years parallels the progress of our understanding of the chemistry of cellulose cross-linking and of the resultant development of more stable cross-linking agents and finishes. For example, the earliest tests to measure formaldehyde release, reported in the late 1950s and early 1960s (3, 4), were actually attempts to assign numbers to subjective evaluation of formaldehyde odor—sniff tests. The first generations of fabrics finished for wash–wear properties were finished with agents such as urea–formaldehyde and melamine–formaldehyde, agents that we now know have poor durability to hydrolysis and an equilibrium position that favors production of free formaldehyde (5). These finishes had formaldehyde release values greater than 4000 μg/g of fabric (6).

During this time, also, the development of polyester fiber allowed the introduction of sharp creases that were durable through laundering and cleaning. For cellulosic textiles to be competitive, a method was devised to impregnate the textile with a solution of cross-linking agent and catalyst, dry the wetted textile, fabricate the garment from uncured fabric, press in desired creases, and then complete the cross-linking reaction via a curing step at the garment stage of manufacturing. Such treatments, called post-cure processes, combined with use of relatively unstable agents, produced fabrics with high tendency to release formaldehyde to workers at the cut-and-sew plants. These postcure fabrics consistently release much higher levels of formaldehyde to workers than their counterparts (precure fabrics) that have undergone the curing step before garment manufacturing (7).

Following the lead of the Southern Regional Research Center (SRRC) (4) and Bacon (3) in development of an objective test for formaldehyde release, the American Association of Textile Chemists and Colorists (AATCC) Research Committee, RA68–Odor Determination, presented two tests for formaldehyde release: the sealed jar method (AATCC test method 112–1965) and the WestPoint Pepperel steam tube method (AATCC test method 113–1965) (8). Both test methods used a colorimetric determination of evolved formaldehyde with pararosaniline. A second generation of cross-linking agents based on the cyclic ureas such as methylolated ethyleneurea, methylolated propyleneurea, and methylolated triazone was developed (9, 10). During this period, the test methods were upgraded by replacement of the Schiff's base reagent with chromotropic acid or Nash reagent. Sulfite titration of formaldehyde was the reference method (11).

In the early 1970s, cross-linking agents such as dimethyloldihydroxyethyleneurea (DMDHEU) and methylolated carbamates (12, 13) were developed and produced more stable finishes on cellulosic textiles. Because of the greater stability of these finishes, attention in testing focused on the influence that the incubation conditions in the accelerated storage portion of the AATCC sealed jar test (by then the test of commerce in the textile industry) had on promoting finish hydrolysis. Scientists recognized that the AATCC test gave an approximation of the potential of the fabric to release formaldehyde, not merely the free formaldehyde present in a finish. In 1975 a modification of the AATCC test method was proposed. It embodied a 4-h sealed jar incubation at 65 °C in place of 20 h at 49 °C. This modification was adopted as an alternative test procedure in 1978 (*14*). Concurrently, other countries saw the need to regulate and test for formaldehyde in textiles (*15*).

As pressures from workers (*16*) and consumer groups (*17*) increased, new finishes and improved techniques were adopted to bring the levels of formaldehyde release from approximately 4000 μg/g, first to 2000 μg/g, second to 1000 μg/g, third to 500 μg/g, and finally to the present levels, often less than 300 μg/g. As the levels measured became lower, more attention was given to the source of formaldehyde measured in these tests. From information derived from the nature of the reaction between a methylolamide and cellulose,

$$RNCH_2OH + CellOH \rightarrow RNCH_2OCell + H_2O \tag{1}$$

it has been established that the source of formaldehyde in finished fabrics is not only the free or unbound formaldehyde, but all the reaction products:

$$HCHO,\ HOCH_2NC(=O)NCH_2OH,\ Cell-OCH_2OH,$$

$$Cell-OCH_2NC(=O)NCH_2OH,\ NC(=O)NCH_2OCH_2NC(=O)N,$$

$$Cell-OCH_2NC(=O)NCH_2O-Cell,\ NC(=O)NCH_2NC(=O)N,$$

$$and\ \ Cell-OCH_2O-Cell$$

For differentiation and measurement of formaldehyde from different sources, test methods with less severe incubation or accelerated storage conditions were developed. A listing of test methods in widespread use and a comparison of the test conditions are in Table I. The earlier methods use extraction techniques and titrimetric analyses; later methods, in particular those under current development, use instrumental techniques.

Vapor Extraction Techniques

AATCC Test Method 112–1982 (Sealed Jar) (*18*).   The most severe of the tests for measurement of formaldehyde release and the most severe of the vapor extraction techniques is the AATCC test method 112, sealed jar test. The test consists of an incubation stage and a colorimetric analysis of the formaldehyde released in the incubation stage. In the incubation stage, the fabric is suspended above water and is in contact only with water in the

Table I. Tests for Formaldehyde in Textiles: Conditions and Sources of Formaldehyde

Method	Liq:Fabric	Test Conditions		Formaldehyde Source
		Time	Temp. (°C)	
Sodium sulfite: $CH_2(NaSO_3)OH$				
Iodine method	50	7 min	0–10	free
Acid-base method	110	0–7 min	0–10	free
$KCN:HOCH_2CN$	25	10 min	50, 0–10	$>NCH_2OH$ + free
MITI test method, JIS–L1041–1960				from hydrolysis
Nash reagent + filtrate	50	1 h	25	(finish + unreacted) + free
Japanese law 112–1973 test				from hydrolysis
Nash reagent + filtrate	100	1 h	40	(finish + unreacted) + free
Shirley Institute test				from hydrolysis
Digestion of filtrate in				(finish + unreacted) + free
22 N H_2SO_4 + chrom. acid,	10	20 min	25	+ digestion of unreacted
20 min at boil				and self-condensed
AATCC sealed jar, 112–1982		4 h	65	from hydrolysis
Nash reagent–chrom. acid		or		(finished + unreacted) +
+ HCHO absorbed in water	—	20 h	49	free

vapor phase under conditions of 100% relative humidity (rh). A 1-g sample of fabric is held over 50 mL of distilled or deionized water in a sealed mason jar. The jar is maintained in an oven at either 49 °C for 20 h or 65 °C for 4 h. After the jar cools to room temperature, the fabric is removed, an aliquot of the water in the bottom of the jar is taken, and color development of the dissolved formaldehyde is done with Nash reagent or chromotropic acid.

An advantage of the Nash reagent over others formerly used with this test or suggested currently as replacements is its specificity for formaldehyde (*19*). Unlike formaldehyde adducts of some other colorimetric reagents, the structure of the Nash–formaldehyde adduct precludes substitution of other aldehydes in the reaction. The methylene group from formaldehyde becomes part of the lutidene ring.

Reactions of formaldehyde with Nash reagent, chromotropic acid, pararosaniline, and 3-methyl-2-benzothiazolinone hydrazone hydrochloride (MBTH) are shown in Reactions 2–5, respectively.

The Nash reagent is the agent most commonly used to develop color in the AATCC-112 test. The range covered by the calibration data is 15 to 60 $\mu g/mL$ of formaldehyde in solution, or 750 to 3000 $\mu g/g$, based on a 1-g sample of fabric. Several techniques broaden the range of analysis and avoid extrapolation from the regression line for lower sample concentrations. These techniques include increasing the amount of sample solution, increasing the ratio of Nash reagent to sample, spiking the unknown solution with a known amount of formaldehyde, and preparing a separate calibration curve to cover the lower concentrations of formaldehyde to be analyzed (*19*). Figure 1 shows the correlation between formaldehyde release values calculated from the same absorbances by using equations derived from a low (0–6) and AATCC-112 prescribed (0–60) calibration curve (*19*). Higher values are obtained with the low calibration curve.

Table II lists several spectrophotometric methods for analysis of formaldehyde, the minimal detectable concentrations, absorbance maxima, and known interferences (*20*). Although the listed minimum detectable

$$CH_3COCH_2COCH_3 + NH_4OAc \xrightarrow[HOAc]{HCHO}$$

NASH REAGENT

Reaction 2.

Reaction 3.

Reaction 4.

concentration for MBTH indicates it would be the reagent of choice in cases in which formaldehyde is the only aldehyde present, analyses of un-treated cotton fabric by the Nash reagent give values closest to zero, the expected formaldehyde level (Table III) (21).

AATCC Test Method 113–1978 (Steam Method) (18). This method differs from test method 112 in that the fabric sample is exposed to low-pressure steam for a very short time (only long enough to collect 10 mL of distillate) in an apparatus designed to promote formaldehyde release. The

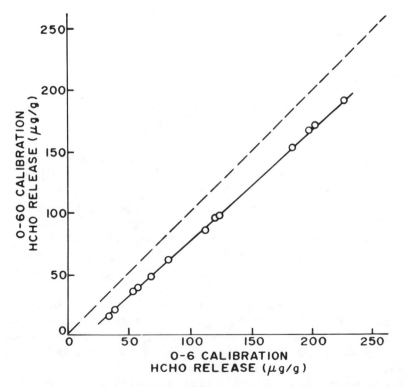

Reaction 5.

Figure 1. *Correlation between formaldehyde release values calculated with regression equations from a 0–60 μg/mL calibration and from a 0–6 μg/mL calibration:* $y = .924x - 14.565$, *and* $r^2 = 0.99954$

Table II. Reagents for Spectrophotometric Analysis for Formaldehyde

Reagent	Minimum Detectable Concentration ($\mu g/mL$)	Absorbance Maximum (nm)	Interferences
Chromotropic acid	0.25	580	nitrogen dioxide, alkenes, acrolein, acetaldehyde, phenol, HCHO precursors
Pararosaniline	0.1	570	sulfur dioxide, cyanide
	0.1	560	almost specific
MBTH	0.05	628	higher aldehydes
Nash reagent	1.4	412	specific

Table III. Analyses of Untreated Cotton Fabric for Formaldehyde

Method	Observed HCHO ($\mu g/g$), Duplicate Determinations	
	A	B
AATCC–112–1982		
Nash (0–60)	– 11	– 12
(0–6)	– 2	– 3
MBTH (0–7.5)	9	9
Na_2SO_3		
10 °C	7[a]	7[a]
25 °C	14	14

[a] One drop of 0.01 N I_2 is equivalent to 7 $\mu g/g$ HCHO.

effluent steam containing the evolved formaldehyde is condensed, and the amount of formaldehyde in the distillate is determined colorimetrically by reaction with phenylhydrazine. By comparison of the color produced with prepared standard dye solutions, a formaldehyde odor potential rating number is obtained for an indication of the degree to which formaldehyde is likely to be evolved in normal storage. The numbers may be assigned qualitative descriptors from "none" to "extremely strong odor potential." This method is most useful in production in which an immediate estimate is needed.

Sun Chemical Company Determination of Formaldehyde Offgassed During Finishing (22).   The Sun Chemical Company method for determination of the offgassed formaldehyde during finishing is only one of several methods that use the principle of scrubbing the formaldehyde vapor from the oven air with water after it passes over the fabric during processing. Other methods are described in the following sections. Water from five scrubbers is combined and analyzed by the AATCC test method 112 with Nash reagent as the colorimetric reagent.

Northern Piedmont Section Determination of Releasable Formalde-hyde in Fabrics (23).   This method is designed to simulate formaldehyde offgassing in a cutting room or retail store. A metered volume of air under pressure is pumped at an exchange rate of one volume of air per hour into a jar containing a weighed fabric sample. A chromotropic acid solution is used to scrub and analyze the exhaust solution. Formaldehyde content is determined hourly over an 8-h period to give an 8-h average.

WestPoint Pepperell Dynamic Chamber Method for Formaldehyde Release (24).   The WestPoint Pepperell dynamic chamber method expands upon the Northern Piedmont principle and a study of formaldehyde offgassing from textile fabrics in a dynamic chamber briefly reported by Wayland et al. (25). In the WestPoint method, fabric samples are suspended vertically from the top of 35.6-L closed glass aquariums. Air maintained at a controlled temperature and relative humidity is used to provide 1.0 and 1.7 air exchanges per hour. Exhaust air is scrubbed with midget impingers via sodium bisulfite solutions. The amount of formaldehyde in the scrubbing solution is determined by the chromotropic acid method. Concurrent analyses of the fabric samples can be carried out for formaldehyde release by the AATCC-112 method and for free formaldehyde or N-methylol groups present via extraction methods that will be discussed later. From periodic analyses, rate constants can be calculated.

Badische Analin Soda Fabrik (BASF) Determination of the Separated Formaldehyde in the Exhaust During Finishing (26).   The BASF method analyzes for the offgassed formaldehyde via scrubbing with either water or bisulfite solution and for free formaldehyde or N-methylol groups present on the fabric via extraction methods that will be discussed later. In the analysis of the offgasses, a dry hot air current, prewarmed to the desired curing temperature in a heat exchanger, is passed through a tube containing the sample at the desired curing temperature. The fabric sample, hung on a thread, is allowed to drop into an extraction flask after the appropriate offgassing analysis time. The offgassed formaldehyde is determined either by iodometric titration of a bisulfite scrubbing solution or colorimetric analysis of a water scrubbing solution.

SRRC Vapor Transport Method (27, 28).   Equipment used in the AATCC test method 112–1982 (sealed jar) is suitable for measuring formaldehyde transport to and from fabrics. Variables that may be controlled are time, temperature, and relative humidity. From determinations of the amounts transported to and from fabrics over a period of time, kinetics of the phenomena can be obtained. Comparisons can be made among finishes and among physicochemical conditions affecting the transport.

Other Vapor Extraction Methods.   Variations and modifications of the methods just discussed have been employed by several research groups attempting to extrapolate results from small, laboratory-scale experiments to estimate or compare release of formaldehyde from various consumer products in an end-use situation. Methods reported by Lovelace Biomedical and Environmental Research Institute (29) in a study for the Consumer Product Safety Commission are representative of such adaptations. Results were reported both from static methods using a large desiccator for the incubation container and dynamic methods using a Laskin-type exposure chamber with controlled temperature, humidity, and one air exchange per hour. Loading is an important factor in all of these tests.

Liquid Extraction Techniques

Cold Sodium Sulfite Method (30, 31). The purpose of this method is to determine only the free or unbound formaldehyde from a textile finished with a formaldehyde-containing reactant by keeping the incubation temperature too low to hydrolyze either reactant or finish. The incubation step and reaction with sodium sulfite are simultaneous; the textile is extracted with aqueous sodium sulfite for 7 min at 0–10 °C to form the bisulfite addition product. The formaldehyde produced can be determined by titrating the NaOH formed (acid–base variation) or by neutralizing and determining unreacted sulfite iodometrically (iodine variation). The sensitivity of the method can be improved somewhat by adjustment of the concentration of the reagents. The improvements are limited in the acid–base method because changing the concentration also changes the pH necessary for reaction. The sodium sulfite methods are not specific for formaldehyde.

Glycolonitrile Formation Method (32). KCN is used in a cold extraction method to assay the free or unbound formaldehyde in a textile and unreacted N-methylol groups. The reaction of KCN with both formaldehyde and N-methylol groups is used; etherified N-methylol groups do not react. Unreacted KCN is titrated with nickel sulfate, and the stoichiometric amount of free formaldehyde plus free N-methylol groups can be determined. Subtraction of free formaldehyde determined separately by the cold sodium sulfite method generates the content of free N-methylol groups.

Shirley Institute Method for Free Formaldehyde (33). The method developed at the Shirley Institute employs a room temperature aqueous extraction procedure with a 10:1 liquor-to-fabric ratio to remove any unreacted finishing agents, water-soluble finish fragments, and unbound formaldehyde from the finished textile. This relatively mild extraction procedure is followed by digestion and color development of the extract filtrate in chromotropic acid and 22 N sulfuric acid for 20 min at a boil. The stated purpose of such a scheme is to give the most appropriate assessment of risks from potentially dermatitic residual formaldehyde derivatives on the textile. It has been hypothesized that levels of formaldehyde greater than 700 ppm (μg/g) produced by this method may give rise to skin irritation during garment manufacture. Again, chromotropic acid is not specific for formaldehyde. Furthermore, residual compounds such as glyoxal can be broken down in the digestion step to analyze as formaldehyde.

Japanese Ministry of International Trade and Industry Test Method JIS-L 1041–1960 (34). Like the Shirley Institute test, water-soluble finish fragments and reactants, as well as free or loosely bound formaldehyde and the easily hydrolyzable methylol groups, are extracted in the 1-h, room temperature, aqueous extraction with a 100:1 liquor-to-fabric ratio. Unlike the Shirley Institute method, the extract is not digested in sulfuric acid–chromotropic acid. Therefore, no further breakdown of extract occurs beyond the mild extraction. Formaldehyde generated in the incubation is analyzed colorimetrically either by Nash reagent or by phloroglucinol. This Ministry of International Trade and Industry (MITI) method was the forerunner of the present Japanese law-112 method used to control textiles that release formaldehyde. The MITI test was used in conjunction with voluntary industry standards in force before law 112–1973 was enacted.

Japanese Test Method Law 112–1973 (35). The Japanese test method law 112–1973 is the quantitative method used to determine the formaldehyde level in textiles for compliance with Japanese Industrial Standard law 112–1973. Japan is the only country to regulate the amount of releasable formaldehyde in fabric. For infants' and children's clothes the allowable amount is 0 μg/g; for underwear, sleepwear, gloves, and hosieries the allowable amount is 75 μg/g. Intermediate garments and outerwear are not regulated by law, but voluntary standards are less than 1000 μg/g for outer garments and 300 μg/g for intermediate garments as per a MITI notice (36).

Japanese test method law 112–1973 uses an aqueous extraction at 40 °C for 1 h with a 100:1 liquor-to-fabric ratio. Formaldehyde in the extract is analyzed colorimetrically by Nash reagent. The aqueous extraction conditions were chosen rather than vapor phase conditions to simulate dissolution in sweat in cases of testing contact dermatitis. The 40 °C temperature was considered close to human body temperature. Modifications to control the pH of the extraction solution have been made for use in other countries (37).

Room Temperature Sodium Sulfite Method (38). The sodium sulfite extraction and titration outlined in the previous section on the cold sodium sulfite method has been carried out at room temperature. As would be expected, the levels of formaldehyde assayed are higher than those from an ice cold method with differences depending on the time of incubation with sodium sulfite. This variation has been used primarily as a research tool.

Instrumental Extraction Techniques

Headspace Gas Chromatography (39, 40). As early as 1978, research workers at SRRC (39) realized the potential for headspace gas chromatography in analyses for releasable formaldehyde in textiles. At that time, headspace chromatography, considered an unconventional technique, had been developed by SRRC to analyze successfully volatiles associated with flavor in raw or processed food products. Qualitative identification of both trimethylamine and formaldehyde with a headspace injection system built in-house and a flame ionization detector was made with routes to quantitative determination offered.

As instrumentation improved, quantification became more achievable. Textile Research Institute (40) used a headspace gas chromatograph (Perkin–Elmer) with the response of a flame ionization detector increased by a reduction technique that converted separated formaldehyde to methanol. The temperature of the headspace vials was 65 °C. Sorption isotherms were established, and kinetics of formaldehyde release into the headspace vial under the conditions of analysis were determined. Formaldehyde measured in the headspace air was considerably less than that generated in either the AATCC-112 test or the Japanese law-112 test. The level was higher than that evolved in the cold sodium sulfite test for free formaldehyde.

CEA Analyzer Method (29). Although this method has not been used specifically for quantitative measurement of formaldehyde release in textiles, it has been called the closest "real-time measurement method for formaldehyde" available. The method involves drawing a metered amount of air from the atmosphere to be monitored through a solution containing pararosaniline. Monitoring can be either continuous or intermittent and thereby permit either dynamic or static chamber analysis.

Other Instrumental Methods. Although other instrumental methods have been suggested for use in analyses for free or released formaldehyde, the methods do not include accelerated storage or end-use incubation steps, but only offer alternatives to the titrimetric or spectrophotometric analyses. Examples of these replacement techniques are alternating current polarography (41) and derivatization by 2,4-dinitrophenylhydrazine for subsequent high-performance liquid chromatography analysis (42), and the more unconventional determination with an enzyme-coated piezoelectric crystal detector (43).

Comparison of Methods for Evolving Formaldehyde from Textiles

As the differences in severity among the tests for formaldehyde release are considered, it is logical to ask what formaldehyde is measured by these tests. A comparison of the amount of formaldehyde released by several test methods from a DMDHEU-treated fabric reveals that widely different levels of formaldehyde are generated by the different test methods (21). The test methods and the amount of HCHO released are as follows: cold sodium sulfite method, 63 μg/g; MITI method, 238 μg/g; Shirley Institute method, 271 μg/g; Japanese law 112–1973, 442 μg/g; and AATCC 112–1982, 908 μg/g.

Most researchers in the field agree that the cold sodium sulfite test measures free or unbound formaldehyde, but that the other tests include contributions from the finish hydrolyzed during the various test conditions (Table I). Although the AATCC-112 test does not involve a water extraction, the fabric is exposed to 100% rh at elevated temperatures in a confined atmosphere. The moisture on the fabric during the long incubation is sufficient to permit extensive hydrolysis of cross-links. Concentrated solutions of acid catalysts may be formed and trapped in unwashed fabrics in the AATCC-112 test in contrast to the very dilute catalyst concentrations from extraction with 50 to 100 mL of water per gram of fabric.

If hydrolysis of cross-links is involved in the incubation step of the test methods, formaldehyde release should vary with catalyst acidity. Table IV illustrates this point (*44*). The cold sodium sulfite test analyzes the free formaldehyde present at any point in time. The amount of formaldehyde built up as a result of storing fabric at relatively high temperatures and humidities is determined by the AATCC-112 test, and to a lesser extent (diluted by a 50:1 or greater liquor-to-fabric ratio) is determined by the Japanese law-112-1973 test.

The AATCC-112 test is much more sensitive to the effects of undercuring; therefore, it is a better indicator of these effects than is the Japanese test. This sensitivity is seen by the increase in formaldehyde release values in the AATCC-112 test with decreasing catalyst-to-agent ratios as shown in Table V (*44*). Scorch test results, a measure of free amide groups present in a finished fabric (*45*), support this conclusion.

Because the cellulose cross-link itself is probably the source of formaldehyde release in a cross-linked and washed fabric, formaldehyde release among finishes should vary according to their stability to hydrolysis and according to the relative humidity in the atmosphere. At 100% rh and 49 °C, the incubation step of the AATCC-112 test promotes hydrolysis of the cellulose cross-link (Compound **A**) to form Structures **I** and **II**. In Figure 2 are formaldehyde release profiles for four finished fabrics after laundering (*46*). Formaldehyde release is plotted versus incubation time in a sealed jar from 4 to 30 h under 100% and 65% rh at 49 °C. Because the rate and extent of release is greater at 100% rh than at 65% rh for all but

Table IV. Effect of Catalyst Type on Formaldehyde Release Tests

	HCHO ($\mu g/g$) *Released in Tests*		
Catalyst	*AATCC*	*Japanese*	*Cold Na_2SO_3*
Lewis acid type			
$Zn(NO_3)_2 \cdot 6H_2O$ (0.5%)	897	413	180
$MgCl_2 \cdot 6H_2O$ (2.7%)	1854	921	150
$Al_2(OH)_5Cl$ (2.0%)	884	530	142
Bronsted acid type			
$MgCl_2 \cdot 6H_2O$–citric acid (0.7%)	2904	638	30
$Al_2(SO_4)_3 \cdot 18H_2O$ (0.5%)	2223	790	123
$NaHSO_4 \cdot H_2O$ (0.75%)	4514	995	191
$Mg(H_2PO_4)_2 \cdot 2H_2O$ (2.2%)	2766	863	131

Table V. Treatments with DMDHEU and $Zn(NO_3)_2 \cdot 6H_2O$

% DMDHEU	% Strength Retained Scorch Test		Free HCHO (µg/g)			
			AATCC		Japanese	
	5.5 %[a]	2.8 %[a]	5.5 %[a]	2.8 %[a]	5.5 %[a]	2.8 %[a]
9	100	24	897	1787	413	460
7	95	16	723	1404	409	425
5	100	40	564	742	350	280
3	100	41	463	531	218	247

[a] Concentration of $Zn(NO_3)_2 \cdot 6H_2O$ catalyst, expressed as a percentage of DMDHEU.

Compound A.

Structures I and II.

the most stable finishes, error can arise from predicting conditions based on 100% rh incubations.

As shown earlier (Table IV), the cold sodium sulfite test does not assess the potential to release formaldehyde. Available formaldehyde at any point in time can be measured. The usefulness of this method, however, is limited by the interferences in the stoichiometry, such as presence of residual catalyst in the acid–base variation, and presence of glyoxal or other aldehydes. Examples of interferences are given in Table VI (31). A 0 °C incubation can be achieved while bypassing the sodium sulfite stoichiometry. A 0 °C incubation in deionized water is followed by Nash color development of the extracted formaldehyde (Table VII) (31).

Conclusion

For prediction of "worst case" storage and for assessment of cross-linking efficiency, the AATCC-112 (sealed jar) test is probably the best method available. For prediction of end-use performance, the incubation conditions should be modified, or a less severe method should be used. Several analytical alternatives are available for handling textiles with low formal-

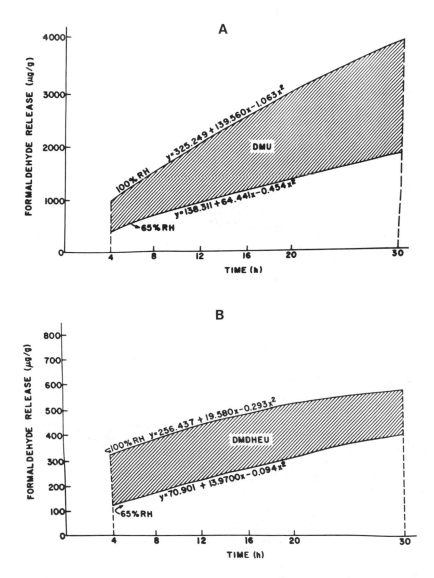

Figures 2A and 2B. *Formaldehyde release profiles: formaldehyde release vs. time at 49 °C and 65% or 100% rh from laundered cotton print cloth finished for durable press with A, dimethylolurea; and B, N,N-methoxymethyldihydroxyethyleneurea.*

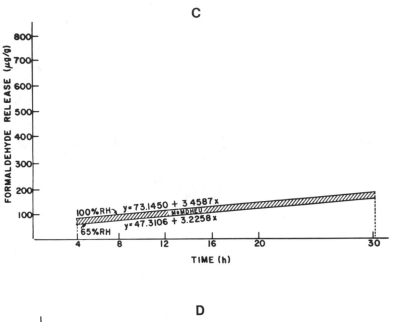

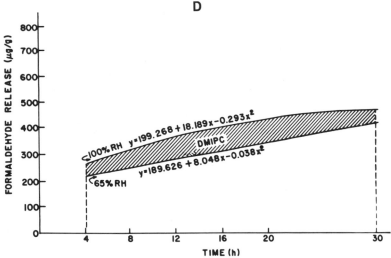

Figures 2C and 2D. Formaldehyde release profiles: formaldehyde release vs. time at 49 °C and 65% or 100% rh from laundered cotton print cloth finished for durable press with C, dimethyloldihydroxyethyleneurea; and D, isopropyl dimethylolcarbamate.

Table VI. Examples of Interferences in Cold Sodium Sulfite Analyses
by Nonformaldehyde Reactants

	Apparent HCHO (μg/g)[a]	
Fabric Finish	Unwashed	Neutral Wash
1,3-Dimethyl-4,5-dihydroxyethyleneurea	233	25
4,5-Dihydroxyethyleneurea	109	28
Acrylamide	—	17
Glyoxal	—	75

[a] Apparent level of HCHO determined due to interferences.

Table VII. Formaldehyde Released During Cold Extractions
of Finished Fabric

	HCHO (μg/g)	
Finish	Cold Na$_2$CO$_3$, pH 9.5	MBTH Development after Cold Extraction, pH 7.0
Dimethyloldihydroxyethyleneurea	17	15
Dimethylolethyleneurea	18	12
Isopropyl dimethylolcarbamate	22	24

dehyde-release characteristics (< 300 μg/g). These include replacement of Nash with MBTH reagent (precise handling is required), Nash analysis modification by increasing the amount of formaldehyde to be analyzed through aliquot manipulation, or use of a separate calibration curve to cover the range of 75–450 μg/g of formaldehyde. The instrumental methods for generating formaldehyde from textiles must, at this time, be considered research tools.

Literature Cited

1. Frick, J. G., Jr. *Chem. Technol.* **1971**, *1*, 100–07.
2. Badische Analin Soda Fabrik, A. G. "BASF Finishing Manual"; Badische Analin Soda Fabrik, A. G.: D–6700 Ludwigshafen, Federal Republic of Germany.
3. Bacon, O. C.; Parker, M. F.; Horn, L. F. *Am. Dyest. Rep.* **1957**, *46*, P933–35.
4. Reid, J. D.; Arceneaux, R. L.; Reinhardt, R. M.; Harris, J. A. *Am. Dyest. Rep.* **1960**, *49 (14)*, 29–34.
5. Petersen, H., presented at the 160th Nat. Meet. Am. Chem. Soc., Chicago, September 1970.
6. Goldstein, H. B. *Text. Chem. Color.* **1973**, *5*, 209–14.
7. Buck, G. S.; Getchell, N. F. U.S. Patent 2 957 746, 1960.
8. Nuessle, A. C. *Am. Dyest. Rep.* **1966**, *55 (17)*, P646–48.
9. Frick, J. G., Jr.; Kottes, B. A. *Text. Res. J.* **1959**, *29*, 314–22.
10. Frick, J. G., Jr.; Kottes, B. A. *Am. Dyest. Rep.* **1959**, *48 (13)*, 23–25.
11. "American Association of Textile Chemists and Colorists Technical Manual"; AATCC: Research Triangle Park, N.C., 1975; Vol. 51.
12. Ruemens, W.; Goetz, N.; Zeidler, Z. *Am. Dyest. Rep.* **1960**, *49*, 752–63.
13. Arceneaux, R. L.; Frick, J. G., Jr.; Reid, J. D.; Gautreaux, G. A. *Am. Dyest. Rep.* **1961**, *50 (22)*, 37–41.

14. Riccobono, P. X.; Ring, R. N.; Roth, A. *Text. Chem. Color.* **1976**, 8, 108–11.
15. Itoh, S.; Tamura, S. *Jpn. Text. News* **1975**, Aug., 66–72.
16. Metropolitan Section "Book of Papers for the American Association of Textile Chemists and Colorists: 1977 National Technical Conference"; AATCC: Research Triangle Park, N.C., 1977; p. 43.
17. Wightman, R. *Daily News Record* **1979**, 9 (66), 1.
18. "American Association of Textile Chemists and Colorists Technical Manual"; AATCC: Research Triangle Park, N.C., 1982–83; Vol. 58.
19. Andrews, B. A. Kottes; Reinhardt, R. M. *Text. Chem. Color.* **1983**, 15, 115–20.
20. Committee on Aldehydes, National Research Council "Formaldehyde and Other Aldehydes"; NAS: Washington, D.C., 1981.
21. Andrews, B. A. Kottes, presented at the Formaldehyde Workshop, Am. Assoc. Chem. Color., Research Triangle Park, N.C., March 1983.
22. North, B. F. *Text. Chem. Color.* **1977**, 9, 223–35.
23. Northern Piedmont Section "Book of Papers for the American Association of Textile Chemists and Colorists: 1980 National Technical Conference"; AATCC: Research Triangle Park, N.C., 1980; p. 29.
24. Roberts, E. C.; Rossano, A. J., Jr. "Book of Papers for the American Association of Textile Chemists and Colorists: 1983 National Technical Conference"; AATCC: Research Triangle Park, N.C., 1983; p. 97.
25. Wayland, R. L., Jr.; Smith, L. W.; Hoffman, J. W. *Text. Res. J.* **1982**, 51 (4), 302–9.
26. Bille, H.; Petersen, H. *Text. Res. J.* **1976**, 57, 155–65.
27. Vail, S. L.; Reinhardt, R. M. *Text. Chem. Color.* **1981**, 13, 131–35.
28. Reinhardt, R. M. *Text. Res. J.* **1983**, 53, 175–80.
29. Pickrell, J. A.; Griffis, L. C.; Hobbs, C. H. "Release of Formaldehyde from Various Consumer Products; Final Report of Lovelace Biomedical Environmental Research Institute to CPSC"; February 1982.
30. DeJong, J. I.; DeJonge, J. *Recl. Trav. Chim. Pays-Bas* **1952**, 71, 643–60.
31. Andrews, B. A. Kottes; Reinhardt, R. M.; Harris, J. A. *Text. Res. J.* **1983**, 53, 688–91.
32. Petersen, H., presented at the Am. Assoc. Text. Chem. Color. Nat. Tech. Conf., Philadelphia, September 1972.
33. *Shirley Inst. Publ.* **1975**, 48, 17.
34. Japanese Ministry of International Trade and Industry, JIS–L 1041–1960, 1960.
35. "Law for Control of Household Goods Containing Harmful Substances," Law No. 112; Welfare Ministry Ordinance: Japan, 1973; No. 34–1974.
36. Matsui, T. *Jpn. Text. News,* **1979**, Nov., 105.
37. D'Angiuro, L. *Melliand Textilber.* **1982**, 63, 522–25.
38. Moran, C. M.; Vail, S. L. *Am. Dyest. Rep.* **1965**, 54 (6), 35–36.
39. Vail, S. L.; Dupuy, H. P. *Text. Chem. Color.* **1979**, 11, 37–38.
40. Weber, R.U.; Kamath, Y. K.; Weigmann, H.-D. "Book of Papers for the American Association of Textile Chemists and Colorists: 1982 National Technical Conference"; AATCC: Research Triangle Park, N.C., 1982; p. 154.
41. Linhart, K. *Melliand Textilber.* **1975**, 56 (3), 240–46.
42. Kuwata, K.; Uebori, M.; Yamasake, Y. *J. Chromatogr. Sci.* **1979**, 17, 264–68.
43. Guilbault, G. G. *Anal. Chem.* **1983**, 55, 1682–84.
44. Andrews, B. A. Kottes; Harper, R. J., Jr.; Vail, S. L. *Text. Res. J.* **1980**, 50, 315–22.
45. Reine, A. H.; Reid, J. D.; Reinhardt, R. M. *Am. Dyest. Rep.* **1966**, 55 (9), 91–94, 153–56.
46. Gulf Coast Section "Book of Papers for the American Association of Textile Chemists and Colorists: 1980 National Technical Conference"; AATCC: Research Triangle Park, N.C., 1980; p. 5.

RECEIVED for review September 28, 1984. ACCEPTED December 18, 1984.

Formaldehyde Release from Pressed Wood Products

BEAT MEYER and KARL HERMANNS

Chemistry Department BG–10, University of Washington, Seattle, WA 98195

During the past decade the residual formaldehyde content of urea–formaldehyde (UF) bonded products has been reduced by a factor of more than 10. This reduction has been achieved by lowering the molar ratio of formaldehyde to urea in the adhesive resin from 1.85 to 1.25 or lower by addition of scavengers after treatment with ammonia or urea, or by other modifications of the resin and panel manufacturing process, and by improved quality control. Thus, formaldehyde emission rates of particle board and plywood paneling have decreased from 10–30 to 0.3 mg/m² day or even less. This chapter explains the chemical equilibrium that regulates formaldehyde release and the relationship between free formaldehyde content, age, temperature, humidity, loading, air exchange rate, formaldehyde indoor air concentration, and other parameters. A comparison of commercial adhesives and panels manufactured in 1983 shows that current generation UF-bonded wood products are capable of meeting 0.1-ppm indoor air standards at currently common load factors as long as temperature, humidity, and ventilation rates remain within a reasonable range.

$\mathbf{M}$EDIUM-DENSITY FIBERBOARD (MDF), particle board, and hardwood plywood paneling are probably the most prominent potential formaldehyde emitters among the currently used consumer products (*1, 2*). All are bonded with urea–formaldehyde resins (UFR). UFR or related hydroxymethylamino reagents such as dimethyloldihydroxyethyleneurea (DMDHEU) are also used on cotton and polyester–cotton fabrics that are used for making upholstery, drapery, and clothing (*3*). Other potential formaldehyde emitters are fiberglass insulation, latex-backed fabrics (*2*), and urea–formaldehyde foam insulation (UFFI). However, UFFI has been withdrawn from the North American market because of quality control problems (*1*). Formaldehyde may also be released from melamine–formaldehyde bonded plywood. In contrast, formaldehyde emission from phenol–formaldehyde bonded particle board or ex-

0065–2393/85/0210/0101$06.00/0
© 1985 American Chemical Society

terior grade plywood is usually very low. This chapter deals only with urea–formaldehyde (UF) bonded wood products (4).

During the last two decades particle board and MDF containing 6–10 wt % UFR and plywood paneling containing 2.5 wt % UFR have increasingly replaced whole wood and are now present in almost every residence and office building. In fact, about 330,000 metric tons are currently used each year to manufacture pressed wood products (3) in the United States. These products are used as construction materials for flooring, wall paneling, cabinet work, and furniture. These products have become so popular that most buildings now contain between 0.2 and 1.2 m^2 of product surface per 1 m^3 of indoor air volume. Under high-load conditions, even traces of residual, unreacted formaldehyde in these products may cause measurable offgassing. This offgassing can lead to noticeable formaldehyde concentrations in indoor air if the air exchange rate is comparatively low, or if air is recirculated to conserve energy that is necessary for heating and air conditioning.

Formaldehyde Complaints

The fact that UF-bonded wood products may release unreacted formaldehyde has been well known since the invention of particle board by Fahrni (5) in 1943, but the formaldehyde exposure risk was initially very moderate because these products were not used in large quantities and thus the odor dissipated rapidly. Consumer complaints about formaldehyde odor appeared only relatively recently, and only after particle board and MDF reached a level of popularity far beyond the most optimistic predictions of their inventors. The first industry conference (6) dealing with formaldehyde emission problems was published in Leipzig, Germany, in 1966. At that time some resin contained between 1 and 6 wt % free formaldehyde. This value has been gradually reduced by a factor of up to one million. The first widely publicized complaint was probably that of public school teachers (7) in Karlsruhe, Germany, in 1973, who refused to work in a classroom building because of irritating odor. The problem was traced to a combination of factors: First, ventilation had been turned off for several weeks during vacation during the hottest summer of the century, and second, the building contained large amounts of new particle board. This episode contained all of the three factors that are involved in most formaldehyde complaints: improper ventilation, large product loads, and an unusually high-emitting product. An almost identical situation in a kindergarten in Wiesbaden in August 1984 led to formaldehyde air levels of 0.7 ppm and caused the introduction of regulations that set a ceiling value of 0.1 ppm for formaldehyde in indoor air in the Federal Republic of Germany (8).

Problems with formaldehyde odor have also been reported in Scandinavia and throughout eastern Europe wherever particle board gained popularity and was used in increasing quantities. In North America, with its much larger forest resources, the introduction of particle board started approximately 10 years later than in Europe (*4*). Here, problems of formaldehyde release often involve the combination of particle board flooring with hardwood paneling. This situation is common in school buildings and in mobile homes. Mobile homes have surface loading factors of UFR-bonded products exceeding 1 m^2/m^3. Also, these buildings have minimal ventilation, are minimally insulated, and are often situated on exposed sites where sunshine and wind can cause UF-bonded panels in the building envelope to reach temperatures above 40 °C or below − 10 °C. High humidity can further compound formaldehyde release problems (*9*). Thus, indoor formaldehyde concentrations have been high in many mobile homes and have occasionally reached occupational safety standard limits. Similarly high levels have been reported when incompletely cured panels were used for shelving in libraries, bookstores, or shoestores. Fortunately, advances in UFR formulations and quality control have made any such elevated formaldehyde air levels unnecessary and, we hope, an experience of the past (*10*). In fact, today, most complaints we receive are due to individual, defective panels.

Formaldehyde Release Mechanism

At room temperature formaldehyde can exist as a dilute gas or as a solid polymer. In the presence of moisture formaldehyde reversibly hydrolyzes to form methylene glycol:

$$CH_2O + H_2O \leftrightharpoons CH_2(OH)_2 \tag{1}$$

The equilibrium is so strongly shifted in favor of the glycol that water is used as an absorber for quantitative analysis. Thus, in the indoor environment, formaldehyde can accumulate wherever moisture condensation occurs, for example, on air conditioning ducts, on cold walls, and on kitchen vents. It is also very effectively absorbed by moisture in the upper respiratory tract of human beings (*1*). The most important problem concerns the interaction of formaldehyde on wood surfaces because wood contains substantial amounts of moisture. The wood moisture content varies from 6 wt% at 30% rh to 27 wt% at 96% rh. This high-moisture adaption capacity makes wood such a comfortable indoor surface material because the transfer is slow. However, this capacity causes a time lag in the transfer of formaldehyde that may conceal the presence of strong formaldehyde-emitting sources. Formaldehyde air concentrations for a new, unoccupied mobile home built in 1981 are shown in Figure 1. This figure shows that formaldehyde indoor air concentrations undergo strong diurnal and seasonal variations with peak concentration differences that can reach a factor of 5 or more (*9*). In an inhabited home the formaldehyde levels are further complicated by occupant activities.

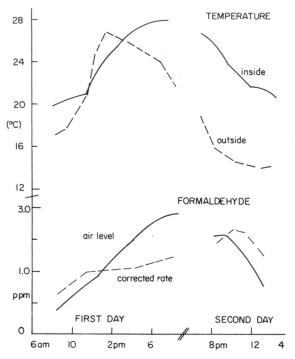

Figure 1. Temperature and indoor air formaldehyde concentration in an unoccupied home. (Adapted from Ref. 11).

Formaldehyde release in pressed wood products is due to latent formaldehyde. UF-bonded wood products are made by hot pressing wood chips that have been resinated with 5–10 wt% UFR. The chips are formed into mats and pressed at 150–190 °C for a period of approximately 10 s/mm of thickness of the finished product. During the press action, hot steam from moist wood particles transfers heat, formaldehyde, and other volatiles from the surface of the mat to the core of the board where unreacted resin components accumulate. Thus, the formaldehyde concentration in the core is approximately twice that of the surface (*12*). In the past, when UF resins with 1 wt% unreacted formaldehyde were used, the content of residual formaldehyde was usually between 0.5 to 1.0×10^{-3} wt%. In the past few years resins having low formaldehyde to urea ratios (F: U) have made it possible to reduce this value by a factor of 10 (*10*). Thus, a modern 16-mm particle board now contains less than 1 g of latent formaldehyde per square meter. Release of formaldehyde from the board into indoor air is diffusion controlled and thus gradually decreases over time. The release rate is strongly affected by the presence of urea because of the equilibrium reaction yielding monomethylolurea (*13*):

$$NH_2CHNH_2 + CH_2(OH)_2 \rightleftharpoons NH_2CONHCH_2OH + H_2O \qquad (2)$$

The equilibrium constant for this reaction is

$$\frac{[NH_2CONHCH_2OH][H_2O]}{[NH_2CONH_2][CH_2(OH)_2]} = k = 2 \times 10^{-2}\, mol/L \qquad (3)$$

Thus, moisture hydrolyzes formaldehyde resins, whereas urea acts as a scavenger. UF resins are manufactured by step-wise reactions. First, urea is reacted with a fourfold excess of formaldehyde to yield a mixture of low-rank chain species with terminal methylol groups. This first reaction product is commercially available as UF concentrate with up to 85 wt% solid content. In the formulation of the resin, more urea is added. In fact, modern resins contain excess urea, as explained later.

Formaldehyde Measurement, Product Characterization, and Exposure Levels

The measurement of formaldehyde air concentrations is normally conducted with the standard NIOSH method (*1*). But this method was designed for measuring occupational threshold limit levels and is not validated for levels below 0.2 ppm. It requires cumbersome air pumps that must be frequently calibrated and glass impingers containing aqueous solutions, and depends on laboratory measurements with time-consuming solution chemistry. Currently, the most convenient field tools are passive samplers that require a 5-day measuring period for collecting formaldehyde on impregnated filters, on solid sorbents, or in solution. These devices can be used to detect 0.01 ppm formaldehyde over 1-week exposure, but a more rapid method for measuring indoor air levels or personal exposure is needed. However, the usefulness of air concentration measurement for identifying defective products in occupied buildings is intrinsically limited because the temperature and humidity of indoor air constantly changes.

Thus, any meaningful prediction of indoor formaldehyde air exposure level depends on knowledge of the formaldehyde emission characteristics of the emitting material. For a reliable prediction of product performance, the total latent formaldehyde content and the formaldehyde release rate under well-defined, standardized conditions must be known. The best method for determining the latent formaldehyde content of pressed wood products is still the European perforator standard method (*14*), even though this test requires careful conditioning and moisture control and is no longer sensitive enough to differentiate among the best of the current commercial products.

Several different methods are currently used to measure formaldehyde release rates. The oldest standard method is the Japanese industrial standard JIS–A5908–1974, which uses a 10-L desiccator (15). In Europe, the most popular test is probably the Wilhelm Klausnitz Institute (WKI) test (16). It is similar to the U.S. textile test (3) American Association of Textile Chemists and Colorists (AATCC) method 112–1978 and suffers from the same problems as all accelerated tests in that it overemphasizes formaldehyde release from those chemical functions that are temperature and moisture sensitive (17). In North America a less sensitive 2-h version of the 24-h JIS test has been adopted (18) for use in production and quality control. The most reliable test involves testing products in air chambers containing real-life load factors, but this test is also the most expensive and time consuming. Currently, some six different large chamber designs are in use (18). The proposed European air chamber standard CEN–N76E (19) uses 1 m^2 board in a volume of 1 m^3 air with 50% rh at 25 °C at two different ventilation rates. Tests in such chambers reveal that formaldehyde release depends on several different interactive parameters. Among these the most prominent are age, temperature, humidity, load factor, and ventilation.

The Effect of Age. If formaldehyde release is due to free formaldehyde in the product, the release rate is proportionate to total formaldehyde in the product, and thus the offgassing decreases exponentially with age. The following experimental formula has been developed (20):

$$C_N = C_o \exp\left[\frac{24N}{P} (0.01 - C_o)y \right] \qquad (4)$$

where C_N is the observed concentration in parts per million; C_o is the original concentration; N designates the air exchange rate [as air change per hour (ach)]; P equals the European perforator value for total latent formaldehyde; and y is the product age expressed in years. The value of C_o depends on the material and adhesive. Figure 2 shows the calculated change in formaldehyde emission rates for two different types of board: one with an initial emission rate of 1 ppm and one for 0.5 ppm. For each board type, the calculated effect of ventilation on aging is shown. The parameters in Equation 4 and Figure 2 were chosen to fit observations from Swedish field studies (20). However, in very new products, the emission rate decreases even more rapidly. It is not uncommon for it to decrease by a factor of two within the first 6 weeks (10).

The Effect of Structure and Porosity of Building Materials. The formaldehyde emission rate is strongly influenced by the nature of the material. Particle board is more porous than MDF or plywood, and thus, for a

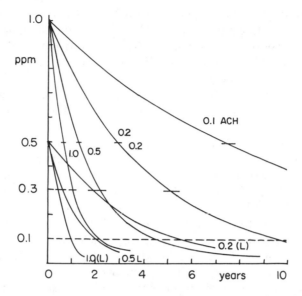

Figure 2. Formaldehyde release from a high-emitting and a low-emitting particle board as a function of age, at different ventilation rates.

given adhesive, the residual formaldehyde effuses at a different rate (*21–24*). This dynamic factor, K, determines formaldehyde emission as follows:

$$C_N = C_o[KL/(N + KL)] \tag{5}$$

where L is the load factor (square meters per cubic meter). The transport coefficient, K (meters per hour), is measured by observing formaldehyde air concentrations in an air chamber as a function of the ratio of load factor to ventilation rate. Typical experimental values of K are 0.4 ± 0.3 for plywood and paneling, 0.5 ± 0.2 for MDF, and 0.8 ± 0.2 for particle board (*21–24*). If the exposed surfaces are treated, the values decrease. Common barrier materials are gypsum board and carpets. Transport coefficients are 0.5 ± 0.1 m/h for nylon carpets with urethane or sponge rubber backing, and 0.62 m/h for 12-mm gypsum wallboard (*25*).

   The Effect of Temperature and Humidity.   Extensive research in Europe and North America has confirmed Japanese work demonstrating that the temperature effect can be predicted for most pressed wood products to ± 10% with the help of the following formula (*21–23*):

$$C_N = C_o[1 + Ah]\exp[-9799(1/T - 1/T_o)] \tag{6}$$

where T_o (original temperature) is 25 °C and standard humidity is 50% rh;

A is the humidity coefficient, and h is the humidity differential. The temperature effect is shown in Figure 3. The increase with temperature is very large: The emission doubles at 32 °C (88 °F) and triples at 35 °C (95 °F). Thus, formaldehyde emission will noticeably increase when building materials are exposed to sun, or when they are used in very warm climates. This effect is significant in poorly insulated mobile homes in the southern belt (9) of the United States where, during summer months, daily temperatures exceed 40 °C.

The humidity coefficient, A, is approximately 0.0175, but calculated humidity values are of very limited value because the moisture exchange between air and wood is so slow that equilibrium is rarely achieved in any occupied buildings. Under equilibrium conditions wood moisture (26) is approximately 9 wt% at 20 °C and 50% rh. At 90% rh this value increases to 20 wt%, and at 10% rh it decreases to 2.5 wt%. Thus, a change of air humidity by 10% in a living room with particle board flooring and plywood paneling at 20 °C can lead to the transfer of 5 L of water between wood surface and indoor air. Thus, wood must be conditioned for 5 or more days before formaldehyde emission can be measured reliably (18). A 10% change in humidity causes about an 18% change in formaldehyde emission.

The Effect of Load Factor and Ventilation Rate.   An increase of the

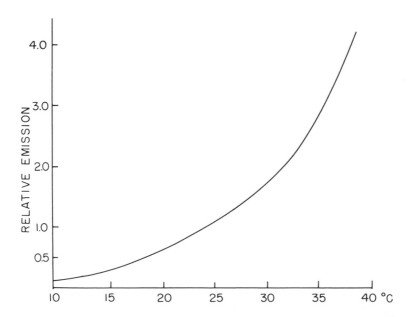

Figure 3. Temperature dependence of formaldehyde emission from pressed wood products (21–24).

product surface, expressed as a ratio of product surface to indoor air volume, normally designated as the load factor, L, increases indoor formaldehyde concentrations. On the other hand, increasing the ventilation rate decreases the formaldehyde concentration. However, the two factors are interconnected, and both factors are time dependent because in normal dwellings the air exchange rate is high enough to prevent full equilibrium formaldehyde concentrations from building up. Thus, formaldehyde measurements are extremely sensitive to air mixing and ventilation rates. If several formaldehyde-emitting materials are present in a room, the emission from all materials must be considered. If comparable adhesives are used in manufacturing the products (23), then

$$C_N = (C_{o1}K_1L_1 + C_{o2}K_2L_2 + C_{o3}K_3L_3)/$$

$$(N + L_1K_1 + L_2K_2 + L_3K_3) \quad (7)$$

unless the room also contains formaldehyde sinks. Low-emitting wood products are highly effective scavengers for high-emitting products because they usually contain unreacted urea functions that can act according to the chemical equilibrium described in Equations 2 and 3. Calculated effects of load factor and ventilation are shown in Figures 4 and 5 for three different adhesive types. Figure 4 represents the load situation in a standard single-wide U.S. mobile home containing approximately 200 m^2 of 5-mm thick hardwood plywood (PW) paneling and approximately 100 m^3 of 16-mm thick particle board (PB) flooring with load factors of 0.95 and 0.45, respectively, to yield a combined load factor of 1.4 m^2/m^3. The top curve is calculated for typical products used during the period of 1970 to 1981; C_{oPB} = 1.5 ppm and C_{oPW} = 0.80 ppm. Curve B assumes C_{oPB} = 0.51 ppm and C_{oPW} = 0.36 ppm to correspond to products marketed from 1981 to 1982. Curve C represents the best commercial products available in spring 1983 with values of 0.20 and 0.12 ppm for C_{oPB} and C_{oPW}, respectively. In conventional housing, loading rates depend on building style and climate region, but loads are always lower than in mobile homes. Typical western-style ramblers usually have plywood loads of 0.2 to 0.6 m^2/m^3 and particle board flooring that yields L = 0.1 to 0.5 m^2/m^3. The corresponding air levels are shown in Figure 5 for the adhesives in Figure 4. These calculated air levels assume good air mixing throughout the building, as is the case in structures that are fully air conditioned and have forced air heating. This high-mixing situation prevails in severe climates such as in central North America.

   In coastal areas with moderate temperatures where heating and cooling is intermittent, air mixing may be significantly lower. This decreased air mixing can lead to high local formaldehyde concentration gradients within buildings and high local concentrations inside areas where cabinets

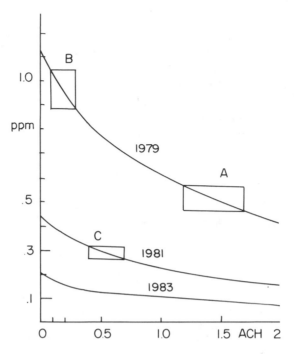

Figure 4. Formaldehyde air levels as a function of air exchange rates for three different UFR adhesives for standard mobile homes.

and shelves or other sources are located. Examples of such situations are bookshelves in libraries and bookstores or kitchen cabinet areas in freshly remodeled houses. In European buildings with hot-water heating, local formaldehyde concentrations also may vary greatly because rooms are individually ventilated and often very intermittently. Thus, bedrooms, living rooms, and kitchens with cabinets may show high local formaldehyde levels.

Control and Reduction of Indoor Formaldehyde Levels

As stated earlier, three factors determine indoor formaldehyde levels: ventilation rates, product loading, and product emission rates. Figures 4 and 5 show that ventilation is only an effective control strategy in buildings with extremely low air exchange rates, such as 0.2 ach or less. Such levels resulted from sudden tightening of homes during the energy crisis of the 1970s. The figures show that a reduction of infiltration from 1.5 ach (area A) to less than 0.3 ach (area B) caused an increase of indoor formaldehyde concentrations by a factor of 2 or more, depending on the product load. The immediate strategy for reducing formaldehyde levels was to restore air exchange rates to approximately 0.5 ach (area C). However, Figures 4 and

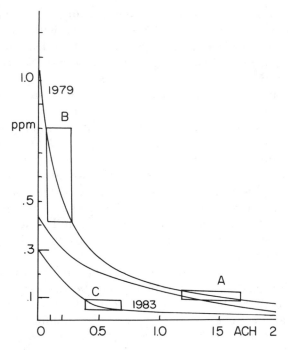

Figure 5. Formaldehyde air levels as a function of air exchange rates for three different UFR adhesives for traditional U.S. homes.

5 show that above 0.5 ach a further increase of ventilation is inefficient and costly. Thus, the best method of formaldehyde control is prevention of its emission. Formaldehyde emission can be controlled by four manufacturing methods: substitution of UFR with other resins, after-treatment of finished products, addition of scavengers to products, and adhesive modification.

Adhesive chemistry is a mature field, and markets are highly competitive. If there were any suitable substitute materials for UFR they would be well known and would have been promptly implemented. The most frequently tested competitors are probably phenolics, isocyanates, and lignosulfonates (4, 27). UF resins are used because they are cheaper, less toxic, more compatible with wood, and, except for quality control problems with formaldehyde, they remain unexcelled in their environmental compatibility. UFR can be used as a plant nutrient, and discarded particle board can be comminuted to yield an excellent soil conditioner. Properly cured UFR is nontoxic, and it has been used in surgical wound dressing (4). If a new adhesive would become available today, it still would take several years to develop any expertise comparable to the well-established UFR application art.

After-treatment of boards to bind residual unreacted formaldehyde is a well-established technology. The process consists of spraying warm, finished board with a urea or similar scavenger solution, fumigating boards with ammonia vapor (28, 29), or sprinkling boards with ammonium carbonate (30). This type of treatment is capable of reducing formaldehyde emission of particle board and MDF to as little as 5–10%. In some cases, the ammonia fumigation method has been successfully used for treating entire homes, but this after-treatment is not nearly as reliable and far more expensive than factory treatment of paneling and boards.

Clearly, the most desirable approach is to eliminate formaldehyde emission at its source. During the past decade the forest products industry and adhesive industry have developed several methods to incorporate scavengers into the product; thus, wood chips are now successfully pretreated with urea (31–33), usually by addition to the wax solution that is applied to impart better water resistance. Other scavengers include ammonia salts (34), lignosulfonate (35), and various types of natural proteins (36). Modification of press time and temperature are effective tools (37), but they require energy expenses.

The most attractive solution consists of improving UFR formulations so that any additional manufacturing steps can be avoided. Formaldehyde emission is directly related to F:U (20, 31, 38). Conventional resins with F:U of less than 1 are feasible, but below a value of approximately 1.5 the internal bond strength of the product is greatly reduced, water swelling increases, and gelation time increases (33). Furthermore, the storage life of such adhesives is short. However, adhesives with F:U between 1.45 and 1.05 are now marketed by many adhesive manufacturers with good success. They are made by a third or fourth addition of urea (39), by high-pressure and temperature condensation (40), by addition of polyfunctional alcohols (40, 41), by cocondensation with small quantities (as little as 0.1 wt%) of melamine or phenol (41–43), or by combining precondensates having different molecular weights (44).

Progress in adhesive formulation has been greatly aided by progress in chemical analysis, such as thin-layer chromatography and, especially, development of C-13 NMR analysis that allows rapid identification of resin components in liquid resin and, recently, even in solid cured resin (45). Furthermore, adhesive chemistry has profited by cross-fertilization with the textile resin industry where similar problems had to be tackled at the same time (3).

Progress in commercialization of low-formaldehyde-emitting adhesives has been rapid. Every European adhesive manufacturer currently sells a low-emitting resin, and North American adhesive formulations are now rapidly improving. We compared seven commercial adhesives that were sold in spring 1983 and found the formaldehyde emission values listed in Table I. These MDF boards were made in a commercial pilot plant (10). The table shows that emissions from currently produced products differ by a factor of approximately 10. Similarly large contrasts have been observed

Table I. Formaldehyde Release from MDF Made with Seven Commercial Adhesives

	2-h Desiccator Value (mg/L)				
Adhesive	3 days	6 weeks	5 months	10 months	F : U
A	—	8.4	—	4.4	1.85
B	4.8	2.3	2.0	1.9[a]	1.65
C	5.6	3.0	2.3	2.0[a]	1.65
D	2.6	1.6	0.86	0.70	1.65
E	2.5	1.4	0.85	0.71	1.26
F	1.4	0.72	0.62	0.59	1.20
G	0.54	0.36	0.38	0.40[a]	1.05

[a] Freshly cut specimen.

by two other groups of investigators (23). However, in spring 1984 already more than 53% of total U.S. plant capacity produced particle board yielding formaldehyde air levels of less than 0.2 ppm at 50% rh, 25 °C, and 0.5 ach, and hardwood plywood paneling with less than 0.3 ppm within 5 days after manufacture (47). In fact, FTM–2 air chamber values of some commercial UF-bonded U.S. particle board is now as low as 0.12 ppm.

The introduction of new resin formulation and scavenger systems is still accelerating. Recently, three companies independently introduced methods that claim to maintain full internal bond strength while reducing formaldehyde emission (48–50) to below 0.1 ppm. One process (50) in full-scale commercial use in Denmark since 1983 applies traditional urea scavengers and wax solutions before the wood ships are dried, prior to resination. This procedure reduces not only formaldehyde emission, but also glue consumption, and thus apparently reduces costs below those of higher formaldehyde-emitting systems (50). As these types of new developments are implemented, the formaldehyde emission from commercial products will clearly decrease beyond the progress shown in Figure 6. The y-axis shows U.S. desiccator values for samples measured in our laboratory. The x-axis lists the corresponding calculated air levels.

As indicated earlier, the translation of laboratory test measurements into field indoor air predictions is a dynamic field, and it is still difficult to make accurate predictions unless the material properties are well known. However, extensive, independent work (46, 51–53) indicates that current state-of-the-art commercial materials, such as the resin system G in Table I, are capable of yielding indoor air levels of 0.1 ppm under appropriate product use conditions within less than 1 month after manufacture except when products are used in hot climates (9) or under other conditions in which building components cannot be kept at a temperature below 30 °C or a humidity below 75% rh. Thus, indoor formaldehyde levels apparently can be kept within the values of the ASHRAE 62–1981 guideline (54), the proposed German legislation (8), and similar indoor air guidelines and

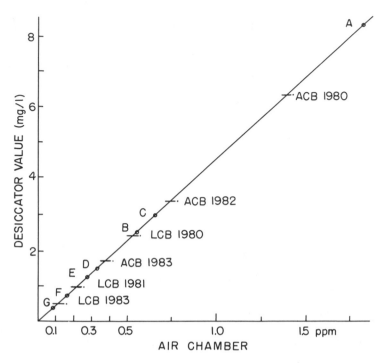

Figure 6. Desiccator values for seven MDFs made with seven adhesives, Table I, and air chamber values FTM-2 for average-emitting (ACB) and low-emitting (LCB) commercial particle board manufactured with resin technologies of 1980 to 1983.

standards (*1*). This situation makes it possible to eliminate the complaints of high formaldehyde levels of the 1970s, and, in fact, it seems that if the current trend continues, formaldehyde levels can be kept at the ambient level.

Acknowledgments

The experimental part of this work was conducted in cooperation with Washington Iron Works in Seattle. The authors thank David C. Smith of Plum Creek Lumber Company, Jack Pommerening of Coe Manufacturing Company, Birger Sundin of AB Casco, and Per Hanetho of Dyno Industries for valuable help and suggestions.

Literature Cited

1. Meyer, B. "Indoor Air Quality"; Addison-Wesley: Reading, Mass., 1983.
2. Pickrell, J. A.; Mokler, B. V.; Griffis, L. C.; Hobbs, C. H.; Bathija, A. *Environ. Sci. Technol.* 1983, *17*, 753.
3. Petersen, H. In "Chemical Processing of Fibers and Fabrics"; Lewin, M.; Sello, S. B., Eds.; Marcel Dekker: New York, 1983.

4. Meyer, B. "Urea–Formaldehyde Resins"; Addison-Wesley: Reading, Mass., 1979.
5. Fahrni, F. French Patent 881 781, 1943.
6. Thomas, M. *Holztechnologie* **1964**, *5 (supplement)*, 79.
7. Deimel, M. In "Organishe Verunreinigungen in der Umwelt"; Aurand, K., Ed.; E. Schmidt: Hamburg, 1978; pp. 416–27.
8. "Formaldehyde"; Bundesgesundheitsamt: Berlin, Federal Republic of Germany, 1984.
9. Meyer, B.; Hermanns, K. "Lawrence Berkeley Laboratory Report"; LBL–18573, 1984.
10. Meyer, B.; Hermanns, K.; Smith, D. *J. Adhes.* **1984**, *17*, 297–308.
11. U.S. Forest Products Laboratory, Department of Agriculture, "Formaldehyde Air Contamination in Mobile Homes: Variation with Interior Location and Time of Day" by Myers, G.; Seymour, J. W.; Khan, T., unpublished data.
12. Meyer, B.; Carlson, N. L. *Holzforschung* **1983**, *37*, 41.
13. de Jong, J. I.; de Jonge, J. *Recueil* **1953**, *72*, 1027.
14. "European Standard N 75 E: Particle Boards—Determination of Formaldehyde Content—Extraction Method Called Perforator Method"; European Committee for Standardization: Brussels, 1982.
15. "Japanese Industrial Standard A–5908: Japanese Standard for Particle Board"; Japanese Standard Association: Tokyo, 1978.
16. Roffael, E. "Formaldehydabgabe von Spanplatten und anderen Werkstoffen"; DRW: Stuttgart, 1982.
17. Vail, S. L.; Reinhardt, R. M. *Text. Chem. Color.* **1981**, *13*, 131.
18. "Tentative Formaldehyde Test Methods FTM–1 and FTM–2, Two-Hour Desiccator Test and Large Air Chamber Test"; National Particle Board Association; Hardwood Plywood Manufacturers Association: Silver Spring, Md., 1983.
19. "European Standard CEN–N 76 E: Situation Report: Particle Boards—Determination of Formaldehyde—Emission under Specified Conditions; Method Called Formaldehyde Emission Method"; European Committee for Standardization: Brussels, 1984.
20. Sundin, B. "Proceedings of the 3rd Medical Legal Symposium on Formaldehyde Issues", Silver Spring, Md., 1982.
21. Berge, A.; Mellegaard, B.; Hanetho, P.; Ormstad, E. B. *Holz Roh. Werkst.* **1980**, *38*, 251.
22. Hoetjer, J. J.; Koerts, F. *Holz Roh. Werkst.* **1981**, *39*, 391.
23. Myers, G. E. *For. Prod. J.* **1983**, *33 (5)*, 27.
24. Andersen, I.; Lundqvist, G. R.; Mølhave, L. *Atmos. Environ.* **1975**, *9*, 1121.
25. Matthews, T. G.; Reed, C. J.; Daffron, C. R.; Hawthorne, A. R. "Proceedings of the International Particle Board Symposium"; 1984, Vol. 18, p. 200.
26. U.S. Department of Agriculture, "Wood Handbook: Wood as an Engineering Material," USDA Handbook 72; Government Printing Office: Washington, 1974.
27. Deppe, H. J.; Ernst, K. "Taschenbuch der Spanplatten Technik"; DRW: Stuttgart, 1977.
28. Verbestel, J. B. European Patent 27 487, 1979.
29. Roffael, E. German Patent Application 2 829 021, 1980.
30. Kaesbauer, F.; Bolz, W.; Diem, H. European Patent 13 372, 1980; U.S. Patent 3 950 472, 1980.
31. Petersen, H.; Reuther, W.; Eisele, W.; Wittmann, O. *Holz Roh. Werkst.* **1974**, *32*, 402.
32. Graser, M.; Hann, E. W. European Patent 1 237, 1981.
33. Gøte-Helgesson, B.; Mannson, B.; Wallin, N. H. British Patent Application 2 019 854, 1979.
34. Neumann, C. European Patent 23 002, 1983.
35. Holmquist, H. W. U.S. Patent 4 186 242, 1980.
36. Tinkelenberg, A.; Vaessen, H. W. L.; Suen, K. W.; Van Doorn, A. J. U.S. Patent 4 282 119, 1981.

37. Ernst, K. *Holz Roh. Werkst.* 1982, *40*, 249.
38. Mayer, J. "Particle Board, Future and Present"; DRW: Stuttgart, 1979; pp. 102–10.
39. Spurlock, H. N. U.S. Patent 4 381 368, 1983.
40. Dudeck, C.; Weber, E.; Diem, H.; Wittmann, O. German Patent Application 3 115 208, 1982.
41. Schittek, H. German Patent 3 027 203, 1982.
42. Diem, H.; Fritsch, R.; Lehnert, H.; Matthias, G.; Schatz, H.; Wittmann, O. German Patent Application 3 145 328, 1983.
43. Diem, H.; Fritsch, R.; Lehnert, H.; Matthias, G.; Schatz, H.; Wittmann, O. German Patent Application 3 125 874, 1983.
44. Brunmueller, F.; Mayer, J.; Lehmann, G.; Wittmann, O.; Bolz, W.; Kaesbauer, F. German Patent Application 2 747 830, 1979.
45. Maciel, G. E.; Szeverenugi, N. M.; Early, T. A.; Myers, G. *Macromol.* 1983, *16*, 598.
46. Sundin, B. "Proceedings of the International Particle Board Symposium"; 1982, Vol. 16, p. 3.
47. "Comparison of Costs Associated with In-Plant Techniques for Reducing Formaldehyde Emissions from Particle Board and Hardwood Plywood Paneling"; Heiden Associates report to the Formaldehyde Institute, submitted to the U.S. Environmental Protection Agency, 1984.
48. Williams, J. H. U.S. Patents 4 409 293 and 4 410 685, 1983.
49. Merkel, D.; Matthias, G., Diem, H.; Wittmann, O. European Patent Application 96 797, 1983.
50. Mansson, B.; Sundin, B.; Sirenius, K. Belgian Patent 896 669, 1983.
51. Lehmann, W. F. "Proceedings of the International Particle Board Symposium"; 1982, Vol. 16, p. 35.
52. Newton, L. R. "Proceedings of the International Particle Board Symposium"; 1982, Vol. 16, p. 45.
53. McVey, D. T. "Proceedings of the International Particle Board Symposium"; 1982, Vol. 16, p. 21.
54. "Standard for Ventilation Required for Minimum Acceptable Indoor Air Quality"; American Society of Heating, Refrigerating and Air-Conditioning Engineers: Atlanta, Ga., 1981; ASHRAE Standard ANSI-ASHRAE 62-1981.

RECEIVED for review September 28, 1984. ACCEPTED December 13, 1984.

Current Status of Measurement Techniques and Concentrations of Formaldehyde in Residences

RICHARD B. GAMMAGE and ALAN R. HAWTHORNE

Health and Safety Research Division, Oak Ridge National Laboratory, Oak Ridge, TN 37831

For measuring concentrations of formaldehyde in residences, scientists are making increasing use of passive integrating monitors that can provide time-weighted average concentrations down to slightly more than 0.01 ppm if the periods of exposure are extended to a few days. The more traditional modified NIOSH method with a 1–2-h sampling time lacks the sensitivity to make accurate measurements at the frequently encountered concentrations of 0.1 ppm or lower. More rigorous intercomparisons of various monitoring systems are required. Marked dependence of formaldehyde concentration on age is observed for different classes of dwellings. As building and furnishing materials that contain urea–formaldehyde resins age, they emit formaldehyde less strongly. Limited studies have revealed diurnal and seasonal within-house fluctuations of two- and tenfold, respectively. Occasional excursions to 0.1 ppm seem to occur in the majority of houses.

Human health problems related to formaldehyde exposure in residences became an increasingly active issue throughout the 1970s. The sectors of the public expressing the most concern were residents of mobile homes and houses insulated with urea–formaldehyde foam insulation (UFFI). For all types of dwellings, the formaldehyde exposure appears to be the highest in mobile homes with a recently reported mean concentration of 0.38 ppm (1). The mean indoor concentration of formaldehyde for several hundred U.S. homes with UFFI, including complaint and noncomplaint homes, has been reported to be 0.12 ppm. Inside UFFI homes in Canada, the mean concentration of formaldehyde was reported recently to be only slightly above 0.05 ppm (2). In about 10% of these homes, however, the formaldehyde concentrations were 0.1 ppm or greater. For comparison, the mean concentrations of formaldehyde inside older conven-

0065–2393/85/0210/0117$06.00/0

tional homes are usually less than 0.05 ppm, and only a few exceed 0.1 ppm (1, 2, 3). The formaldehyde ceiling concentration for personal comfort established by the American Society of Heating, Refrigeration, and Air Conditioning Engineers (ASHRAE) is 0.1 ppm (4).

The primary sources of the airborne formaldehyde are urea–formaldehyde (UF) resins used in pressed-wood products such as particle board, fiberboard, plywood, insulation, and other building material such as decorative paneling. Degradation of the UF polymeric structure by moisture-induced reactions leads to a chronic release of formaldehyde (5). Usually lesser amounts of formaldehyde can also be emitted by combustion sources and tobacco smokers.

To place some perspective on the scale of formaldehyde production, approximately 6 billion lb was produced in 1984 (6). Resins made with urea or phenol account for half of the total formaldehyde consumption in the United States, and most of these resins go into housing materials. It should, therefore, be of little surprise that formaldehyde is ubiquitous to modern living environments.

This review is intended to be read with two important questions in mind: How well are the available monitoring devices able to cope with current and future demands? What is satisfactory, lacking, or amiss in our current state of knowledge about the levels and behavior of formaldehyde in residences?

Commonly Used Monitors

Most of the methods that have been used for measuring formaldehyde levels in air have recently been reviewed by Balmat (7) for the Formaldehyde Institute. Only those monitoring techniques that have seen or are seeing extensive use in residential monitoring will be considered here.

Modified NIOSH Method (8).   The monitoring technique used with the greatest frequency has been the midget impinger sampler containing 1% sodium bisulfite solution instead of the previously advocated pure water. Subsequent colorimetric analysis is limited to the chromotropic acid method. Sampling is usually conducted for 1 or 2 h at an air flow rate of approximately 1 L/min. Lower limits of detection of 0.04 and 0.1 ppm have been published for the method (9, 10). More recently a minimum detectable concentration of 0.1 ppm in field work has been quoted by Dally (11) for a 1-h sampling time. The sensitivity could, of course, be improved by increasing the time of air sampling. In most instances this option is unattractive because a technician usually attends the sampler during operation, and expenses escalate with longer sampling times. Losses of formaldehyde via evaporation can introduce additional difficulties.

The NIOSH method was developed with the monitoring of the workplace atmosphere in mind, as the name of the parent organization implies. The OSHA standard for the workplace is 3 ppm of formaldehyde averaged over 8 h. In residences, however, one is usually concerned with measuring much smaller concentrations of formaldehyde. Nevertheless, during the latter 1970s, the modified NIOSH method, probably because it was the best available at the time, became the standard method for measuring formaldehyde in residences.

Very recently an error analysis has been reported (2) for results obtained in large-scale residential studies in Canada. The aim of an absolute, total error limit (sampling plus analysis) of 15% at the 90% confidence level was not achieved. The observed uncertainty at a concentration of 0.1 ppm was 38%. The magnitude of this uncertainty increased dramatically for concentrations below 0.1 ppm and was caused largely by an increase in the coefficient of variation associated with the sample analysis. Because of the large uncertainties in results at the formaldehyde concentrations commonly encountered in residences (<0.1 ppm), a more sensitive technique than the NIOSH chromotropic acid method was required (2).

The modified NIOSH method may be adequate for testing compliance with the 0.4 ppm of formaldehyde indoor ambient air quality standard for mobile homes sold in Wisconsin (12). Its appropriateness for making accurate measurements at levels of 0.1 ppm or less is in serious question in our judgment because of the Canadian findings (2). Most measurements using this technique have indeed been reported for such low formaldehyde concentrations. Quite often one sees formaldehyde concentrations of 0.1 ppm quoted to two and even three decimal places; such a practice implies an accuracy that cannot possibly exist.

Passive Samplers.   Passive sampling devices are rapidly becoming more popular. The main reasons are cost effectiveness, small size, and ability to provide time-weighted average concentrations of formaldehyde over sampling periods of 1 or more days. These devices have the added advantage that several companies offer the badge-type monitors with mail delivery and return for analysis. This feature, together with a relatively low cost, allows individual homeowners to measure formaldehyde levels in their own residences. Large-scale field studies are also turning to more extensive use of these passive monitors (2, 13, 14). A corresponding decline is occurring in the use of the modified NIOSH method.

The mail service, passive badge, or tube-type devices are exemplified by 3M's series 3750 and Air Quality Research's (AQR) PF–1 formaldehyde monitors. Each type contains a sorbent of sodium bisulfite in solid form, and each samples formaldehyde vapor at a rate that is diffusionally controlled. The lower limits of detection are of the order of 1 ppm·h, and exposure times between 1 and 7 days are recommended by the suppliers. Anders (14) has reported on the results of more than 10,000 3M badge-type analyses done for homeowners throughout the United States. Several tens of thousands of the AQR devices are currently being used in the Canadian National Testing Survey (2). It has been reported privately to us that in these large-scale studies in the field with the AQR monitors, the uncertainty at a formaldehyde concentration of 0.1 ppm is ±20% at the 80% confidence level (15). In other field studies conducted with a similar passive sampler developed at Lawrence Berkeley Laboratory, a considerably better overall accuracy was reported equal to ±14% at the 95% confidence interval (16). These two disparate field experiences indicate that more extensive field validation is necessary. After such evaluation, it could be said with more certainty whether or not a particular type of passive sampling device is superior to the modified NIOSH monitoring technique.

In fact, our principal criticism of the use of these and other passive devices is that they were introduced into residential service before critical assessment of the methodologies and round-robin and intercomparison testing had been carried out by investigators other than the developers or vendors themselves. The introduction of these devices into widespread use without complete validation under field conditions has probably resulted in measurements with less than the claimed precision and accuracy. In the last half of 1983, we conducted some limited intercomparison testing of different devices both inside residences (17) and in the laboratory (18). The evaluations were not extensive and serve primarily to indicate the need for more rigorous evaluations.

For the field intercomparison (17), five homes were selected in east Tennessee to provide a wide range in house age, type of construction, and level of formaldehyde (19). Results are given in Table I. Monitoring was generally for 24 h (nonstandard times of exposure are shown in parentheses) and was carried out in the manner shown in Figure 1. Measurements with each of the four monitoring devices were carried out in quadruplicate. The LBL reference method (20) used a refrigerated train of impingers containing pure water with subsequent colorimetric analysis with pararosaniline (20). The Oak Ridge National Laboratory (ORNL) passive monitor (21) also used pure water and the pararosaniline method of analysis. Several important ancillary parameters that could affect the levels of formaldehyde, or the performance of the passive monitors (such as face velocity or relative humidity), were also measured (Table II). If, for example, the air drift velocity is too low, the effective sampling rate for formaldehyde is decreased. The dimensions of the sampler determine the limiting air velocity below which the sampling rate is reduced. The sampling rate can also be affected adversely by operation at relative humidities that are either too high (16) or too low (18).

Specific problems in the field tests (Table I) were encountered with each type of passive formaldehyde monitor. Unexposed controls gave high blank readings, and field exposures were low by as much as 50%. The reusable ORNL monitors, which had received several prior field exposures and whose semipermeable membranes had discolored, failed to record a measurable level of formaldehyde in house Number 3. A remeasurement, how-

Table I. Field Testing of Passive Formaldehyde Monitors Inside Five East Tennessee Houses

	House Number				
Monitor	2	3	4	5	7
LBL reference	0.03 ± 0.01	0.31 ± 0.03	0.06 ± 0.01	0.18 ± 0.02	0.38 ± 0.01
Passive commercial (A)	0.12 ± 0.03	0.30 ± 0.01 (60 h)	0.09 ± 0.05	0.19 ± 0.03	0.42 ± 0.06 (0.14 ± 0.00) (120 h)[a]
Passive commercial (B)	0.05 ± 0.02	0.16 ± 0.01 (60 h)	0.03 ± 0.01	0.13 ± 0.01	0.17 ± 0.02
ORNL passive	<0.03	<0.03 (0.36 ± 0.02)[b]	0.03 ± 0.01	0.22 ± 0.02	0.27 ± 0.03

NOTE: All measurements are expressed as parts per million.
[a] Additional measurement for extended time.
[b] Repeat measurement with new monitors.

Figure 1. Field intercomparison of passive formaldehyde monitors in house No. 5; the air drift velocity is being measured by observing the drift of smoke powder along a meter stick.

Table II. Age of Homes and Other Environmental Parameters During Testing of Formaldehyde Monitors

House No.	Air Exchange Rate (h^{-1})	Air Drift Velocity (cm/s)	Indoor Relative Humidity $(\%)$	Indoor Temperature $(°C)$
2 (10 years old)	0.6	1–20	55–60	26
3 (3 years old, UFFI prefit)	0.7	5–15	70–80	24
4 (15 years old, UFFI retrofit)	0.2–0.5	5–15	65–75	23
5 (1 year old)	0.5	1–10	60–65	24
7 (2 years old, energy supersaver)	0.9	5–10	60–65	23

ever, with new, unused ORNL monitors gave a more acceptable result for the formaldehyde concentration. This adverse experience points to a need for a quality control program to establish the criteria for reusability of passive monitors of the ORNL type.

Because large numbers of passive formaldehyde monitoring devices are being used today, an intercomparison project to achieve adequate standards of reliability and quality assurance is needed. We suggest a project organized something along the lines of the highly successful series of international intercomparisons of integrating radiation dosimeters for environmental monitoring (22).

The laboratory intercomparisons that we conducted (18) toward the end of 1982 revealed more problems, especially for the commercially available passive monitors that were tested. The formaldehyde generation facility (21) was designed to provide low-concentration atmospheres of formaldehyde vapor down to 0.005 ppm. Dilute formalin solution was injected into a heated stream of air through a mechanically driven syringe. The humidity of the stream of formaldehyde-containing air was controlled before it entered the exposure chamber. The exposure chamber was Teflon lined and 0.2 m^3 in volume and continued multiple sampling ports.

One of the intercomparison tests, in which both active and passive sampling devices were employed, was conducted at a nominal formaldehyde concentration of 0.300 ppm for 24 h. The results are shown in Table III. The commercial passive monitors read abnormally low. The problems may have been related to a shelf storage time that was longer than 3 months, or to a low relative humidity (20%) inside the exposure chamber. The intercomparison project, whose desirability was alluded to earlier, should contain a laboratory exposure component. The protocol for such testing would define ranges of formaldehyde concentration, length of

Table III. Laboratory Testing of Active and Passive Formaldehyde Monitors Exposed for 24 h at a Nominal Concentration of 0.300 ppm

Method	Measured Concentration (ppm)	Comment
CEA 555 (23)	0.315 ± 0.010	avg. of five readings during working hours
Pumped molecular sieve (24)	0.290 ± 0.010	avg. of eight 15-min samples during working hours
Refrigerated sampler (20)	0.335 ± 0.010	avg. of three 24-h samples
Passive membrane (ORNL) (21)	0.340 ± 0.050	avg. of three 24-h samples
Passive commercial (A)	0.180 ± 0.010	avg. of three 24-h samples
Passive commercial (B)	no response	avg. of three 24-h samples
Mean response (excluding commercial monitors)	0.320 ± 0.025	

exposure, temperature, humidity, velocity of air movement, and storage time before and after exposure. Evaluation protocols for diffusive samplers were discussed recently by Brown et al. (25). Performance in both laboratory and field trials was considered, but only from the viewpoint of occupational environments.

Concentrations of Formaldehyde

Housing Groups. A panel evaluated exposures at a recent EPA-sponsored workshop on formaldehyde (26). The panel members judged that the range and mean concentrations of formaldehyde for various types of housing are known fairly well, especially when large sample sizes are involved. Table IV summarizes data from studies of formaldehyde concentrations for residences in different parts of the United States, Canada, and the United Kingdom.

The older conventional home is the class of dwelling in which the lowest mean concentrations (approximately 0.03 ppm) of formaldehyde are encountered. The age of the conventional house is very important because the formaldehyde concentration is elevated significantly in newer homes (34). This dependency of the mean concentration of formaldehyde on the age of the home is depicted in Figure 2; of the 40 houses that were monitored, about the same number of houses (approximately six houses) were in each of the age groups that compose the abscissa. The age–formaldehyde concentration relationship for conventional homes has not received the same scrutiny that it has for mobile homes (32) and UFFI (3). Nevertheless, it seems to be a parameter that is just as important for modern conventional houses as it is for mobile homes.

The factor of house age can cause problems when the influence of an additional formaldehyde emitting agent, such as UFFI, is being sought. For example, when formaldehyde concentrations in sets of UFFI and non-UFFI houses are being compared, the ages of the houses in each set should have the same balance to be able to distinguish any effects on formaldehyde concentrations due to the UFFI alone. In general, however, this age balance has not been considered. When examining the field studies listed in Table IV or elsewhere, we suggest that more credence in the mean concentrations of formaldehyde in UFFI and non-UFFI homes should be placed on the larger scale studies in which the effect of the house age factor will tend to be balanced by the large numbers of houses involved. Large numbers of houses monitored will also have the advantage of tending to reduce the impact of diurnal and seasonal fluctuations in formaldehyde concentrations. The mean concentration of formaldehyde in a relatively small study group of non-UFFI houses has, for example, received criticism (30); the inclusion of a brand new, unfoamed house with a high formaldehyde concentration of 0.34 ppm caused the data to be skewed significantly.

Houses that have been insulated with UFFI have, when considered as

Table IV. Reported Levels of Formaldehyde in the Indoor Air of Classes of Private Residences

| Type of Residence | No. of Residences | Formaldehyde (ppm) | | Ref. |
		Range	Mean	
U.S. homes without UFFI	41	0.01–0.1	0.03	1
U.S. homes with UFFI (complaint and noncomplaint)	636	0.01–3.4	0.12	1
U.S. mobile homes	431	0.01–3.5	0.38	1
Canadian houses without UFFI	383	(3% >0.1 ppm)	0.036	2
Canadian houses with UFFI	~1850	(10% >0.1 ppm)	0.054	2
U.S. houses without UFFI and without particle board	17	—	0.025	27
U.S. houses with UFFI but without particle board subfloors	~600	—	0.050	27
U.S. mobile homes	several hundred		0.12	27
U.K. buildings without UFFI	50	<0.02->0.3 (3% >0.1 ppm)	0.047	28
U.K. buildings with UFFI	128	~0.01->1 (7% >0.1 ppm)	0.093	28
U.S. houses without UFFI	42	0.03–0.17	0.06	29
U.S. houses without UFFI	31	—	0.07	30
U.S. houses with UFFI	—	—	0.06	30
Mobile homes (Minnesota complaint)	~100	0–3.0	0.4	31, 32
Mobile homes (Wisconsin complaint)	—	0.02–4.2	0.9	31
Mobile homes (Wisconsin)	65	<0.10–3.68	0.47	11
Mobile homes (Washington complaint)	—	0–1.77	0.1–0.44	31
U.S. mobile homes				
Never occupied	260	—	0.86	33
Older, occupied			0.25	33
East Tennessee homes	40	<0.02–0.4	0.06	34
Age 0–5 years	18	—	0.08	34
Age 5–15 years	11	—	0.04	34
Age >15 years	11	—	0.03	34
Conventional California, Colorado, and S. Dakota homes	64	0.02–0.11	0.05	35
Specialized U.S. housing	52	0.03–0.3	0.1	35

a class, only a modest elevation of formaldehyde concentrations compared to non-UFFI houses. This statement, however, needs to be counterbalanced by the circumstance that a considerably higher percentage of UFFI houses, as opposed to non-UFFI houses, have formaldehyde concentrations at or exceeding 0.1 ppm.

As a class, mobile homes are the dwellings with the apparently highest concentrations of formaldehyde; the mean concentrations are close to 0.4 ppm in most of the studies listed in Table IV, and concentrations for individual mobile homes have been recorded as high as 4 ppm.

Individual Houses: Temporal Considerations. Individuals are usu-

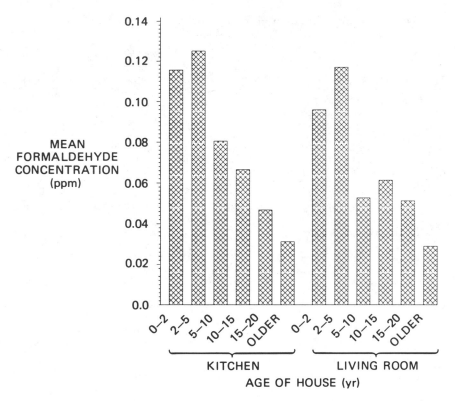

Figure 2. Mean concentrations of formaldehyde in 40 homes in east Tennessee measured biweekly with passive formaldehyde monitors during the warmer seasons of the year; a more complete set of data is provided in Ref. 34. (Reproduced with permission from Ref. 38. Copyright 1984 CRC.)

ally more concerned with the current levels of formaldehyde inside their homes than with average concentrations of formaldehyde for any group of housing. The first question to answer is what is the most appropriate type(s) of formaldehyde measurement to make? This decision depends on whether one's aim is trying to determine an individual's short-term or longer term exposure profile.

The passive formaldehyde monitor provides a time-weighted average concentration. The monitor is exposed for a period of time that is usually between 1 day and 1 week. Such measurements of integrated exposure must be made during the different seasons of the year to provide an adequate profile of the homeowner's potential exposure. A rather extreme example of seasonal fluctuations of formaldehyde levels inside a prefit UFFI home (36) is shown in Figure 3. This home was one of 40 houses that were studied, most of which showed more moderate seasonal fluctuations in formaldehyde levels (34). Most of the individual measurements were made

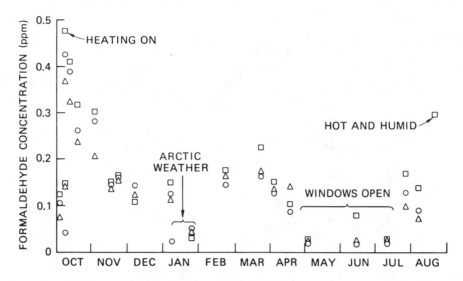

Figure 3. Seasonal fluctuations of formaldehyde concentrations inside the small, south-facing study of a 3-year-old UFFI-prefit house. Key: □, study; ○, living room; and △, kitchen. (Reproduced with permission from Ref. 38. Copyright 1984 CRC.)

over 24-h periods with passive formaldehyde monitors (*21*). The suggested reasons for the marked fluctuations in concentration of formaldehyde vapor are given in Reference 36. The marked increase in formaldehyde concentration in October was, for example, speculatively attributed to the onset of the heating season; heated and low-humidity air issuing from floor vents located by the interior walls was possibly causing evaporation of formaldehyde-bearing pools of moisture within the walls. Temperature and humidity both can have strong influences on the emission rates of formaldehyde (*1, 37*).

Another important finding of the east Tennessee 40-home study (*34*) was that in a majority of the homes, the formaldehyde concentration exceeded 0.1 ppm on 1 or more days of the year. Unfortunately, very few measurements of seasonal variability in formaldehyde levels have been made in other studies (*38*).

A profile of fluctuations in formaldehyde concentration may be needed within a time frame shorter than 24 h. Short-term peak exposures might, for example, be a triggering event for asthma (*39*). The passive integrating formaldehyde monitor is unsuitable for such a task. The modified NIOSH method (*8*) is generally too insensitive for making accurate measurements of diurnal fluctuations in formaldehyde levels. Consequently, little information is available about diurnal or other short-term fluctuations.

One example of a study of short-term variations made by the Oak Ridge group (*36*) is shown in Figure 4; diurnal fluctuations observed re-

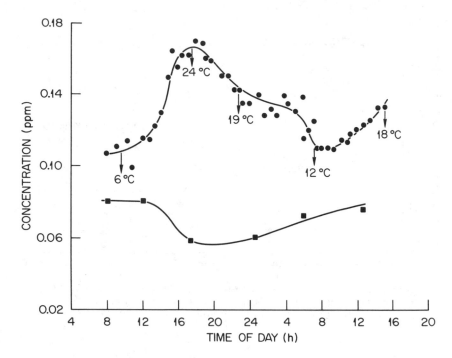

Figure 4. Diurnal fluctuations in formaldehyde concentrations inside a 3-year-old UFFI-prefit house and a 10-year-old non-UFFI house; the outdoor temperatures are indicated. Key: ●, UFFI-prefit house; and ■, non-UFFI house. (Reproduced with permission from Ref. 38. Copyright 1984 CRC.)

sulted in a near doubling of formaldehyde concentrations. The sampling technique uses molecular sieve sorbent (24) and a 15-min collection time and has a detection limit of 0.025 ppm. An empirical equation has been developed to predict breakthrough as a function of sampling rate, relative humidity, and sorbent mass. The formaldehyde collections in this instance were labor intensive and required round-the-clock involvement of a technician. An automated sampling system of the type developed by Dietz (40) would be much better for short-term repetitive sampling.

Available methods are not altogether suitable for measuring, in a cost-effective manner, changes in formaldehyde levels over the short term (hour by hour). This situation, together with the sparsity of reported field studies, led the exposure panel of the "Consensus Workshop on Formaldehyde" (26) to conclude that data were inadequate to characterize the frequency or magnitude of short-term peak (acute) exposures of various groups within the population.

Acknowledgments

Oak Ridge National Laboratory is operated by Martin Marietta Energy Systems, Inc., for the U.S. Department of Energy under Contract No. DE–AC05–840R21400.

Literature Cited

1. Gupta, K. C.; Ulsamer, A. G.; Preuss, P. W. *Environ. Int.* 1982, *8*, 349–58.
2. Urea–Formaldehyde Foam Insulation Information and Coordination Center "Final Report of the Canadian National Testing Survey"; UFFI Centre: Quebec, 1983.
3. U.S. Consumer Product Safety Commission "Revised Carcinogenic Risk Assessment of Urea–Formaldehyde Foam Insulation: Estimates of Cancer Risk Due to Inhalation of Formaldehyde Released by UFFI" by Cohen, M. S.; Government Printing Office: Washington, 1981; also Hileman, B. *Environ. Sci. Technol.* 1982, *16*, 543A.
4. American Society of Heating, Refrigeration, and Air Conditioning Engineers "Guideline for Indoor Formaldehyde in Ventilation for Acceptable Indoor Air Quality"; ASHRAE: New York, 1981; Standard 62.
5. Allen, G. G.; Dutkiewicz, J.; Gilmartin, E. J. *Environ. Sci. Technol.* 1980, *14*, 1235–40.
6. "Key chemicals—Formaldehyde"; *Chem. Eng. News* 1984, p. 14.
7. Balmat, J. L. "Formaldehyde Institute Formaldehyde Methods Manual"; Formaldehyde Institute: Scarsdale, N.Y., 1983.
8. "National Institute for Occupational Safety and Health Manual of Analytical Methods," 2d ed; Taylor, D. G., Ed.; NIOSH: Cincinnati, 1977; Vol. 1, DHHS(NIOSH) Publication No. 77–157–A, P&CAM 125; also "National Institute for Occupational Safety and Health Manual of Analytical Methods," Taylor, D. G., Ed.; NIOSH: Cincinnati, 1981; Vol. 7, DHSS(NIOSH) Publication No. 82–100, P&CAM 354.
9. Altshuller, A. P.; Miller, D. L.; Sleva, S. F. *Anal. Chem.* 1961, *33*, 621–25.
10. U.S. Department of Health, Education and Welfare, Public Health Service "National Institute for Occupational Safety and Health Manual of Analytical Methods," 2d ed.; Government Printing Office: Washington, 1973; Vol. 1, pp. 125–29.
11. Dally, K. A.; Hanrahan, P. P.; Woodbury, M. A. *Arch. Environ. Health*, 1981, *36*, 277–84.
12. Wisconsin Register, 1982, No. 316, Mobile Homes.
13. Girman, J. R.; Geisling, K. L.; Hodgson, A. T., "Sources and Concentrations of Formaldehyde in Indoor Environments," Lawrence Berkeley Laboratory Report LBL–14574, June 1983.
14. Anders, L. W.; Shor, R. M., presented at the Am. Ind. Hyg. Conf., Philadelphia, Pa., 1983.
15. Richardson, G. M., personal communication.
16. Hodgson, A. T.; Geisling, K. L.; Remiju, B.; Girman, J. R. "Validation of a Passive Sampler for Determining Formaldehyde in Residential Indoor Air," Lawrence Berkeley Laboratory Report LBL–14626, 1982.
17. Gammage, R. B.; Hingerty, B. E.; Womack, D. R., Hawthorne, A. R., presented at the Am. Ind. Hyg. Conf., Philadelphia, Pa., 1983.
18. Hawthorne, A. R.; Matthews, T. G.; Gammage, R. B.; Westley, R. R.; Morris, S. A., presented at the Am. Ind. Hyg. Conf., Philadelphia, Pa., 1983.
19. Hawthorne, A. R.; Gammage, R. B.; Dudney, C. S.; Womack, D. R.; Morris, S. A.; Westley, R. R.; Gupta, K. C. *Proc. Spec. Conf. Meas. Monit. Non-Criter. (Toxic) Contam. Air* 1983, 514–26.
20. Miksch, R. R.; Anthon, D. W.; Fanning, L. Z. *Anal. Chem.* 1981, *53*, 2118–23.

21. Matthews, T. G.; Hawthorne, A. R.; Howell, T. C.; Metcalfe, C. E.; Gammage, R. B. *Environ. Int.* **1982**, *8*, 143–51.
22. de Planque, G.; Gesell, T. F. *Health Phys.* **1979**, *36*, 221–34.
23. Matthews, T. G. *Am. Ind. Hyg. Assoc. J.* **1982**, *43*, 547–52.
24. Matthews, T. G.; Howell, T. C. *Anal. Chem.* **1982**, *54*, 1495–98.
25. Brown, R. H.; Harvey, R. P.; Purnell, C. J.; Saunders, K. J. *Am. Ind. Hyg. Assoc. J.* **1984**, *45*, 67–75.
26. "Exposure Panel Consensus Report"; Consensus Workshop on Formaldehyde, Little Rock, Ark., October 1983.
27. Orheim, R. M., letter to the editor *Chem. Eng. News* **1982**, *60*, 2; also personal communication.
28. Everrett, L. H., presented at the Consensus Workshop on Formaldehyde, Little Rock, Ark., October 1983.
29. Godish, T. *J. Environ. Health* **1981**, *44*, 116–21.
30. Frank, C. *Fed. Reg.* **1982**, *47*, 14386–87.
31. Calvert, J. G. "Formaldehyde and Other Aldehydes"; NAS: Washington, D.C., 1981; Chap. 5.
32. Garry, V. F.; Oatman, L.; Plens, R.; Gray, D. *Minn. Med.* **1980**, *63*, 107–11.
33. Clayton Environmental Consultants *Fed. Reg.* **1983**, *48*, 37139.
34. Hawthorne, A. R.; Gammage, R. B.; Dudney, C. S.; Hingerty, B. E.; Schuresko, D. D.; Parzyck, D. C.; Womack, D. R.; Morris, S. A.; Westley, R. R.; White, D. A.; Schrimsher, J. M. "An Indoor Air Quality Study of Forty East Tennessee Homes," Oak Ridge National Laboratory Report, ORNL–5965, 1985.
35. Nero, A. V.; Grimsrud, D. T. "The Dependence of Indoor Pollutant Concentrations on Sources, Ventilation Rates, and Other Removal Factors," Lawrence Berkeley Laboratory Report, LBL–16525, 1983.
36. Gammage, R. B.; Hingerty, B. E.; Matthews, T. G.; Hawthorne, A. R.; Womack, D. R.; Westley, R. R.; Gupta, K. C. *Proc. Spec. Conf. Meas. Monit. Non-Criter. (Toxic) Contam. Air* **1983**, 453–62.
37. Matthews, T. G.; Hawthorne, A. R.; Daffron, C. R.; Reed, T. J.; Corey, M. D. "Proceedings of the 17th International Washington State University Particle Board–Composite Materials Symposium," Pullman, Wash. 1983.
38. Gammage, R. B.; Gupta, K. C. "Formaldehyde"; Walsh, P. J.; Dudney, C. S.; Copenhaver, E. D., Eds.; CRC: Boca Raton, Fla., 1984; pp. 109–42.
39. Hendrick, D. J.; Lane, D. J. *Br. J. Ind. Med.* **1977**, *34*, 11.
40. Dietz, R. N. "Brookhaven Air Infiltration Measurement System (BNL–AIMS) Manual for Field Deployment," Brookhaven National Laboratory Report, BNL–31544, 1984.

RECEIVED for review September 28, 1984. ACCEPTED January 10, 1985.

Formaldehyde Emission from Combustion Sources and Solid Formaldehyde-Resin-Containing Products

Potential Impact on Indoor Formaldehyde Concentrations

T. G. MATTHEWS, T. J. REED, B. J. TROMBERG, C. R. DAFFRON, and ALAN R. HAWTHORNE

Health and Safety Research Division, Oak Ridge National Laboratory, Oak Ridge, TN 37831

The formaldehyde (CH_2O) emission rates of combustion sources and solid CH_2O-resin-containing products commonly found in domestic environments are surveyed. The potential impact of these sources on indoor CH_2O concentrations is estimated with simple, steady state, indoor pollutant concentration models. Source emission rates, product loadings for solid emission sources, duty cycles for combustion sources, and potential permeation barriers are considered in the model. The strongest contributors to indoor CH_2O are pressed-wood products and foam insulation containing urea–formaldehyde resins. Combustion sources and phenol–formaldehyde resin bonded products are generally weak emitters.

NONOCCUPATIONAL INDOOR ENVIRONMENTS can contain a variety of CH_2O emitters including both combustion sources and solid CH_2O-resin-containing products (Figure 1). Common combustion sources include gas burners and ovens, kerosene heaters, and cigarettes. Consumer and construction products incorporating CH_2O resins include urea–formaldehyde foam insulation (UFFI), pressed-wood products, fibrous glass insulation, ceiling panels, and textiles. The potential impact of each source on indoor CH_2O concentrations depends on the CH_2O emission characteristics of the product, product use parameters, and the ventilation characteristics of the indoor environment. For combustion appliances, wide variation in the CH_2O emission rates can be expected on the basis of product design, tun-

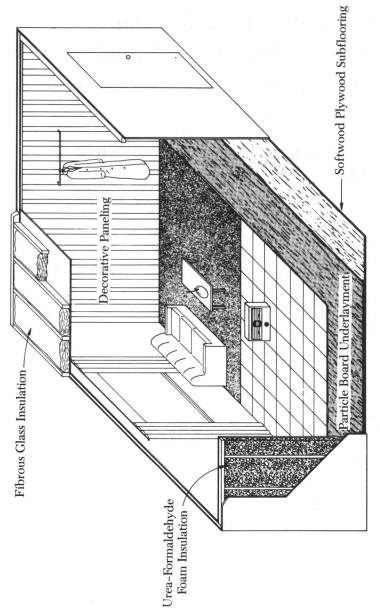

Figure 1. Single indoor compartment model.

Softwood Plywood Subflooring

Particle Board Underlayment

Decorative Paneling

Fibrous Glass Insulation

Urea–Formaldehyde
Foam Insulation

ing, and duty cycle (1). The efficiency of the combustion process and thus the CH_2O generation rate may also be a function of a warm-up cycle or oxygen concentration in the indoor atmosphere (2). The CH_2O emission from CH_2O-resin-containing products is generally a function of the type and amount of resin it contains, environmental parameters such as temperature, relative humidity (rh), room CH_2O concentration, and age of the material (3). Product use parameters include the loading of the material (i.e., surface area [square meters] per air volume [cubic meters])[1] and the presence of intervening permeation barriers, such as resilient flooring or plastic vapor barriers, between the CH_2O emitter and the indoor atmosphere. In indoor environments that depend on natural ventilation for pollutant removal, the air infiltration rate for the dwelling, distribution of emission sources, and intercompartment mixing efficiencies are all important parameters. Formaldehyde concentrations outside the home are generally very low (i.e., <0.01 ppm) and thus have a minimal contribution to indoor concentrations (4).

Formaldehyde Emission Rates from Common Combustion and Formaldehyde-Resin-Containing Products

Most CH_2O emission sources can be conveniently divided into three categories: combustion sources, CH_2O-resin-containing products with direct exposure in the indoor atmosphere, and CH_2O-resin-containing products separated from the indoor atmosphere by an intervening permeation barrier. Combustion sources that are commonly used throughout the United States include gas stoves, kerosene heaters, and cigarettes. Formaldehyde emissions from all three sources have been measured in room-size environmental chambers under approximate consumer use conditions. These experimental results (and references) are summarized in Table I. Average duty cycles for each combustion source that are considered to be crude estimates of highly variable consumer use patterns are also reported. The average duty cycles for a single gas burner and gas oven were calculated for a typical U.S. family of four with average income and accounted for contemporary pilotless burner designs (5). The average gas–energy consumption for cooking, 19,000 Btu/day, was divided into two roughly equal portions: 9000 Btu/day for gas burners and 10,000 Btu/day for gas ovens (5). For a gas burner, an average energy consumption rate of 9000 Btu/h yielded an approximate duty cycle of 1.0 h/day. For a gas oven, an average energy consumption rate of 15,000 Btu/h yielded a duty cycle of 0.7 h/day. The average duty cycle for kerosene heaters (i.e., 8 h/day) during the heating season has been estimated by the U.S. Consumer Product Safety Commission (6). The average cigarette consumption rate was estimated from U.S. Department of Agriculture statistics (7). In 1978, 53 million U.S. smokers

---

[1]A units and conversions listing is given in Appendix 1.

Table I. Formaldehyde Emission Rates of Combustion Sources

Product	Duty Cycle[a]	Formaldehyde Emission Rate (mg/h)	
		Measured	Average over 24 h
Gas burner	1.0 h/day	16 ± 10 (18)	0.67 ± 0.41
		4.3 ± 2.6 (19)	0.18 ± 0.11
Gas oven	0.7 h/day	23 ± 3.4 (18)	0.67 ± 0.10
Kerosene heater			
Convective	8 h/day	1.0 ± 0.66[b]	0.33 ± 0.22
Radiant	8 h/day	4.0 ± 2.0[b]	1.3 ± 0.66
Cigarettes	10 cig/day	0.97 ± 0.06 (21)	0.40 ± 0.03
		1.44 (22)	0.60

[a]Estimation of duty cycle is described in text.
[b]Data for new heaters (i.e., ≤1 yr old) (20).

consumed 615 billion cigarettes. By assuming a single smoker per family of four, this result corresponds to a cigarette consumption rate of 32 cigarettes per smoker per day. By assuming these cigarettes are smoked in a minimum of three different locations in compartments inside the home or outdoors, the average cigarette consumption rate is approximately 10 cigarettes per indoor compartment per day. In practice, the actual use of any of these sources will vary considerably.

For each combustion source, the average CH_2O emission rate over a 24-h period has been calculated (Table I). These data are used in subsequent modeling to predict the potential steady state impact of each source on indoor CH_2O concentrations. The average emission rates for all three categories of combustion sources are fairly similar and constitute a range of approximately 0.3–1.3 mg/h.

Wide varieties of CH_2O-resin-containing products with direct exposure in the indoor environment are found (Table II). The significant emitters are typically pressed-wood products incorporating urea–formaldehyde resins. Textiles, carpeting, ceiling tiles, and resilient flooring are generally weak emitters. UFFI is a potentially strong CH_2O emitter that is separated from the indoor environment by gypsum board in common exterior wall construction. However, in this report UFFI is treated analogously to emitters with direct exposure because the CH_2O emission rate data were taken from simulated wall panels with an interior surface of painted gypsum wallboard (8). The remaining CH_2O-resin-containing products with indirect exposure in the indoor environment are listed in Table III. With the exception of particle board underlayment, all are weak CH_2O emitters. Softwood plywood subflooring is typically among the weakest pressed-wood emitters because it is fabricated with phenol–formaldehyde resins (9). In contrast, urea–formaldehyde resins are used in most particle board (except waferboard and oriented strand board), hardwood plywood paneling, and medium-density fiberboard products (9).

Table II. Formaldehyde Emission Rates from CH_2O-Resin-
Containing Products with Direct Exposure in Indoor Environments

	Formaldehyde Emission Rate (mg/m^2h)	
Product	Average	Range
Nonapparel textiles[a]		
Drapery	<0.01	<0.01–0.02
Upholstery	<0.01	≤0.01
Apparel textiles (unwashed)[a]	0.01	<0.01–0.02
Carpeting[a]	<0.01	<0.01
Ceiling tiles (12)	0.01 ± 0.01	<0.01–0.02
Resilient flooring (23)	<0.01	<0.01
Furniture (uncovered boards)		
Industrial particle board (24)	0.31 ± 0.14	0.15–0.62
Medium-density fiberboard (24)	1.5 ± 0.51	0.57–2.3
Decorative hardwood plywood paneling (24)		
Print overlay	0.28 ± 0.20	0.05–0.63
Paper overlay	0.11 ± 0.07	0.03–0.27
Domestic veneer overlay	0.12 ± 0.04	0.07–0.24
UFFI[b]	0.23 ± 0.19	0.05–0.80

[a]Average data are calculated from the midpoint in the range for all product subgroups in each product category (10).
[b]UFFI is normally encased behind gypsum wallboard. It is treated as an emitter with direct indoor exposure because CH_2OER' data were taken from simulated exterior wall panels (8).

Table III. Formaldehyde Emission Rates from CH_2O-Resin-
Containing Products with Indirect Exposure in Indoor
Environments Through Permeation Barriers

	Formaldehyde Emission Rate (mg/m^2h)	
Product	Average	Range
Carpet cushion (23)	0.01 ± 0.01	<0.01–0.02
Fibrous glass insulation (12)	0.02 ± 0.01	<0.01–0.02
Softwood plywood subflooring (11)	0.02 ± 0.01	0.01–0.03
Particle board underlayment (24)	0.30 ± 0.22	0.11–0.78

The product conditioning and CH_2O emission rate measurement techniques used to characterize the CH_2O emission sources listed in Tables II and III are varied. The CH_2O emission rate data for textiles and carpeting have been reported by Pickrell et al. (10) for purposes of product ranking. Product conditioning and testing were performed at 20 °C and 100% rh in sealed desiccators. The quantitative relationship between the CH_2O emission rate measurements under these test conditions and environmental chamber tests under simulated real-world conditions (e.g., 23 °C, 50% rh) was not determined. However, the CH_2O emission rates of the textile and carpet products are sufficiently low that such intermethod correlations are presumably unimportant.

The data for pressed-wood products, fibrous glass insulation, carpet cushion, ceiling tiles, and resilient tile flooring were obtained with the Oak Ridge National Laboratory (ORNL) developed formaldehyde surface emission monitor (FSEM) and environmental chamber test methods. Product conditioning was at ~23 °C and 50% rh. The FSEM is a passive, non-destructive flux monitor that uses $13\times$ molecular sieve to collect emitted CH_2O and maintain low CH_2O concentrations (typical of indoor environments) near the surface of the emitter. A strong, approximately one-to-one correlation between CH_2O emission rate data taken with the FSEM and small-scale environmental chamber tests (at 23 °C, 50% rh, and 0.1 ppm CH_2O) has been measured for a variety of pressed-wood products and UFFI encased in simulated wall panels (11). A similar, but more qualitative, intermethod correlation has been observed for measurements of fibrous glass insulation and ceiling tiles (12).

In this review, a simple steady state model for indoor pollutant concentration is used to estimate the potential impact of a variety of CH_2O emission sources on the CH_2O concentration in a single compartment. The indoor air quality model is based on the pollutant mass balance equation of Wadden and Scheff (13). Separate emission models are also developed for combustion sources, CH_2O-resin-containing products with direct exposure in the indoor compartment, and solid emission sources separated from the indoor compartment by intervening permeation barriers. The emitter models are incorporated into the pollutant concentration model to account for any CH_2O concentration dependence of the emitters. All modeling is restricted to standard temperature and relative humidity conditions (i.e., 23 °C, 50% rh). The steady state modeling will not consider the impact of short-term changes in emission sources, environmental or ventilation parameters, and the resulting time-dependent CH_2O concentration behavior. Assumptions for each step in the model development are identified.

Steady State Formaldehyde Concentration Model for a Single Compartment

At steady state, the CH_2O concentration in a single compartment may be expressed as

$$[CH_2O]_{SS} = [CH_2O]_O + CH_2OER/(C \cdot ACH \cdot V) \qquad (1)$$

where $[CH_2O]_{SS}$ is the steady state CH_2O concentration inside the compartment (mg/m^3), $[CH_2O]_O$ is the steady state CH_2O concentration outside the compartment (mg/m^3), CH_2OER is the CH_2O emission rate of CH_2O

sources inside the compartment under existing environmental conditions (mg/h), C is the fraction of air coming into the compartment that completely mixes within the compartment volume (where the remaining fraction of air, $1 - C$, does not mix with the compartment volume), ACH is the volumetric flow rate of air in and out of the compartment in units of compartment volume per hour (h^{-1}), and V is the volume of the compartment (m^3).

Several simplifying assumptions are required to reduce the temporally dependent, mass balance equation of Wadden and Scheff (*13*) to the steady state model for CH_2O given in Equation 1. (A derivation of Equation 1 and a discussion of the probable impact of these assumptions are given in Appendix 2.) These assumptions are the following: a steady state condition exists, all parameters in Equation 1 are constant, uniform mixing exists and results in no CH_2O concentration gradients in the atmosphere inside the compartment, no mechanical or natural air filtration systems exist, and no permanent losses of CH_2O due to sinks occur. By noting that $C \cdot ACH$ is the effective air infiltration rate [$AIF(h^{-1})$] for the compartment,

$$[CH_2O]_{SS} = [CH_2O]_O + CH_2OER/(AIF \cdot V) \tag{2}$$

For area-dependent sources,

$$[CH_2O]_{SS} = [CH_2O]_O + CH_2OER' \cdot area/(AIF \cdot V) \tag{3}$$

where CH_2OER' is the CH_2O emission rate of the solid, CH_2O-resin-containing products (mg/m^2h) under existing environmental conditions, and area is the area of the solid emission sources (m^2).

Source Emission Models for Combustion and Solid Emission Sources

The CH_2OER of combustion sources at steady state is assumed to be a constant, determined from the mean emission levels measured in laboratory chamber studies averaged over a 24-h period with estimated duty cycles. The model does not account for potential short-term escalations in indoor CH_2O concentrations due to intermittent use of combustion sources. In addition, the model does not account for the effects of any environmental, tuning, or warm-up parameters on the CH_2OER of combustion sources.

For solid emission sources with direct exposure, an inverse linear dependence of CH_2OER' on CH_2O concentration is assumed. This assumption is consistent with the application of Fick's law to the bulk–vapor interphase at the surface of a solid emission source (*14*).

$$CH_2OER' = m([CH_2O]_B - [CH_2O]_V) \tag{4}$$

where m is the mass transport coefficient for the CH_2O emitter (m/h), $[CH_2O]_B$ is the CH_2O concentration in the bulk phase (mg/m^3), and

$[CH_2O]_V$ is the CH_2O concentration in the adjoining vapor phase (mg/m^3). By assuming that $[CH_2O]_B$ and m are insensitive to changes in $[CH_2O]_V$,

$$CH_2OER' \simeq -m[CH_2O]_V + b \tag{5}$$

where b is a constant, the CH_2OER' (mg/m^2h) at zero CH_2O concentration in the vapor phase. A general inverse linear dependence of the CH_2OER' on the CH_2O vapor concentration has been experimentally observed for a variety of pressed-wood products, UFFI, and fibrous glass insulation. A limiting condition is also applied to Equation 5 that is consistent with the assumption for Equation 1 that there are no permanent losses of CH_2O due to sinks. At steady state, the CH_2OER' of all sources must be ≥ 0. However, the product age dependence of the CH_2OER' is not considered.

For solid CH_2O emission sources covered by permeation barriers, two linear equations are used to describe the CH_2OER of the source–permeation barrier combination. Fick's law is used to describe CH_2O transport across the permeation barrier.

$$CH_2OER' = K([CH_2O]_{BB} - [CH_2O]_{SS}) \tag{6}$$

where K is the mass transport coefficient for the permeation barrier (m/h), $[CH_2O]_{BB}$ is the CH_2O concentration below the barrier (mg/m^3), and $[CH_2O]_{SS}$ is the CH_2O concentration above the permeation barrier (mg/m^3). It is assumed that no CH_2O concentration gradients exist below or above the permeation barrier. This treatment may tend to underestimate the effectiveness of the permeation barriers in comparison to real-world conditions. The reduced air movement over the surface of the primary emission source (e.g., particle board underlayment) caused by the presence of the permeation barrier may enhance CH_2O concentration gradients near the surface of the emitter and thus reduce the CH_2OER'. A simple inverse linear dependence of the CH_2OER' on the CH_2O concentration below the barrier is also assumed.

$$CH_2OER' = -m[CH_2O]_{BB} + b \tag{7}$$

Analogous to Equation 5, the CH_2OER' at steady state is assumed to be ≥ 0. By combining Equations 6 and 7, the resultant emission model for a combination of CH_2O source and permeation barrier is

$$CH_2OER' = (K \cdot b - m \cdot K \cdot [CH_2O_{SS}])/(m + K) \tag{8}$$

Recent laboratory studies of CH_2O emissions from particle board underlayment covered with carpet and tile flooring barriers generally support this model (3).

Steady State Formaldehyde Concentration Models Incorporating Source Emission Models

The potential impact of CH_2O emission sources on indoor CH_2O concentrations ($\Delta[CH_2O]$) is estimated from the difference between the steady state CH_2O concentration with various sources contained inside the compartment ($[CH_2O]_{SS}$) and the concentration outside the compartment ($[CH_2O]_O$).

$$\Delta[CH_2O] = [CH_2O]_{SS} - [CH_2O]_O \qquad (9)$$

$[CH_2O]_O$ is assumed to be a constant reflecting the CH_2O concentration outdoors or in neighboring compartments. $[CH_2O]_{SS}$ is estimated from pollutant concentration models (i.e., Equations 2 and 3) incorporating emitter models (i.e., Equations 5 and 8) for various CH_2O emission sources.

For combustion sources $[CH_2O]_{SS}$ is calculated directly with the single compartment CH_2O concentration model (Equation 2); the CH_2OER of combustion sources is assumed to be constant, independent of the CH_2O vapor concentration. For solid emission sources with direct exposure inside the compartment, $[CH_2O]_{SS}$ is calculated from a combination of the indoor CH_2O concentration (Equation 3) and emitter (Equation 5) models.

$$[CH_2O_{SS}] = (A \cdot b + [CH_2O]_O)/(1 + mA) \qquad (10)$$

where A is the area/$(AIF \cdot V)$. For solid sources separated from the indoor compartment by an intervening permeation barrier, $[CH_2O]_{SS}$ is calculated from a combination of the CH_2O concentration model (Equation 3) and the emitter and permeation barrier model (Equation 8).

$$[CH_2O]_{SS} = \frac{A \cdot b + [1 + (m/K)] \cdot [CH_2O]_O}{1 + (m/K) + mA} \qquad (11)$$

In the case of a very inefficient permeation barrier (i.e., $K \gg 1$) Equation 11 reverts back to Equation 10. With a highly effective permeation barrier (i.e., $K \simeq 0$), Equation 11 reduces to

$$[CH_2O]_{SS} \approx [CH_2O]_O \qquad (12)$$

to indicate that the CH_2O emitter has little impact on the indoor CH_2O concentration.

For a single compartment with multiple CH_2O point sources (u) and area sources (v), $[CH_2O]_{SS}$ is determined with the following expression:

$$[CH_2O]_{SS} = [CH_2O]_O + \sum_{i=1}^{u} (CH_2OER_i/(AIF \cdot V))$$

$$+ \sum_{i=1}^{v} (CH_2OER_i' \cdot area)/(AIF \cdot V) \qquad (13)$$

For solid emission sources, appropriate emitter models are substituted for CH_2OER .

Potential Impact of Formaldehyde Emission Sources on Steady State Indoor Formaldehyde Concentrations

By substituting the average CH_2OER' data for combustion and solid emission sources into appropriate steady state CH_2O concentration models, the potential impact of individual sources on indoor CH_2O concentrations (i.e., $\Delta[CH_2O]$, Equation 9) can be estimated. However, further assumptions concerning the general design of the indoor compartment are required. The length, width, and height of the compartment are assumed to be 4, 4, and 2.5 m, respectively. The compartment volume is thus 40 m³. The AIF is assumed to be 0.5 h⁻¹. This AIF is consistent with the results taken in the 40-home study in the Oak Ridge–Knoxville (Tenn.) area in 1982–83 (15). The CH_2O concentration outside of the compartment ($[CH_2O]_O$) is assumed to be 0.03 mg/m³. This concentration reflects a compromise between anticipated near-zero outdoor levels and the potential for residual CH_2O levels in adjacent indoor compartments.

For combustion sources, the CH_2OER' data (Table I) are substituted into Equation 2. The $\Delta[CH_2O]$ values calculated from Equation 9 are listed in Table IV. The results indicate that, for each combustion source, a small impact on the steady state CH_2O concentration is anticipated. This hypothesis may be contingent, however, on several factors: consumer use patterns for cigarettes; duty cycle and local ventilation for gas stoves; and the duty cycle, design, and tuning of kerosene heaters. Certain kerosene

Table IV. Potential Contribution of Combustion Sources to Indoor CH₂O Concentrations

Product	$\Delta[CH_2O]$ (ppm)	Ref.
Gas stove		
Burner	0.03	18
Burner	0.01	19
Oven	0.03	18
Kerosene heater		
Convective	0.01	20
Radiant	0.05	20
Cigarettes	0.02	21, 22

heaters have been found to be strong CH_2O emitters under various tuning conditions (2).

For solid emission sources with direct exposure in the indoor atmosphere, CH_2OER' data (Table II) are substituted into either Equation 3 or Equation 10. Equation 3 is typically used for weak-emitting products for which the CH_2O concentration dependence of the CH_2OER' has not been measured. The CH_2O emission rate is assumed to be constant, independent of CH_2O concentration, and thus maximizes the impact of the emitter on the modeled CH_2O concentration. Equation 10 is used for pressed-wood products and UFFI for which CH_2O concentration dependent emission models (i.e., Equation 5) can be estimated. The slope and intercept values (i.e., m and b of Equation 5) of the emitter model must be empirically determined. Estimates of these constants are listed in Table V. The pressed-wood product constants were determined from measurements of boards sampled primarily from U.S. industry in 1983. The measurements were taken when the boards were typically less than 1 year old. The UFFI constants were determined from measurements of 1.3-year-old simulated wall panels (8) foamed in 1980 (16). However, the specific age of the CH_2O emitters or the age dependence of the CH_2OER values from the CH_2O sources are not considered in the modeling. Calculations used to determine the average slope and intercept values for each product category are detailed in Appendix 3.

To use Equations 3 and 10, additional assumptions concerning the area (square meters) of the solid CH_2O emission sources are required. Unfortunately, available annual survey data for product sales were difficult to relate to cumulative average quantities of individual furnishings actually contained in "typical" detached U.S. housing. As a result, the selected area for each product was determined whenever possible according to common product use characteristics in an individual compartment. For example,

Table V. Summary of Slope and Intercept Constants for Linear CH_2O Emission Models for Pressed-Wood Products and UFFI

Product	Slope (m/h)[a]	Intercept (mg/m²h)
Particle board underlayment	−0.60 ± 0.23	0.38 ± 0.03
Industrial particle board	−0.47 ± 0.18	0.37 ± 0.03
Hardwood plywood paneling		
Print overlay	−0.40 ± 0.09	0.33 ± 0.01
Paper overlay	−0.22 ± 0.10	0.14 ± 0.01
Domestic veneer overlay	−0.27 ± 0.13	0.15 ± 0.02
Medium-density fiberboard	−0.94 ± 0.06	1.6 ± 0.1
UFFI	−0.46 ± 0.47	0.34 ± 0.34

NOTE: Calculations for the determination of the slope and intercept values are described in Appendix 2.

[a]Substitute positive slope values into Equations 5, 7, 8, 10, or 11 because the negative sign is incorporated into the equations.

the area of carpeting and resilient flooring and the area of ceiling tiles were chosen to cover the entire floor and ceiling, respectively. For other home furnishings such as textiles and furniture, two different areas were chosen that vary by a factor of 5 to attempt to account for widely variable product usage. For furniture constructed with pressed-wood products, it is particularly difficult to estimate the effective emitting area because much of the surface is sealed with a variety of decorative permeation barriers.

For decorative paneling, definitive survey data for about the last decade are available for the approximately 25% of new home constructions that contain paneling (17). However, remodeling and repair applications generally account for >20 times the sales of paneling for new home construction (17). Unfortunately, the average paneling use for remodeling and repair applications in homes that contain paneling is unavailable. As a result, it is difficult to estimate the cumulative quantity of decorative paneling that is present in an average home. In addition, the CH_2O emissions from existent paneling will depend on the decay rate for CH_2O emissions from each board, which could be highly variable from home to home. As a consequence, one wall of decorative paneling was chosen as the minimum quantity that would normally be present in a room for decorative purposes.

For UFFI, a surface area of 1.5 walls was selected as a compromise between corner and side rooms in a home that would contain two and one insulated exterior walls, respectively. A reduction in surface area of 1 m^2 was also made for a window.

The specific areas assumed for each category of solid emission sources with direct indoor exposure and the $\Delta[CH_2O]$ values calculated with Equation 9 are listed in Table VI. The results indicate that the impact of carpeting, resilient flooring, textiles, and ceiling tiles is anticipated to be minimal. The impact of furniture is expected to be highly variable, depending upon the category and quantity of pressed-wood products and the presence of permeation barriers. Medium-density fiberboard is a particularly strong CH_2O emitter in uncovered form. The impact of one wall of decorative hardwood plywood paneling on indoor CH_2O concentrations may be significant, especially if printed paneling is used. The impact of UFFI, on the basis of the simulated wall panel data (8), appears to be significant.

To evaluate the impact of individual emission sources covered by permeation barriers, Equation 11 and, subsequently, Equation 9 are used. For Equation 11, assumptions concerning the area of each CH_2O emission source, the CH_2O concentration dependence of each source (i.e., Table V), and the CH_2O transport coefficient, K, for individual permeation barriers are required. Data for each parameter are only available for particle board underlayment and specific decorative flooring materials. A comparison of the modeling results for the underlayment with no barrier, a carpet and carpet cushion barrier, and a resilient flooring barrier is shown in Table

Table VI. Potential Contribution of Solid Emission Sources with
Direct Exposure in Indoor Environments to Indoor CH_2O
Concentrations

Product	Area (m^2)	$\Delta[CH_2O]$ (ppm)
Textiles		
Nonapparel	5	<0.01
	25	0.01
Apparel	5	<0.01
	25	0.01
Carpeting (10)	16	<0.01
Ceiling tiles (12)	16	<0.01
Resilient flooring (23)	16	<0.01
Furniture (uncovered board) (24)		
Industrial particle board	1.0	0.01
	5.0	0.06
Medium-density fiberboard	1.0	0.06
	5.0	0.25
Decorative paneling (24)		
Print overlay	10	0.11
Paper overlay	10	0.05
Domestic veneer overlay	10	0.05
UFFI (8)	14	0.14

Table VII. Potential Contribution of Solid Emission Sources Separated from the Indoor
Atmosphere by Permeation Barriers to Indoor CH_2O Concentrations

Product	Barrier	K (m/h)	Area (m^2)	$\Delta[CH_2O]$ (ppm)
Particle board	none	∞	16	0.16
Underlayment	carpet and cushion	0.43 (23)	16	0.08
	tile	<0.003 (23)	16	<0.01
Softwood plywood	particle board			
subflooring	underlayment	0.002	16	<0.01
Carpet cushion	carpet	—	16	<0.01[a]
Fibrous glass ceil-				
ing insulation	gypsum board	0.57 (8)	16	≤0.01[a]
Fibrous glass wall				
insulation	gypsum board[b]	<0.57 (8)	14	≤0.01[a]

[a]$\Delta[CH_2O]$ was calculated with the worst-case assumption that $K = \infty$ for the barrier.
[b]The effective K values for noncontinuous barriers such as kraft paper or foil behind gypsum board are unavailable.

VII. Permeation barriers such as resilient tile flooring are anticipated to
cause marked reductions in the impact of particle board underlayment on
indoor CH_2O concentrations.

Other products that are covered by permeation barriers, such as
softwood plywood subflooring, carpet cushion, and fibrous glass insulation, are generally weak emitters; the impact on indoor CH_2O concentrations is anticipated to be small, independent of the presence of permeation
barriers. Worst-case calculations without permeation barriers with Equa-

tion 3 indicate a $\Delta[CH_2O]$ of ≤ 0.01 ppm for all products (except particle board underlayment) that are listed in Table VII. The low permeability of particle board underlayment provides additional evidence for a minimal impact of softwood plywood subflooring on indoor CH_2O concentrations. Additional calculations have also been made for fibrous glass attic insulation with a two-compartment model with an attic temperature of 38 °C. A maximum $\Delta[CH_2O]$ of approximately 0.02 ppm was estimated (*12*).

Potential Impact of Multiple Formaldehyde Emitters on Indoor Formaldehyde Concentrations

To simulate the interaction of multiple CH_2O emission sources in an indoor compartment and their combined effect on steady state indoor CH_2O concentration, Equation 13 and Equation 9 are used. To simplify the multiple emitter treatment, only solid emission sources with significant emission potential are incorporated in the model. The sources, assumed source areas, and calculated $\Delta[CH_2O]$ values as a function of AIF are summarized in Table VIII. A wide range of CH_2O concentrations (i.e., 0.01–0.38 ppm) is achieved as a function of air infiltration rates and emission sources that might be anticipated in domestic environments. The CH_2O concentration range is consistent with that observed in 1–5-year-old homes investigated in the ORNL 40-home study (Table IX). The closest correlation between the modeled and measured CH_2O concentrations would be anticipated for recently built homes in which the pressed-wood products and UFFI would be most similar to those evaluated in laboratory studies. Older homes (i.e., >5 year) in the 40-home study had consistently lower CH_2O concentrations (Table IX). This result is presumably evidence for the aging of CH_2O-resin-containing emission sources. Additional factors such as tighter home construction in newer homes may also be involved.

Table VIII. Potential Impact of Multiple CH_2O Emission Sources on Indoor CH_2O Concentrations ($\Delta[CH_2O]$, ppm) as a Function of Air Infiltration Rate and Sequential Reductions in Emission Sources

AIF (h^{-1})	All Emitters[a]	UFFI Removed	Tile Flooring Installed	Replacement Low-Emission Paneling[b]	MDF Removed
0.25	0.38	0.32	0.28	0.20	0.13
0.50	0.28	0.21	0.17	0.12	0.07
1.0	0.18	0.13	0.10	0.07	0.04
2.5	0.09	0.06	0.04	0.03	0.02
5.0	0.05	0.03	0.02	0.01	0.01

[a]UFFI: 1.5 walls, 14 m²; carpet-covered particle board underlayment: 16 m²; decorative hardwood plywood paneling: one wall, 10 m² print overlay; cupboards, industrial particle board: 2 m²; and furniture, medium-density fiberboard: 1 m².
[b]Replacement paneling with domestic veneer overlay.

Table IX. Frequency of Room-Average CH_2O Concentrations Determined from Measurements Taken During Multiple Visits to Homes in the ORNL 40-Home Study (25)

Average Formaldehyde Concentration (ppm)	Homes with UFFI		Homes Without UFFI	
	<5 yrs old	>5 yrs old	<5 yrs old	>5 yrs old
<0.025	85	86	67	133
0.025–0.075	71	45	64	143
0.075–0.125	91	18	63	55
0.125–0.175	57	11	34	8
0.175–0.225	32	1	9	4
0.225–0.275	8	1	22	0
0.275–0.325	3	0	9	0
0.325–0.375	1	3	13	0
>0.375	0	0	12	0

Acknowledgments

The opinions expressed in this chapter are those of the authors and do not necessarily represent the views of the U.S. Consumer Product Safety Commission.

This research was sponsored jointly by the U.S. Consumer Product Safety Commission under Interagency Agreement CPSC–IAG–82–1297 and the Office of Health and Environmental Research, U.S. Department of Energy, under contract W–7405–eng–26 with the Union Carbide Corporation.

Appendix 1. Units and Conversions

Formaldehyde concentration at 23 °C: 1 ppm = 1.24 mg/m^3 = 1240 μg/m^3 = 7.74 × 10^{-8} lb/ft^3

Emission rate: 1 mg/m^{2}h = 0.28 μg/m^{2}s = 2.05 × 10^{-7} lb/ft^{2}h

Transport rate: 1 m/h = 0.017 m/s = 5.08 × 10^{-3} ft/s

Air exchange rate: 1 compartment volume/h = 0.017 compartment volume/s

Energy: 1 Btu = 1.06 kJ

Loading: 1 m^2/m^3 = 0.30 ft^2/ft^3

Appendix 2. Application and Impact of Simplifying Assumptions Used to Derive Steady State CH_2O Concentrations Models

The steady state CH_2O concentration model for a single compartment that is used in this report (i.e., Equation 1) is developed from the pollutant mass balance model of Wadden and Scheff (13).

$$\frac{V \, d[CH_2O]_i}{dt} = Cq_o[CH_2O]_o(1 - F_o) + Cq_1[CH_2O]_i(1 - F_1)$$

$$+ Cq_2[CH_2O]_o - C(q_o + q_1 + q_2)[CH_2O]_i$$

$$+ CH_2OER - CH_2OR \tag{A1}$$

where

C	=	mixing efficiency factor
$[CH_2O]_i$	=	CH_2O concentration inside the compartment
$[CH_2O]_o$	=	CH_2O concentration outside the compartment
t	=	time
q_o	=	volumetric flow rate for make-up air
q_1	=	volumetric flow rate for recirculation
q_2	=	volumetric flow rate for infiltration
F_o	=	filter efficiency for make-up air
F_1	=	filter efficiency for recirculation air
V	=	room volume
CH_2OER	=	indoor source emission rate
CH_2OR	=	indoor sink removal rate

The mixing efficiency factor, C, is the fraction of incoming air to the compartment that completely mixes with the compartment volume. Uniform mixing of air and the absence of CH_2O concentration gradients are assumed throughout the compartment.

By solving Equation A1 for $[CH_2O]_i$ as a function of t, by holding all variables other than $[CH_2O]_i$ and t constant, and by using the boundary condition $[CH_2O]_i = [CH_2O]_s$ at $t = 0$,

$$[CH_2O]_i = \frac{C[q_o(1 - F_o) + q_2][CH_2O]_o + CH_2OER - CH_2OR}{C(q_o + q_1F_1 + q_2)}$$

$$\times [1 - e^{-(C/V)(q_o + q_1F_1 + q_2)t}]$$
$$+ [CH_2O]_s e^{-(C/V)(q_o + q_1F_1 + q_2)t} \tag{A2}$$

By assuming t is equal to infinity, the steady state CH_2O concentration (i.e., $[CH_2O]_{ss}$) is

$$[CH_2O]_{ss} = \frac{C[q_o(1 + F_o) + q_2][CH_2O]_o + CH_2OER - CH_2OR}{C(q_o + q_1F_1 + q_2)} \tag{A3}$$

Further simplifying assumptions include the absence of mechanical or natural air filtration systems:

$$F_o = F_1 = 0 \tag{A4}$$

the absence of permanent loss mechanisms due to sinks:

$$CH_2OR = 0 \qquad (A5)$$

and a single expression for the compartment air exchange rate (ACH):

$$V \cdot ACH = q_o + q_2 \qquad (A6)$$

The final expression is

$$[CH_2O]_{ss} = [CH_2O]_o + CH_2OER/(C \cdot ACH \cdot V) \qquad (A7)$$

The influence of these simplifying assumptions on the results of the modeling can be qualitatively evaluated in terms of their potential positive or negative impact on the modeled CH_2O concentrations. In addition, generic comparisons can be drawn between the predictions of the simplified model that incorporates the assumptions (i.e., Equation A7) and more fundamental mass balance equations (i.e., Equation A1). The assumption of a steady state condition, holding all parameters constant at time equal to infinity, restricts the application of the simplified model to compartments with temporally invariant environmental and ventilation conditions. No consistent positive or negative bias is anticipated between the results for steady state and time-dependent models (i.e., Equations A1 and A3), provided the time-dependent model parameters average the same magnitude as those for the steady state model. The assumptions of uniform mixing and no CH_2O concentration gradients in the air throughout the indoor compartment are applied to both the temporally dependent model (i.e., Equation A1) and simplified steady state models (i.e., Equation A7). These assumptions are expected to cause a positive bias of uncertain magnitude on the modeled CH_2O concentrations. Elevated CH_2O concentrations near the surface of solid CH_2O emission sources due to nonuniform mixing throughout the compartment are expected to suppress the CH_2O emission rates in comparison to uniformly mixed conditions. The assumption of no mechanical and natural ventilation systems and CH_2O sinks should also cause a positive bias on the CH_2O concentration modeling. All three model terms represent finite CH_2O loss mechanisms of uncertain and probably variable magnitude in homes. Water-bearing sorbent materials (i.e., sinks) such as gypsum board are assumed to buffer the attainment of steady state CH_2O concentrations but are not a permanent CH_2O loss mechanism. Recent studies of the CH_2O sorption and desorption characteristics of gypsum board generically confirm this assumption (3). A strong time-dependent buffer to sudden changes in CH_2O vapor concentration and a weak (i.e., < 10%) permanent CH_2O loss mechanism were observed.

Appendix 3: Determination of Slope and Intercept Values for Pressed-Wood Product and UFFI Linear CH₂O Models

A simple inverse linear dependence of the CH_2O emission rate of pressed-wood products and UFFI on CH_2O concentration is assumed (Equation 5). For each product this linear model is described by two constants: the slope and intercept value.

The slope values for pressed-wood products and UFFI (Table V) were determined from linear least squares analyses of CH_2O emission rate data (i.e., dependent variable) as a function of CH_2O concentration (i.e., independent variable). For pressed-wood products, the CH_2O emission rate data at a variety of CH_2O concentrations were determined from environmental chamber experiments in which the CH_2O concentration was indirectly controlled through variation in product loading and air exchange rate (23). The results for each category of pressed-wood products listed in Table V represent the average slope for typically three or more boards sampled from U.S. industry in 1982–83. With UFFI, CH_2O emission rate data were measured from insulation encased in simulated exterior wall panels. The air exchange rate through a chamber on the gypsum board side of the wall panels was varied to determine the CH_2O emission rate at different CH_2O concentrations. The results listed in Table V represent the average slope for measurements of nine wall panels (8).

The intercept values (i.e., of the dependent variable) for pressed-wood products were calculated with the slopes determined from the linear regression analysis of the environmental chamber data mentioned earlier and the results of a pressed-wood survey involving more than 100 boards sampled from U.S. industry in 1983 (24). The survey was conducted with the ORNL-developed FSEM. A strong, approximately one-to-one correlation between CH_2O emission rate data taken with the FSEM and with the environmental chamber at 23 °C, 50% rh, and 0.1 ppm CH_2O has been observed for a variety of pressed-wood products (11, 23).[2] By substituting the average FSEM data from Table II for the assumed CH_2O emission rate at 0.1 ppm CH_2O (i.e., $CH_2OER'_{0.124\ mg/m^3}$) and the average slope (m) determined from environmental chamber experiments (i.e., Table V) into a rearranged form of Equation 5, the intercept (b) for the dependent variable can be calculated.

$$b(mg/m^2h) = CH_2OER'_{0.124\ mg/m^3} + m\,(m/h) \cdot 0.124\ (mg/m^3) \quad (A1)$$

For UFFI, independent survey data of a large number of simulated wall panels were unavailable. The intercept for the dependent variable was cal-

_____

[2]The intermethod correlation between the FSEM and environmental chamber is strongly dependent on equal conditioning of the pressed-wood specimens used in both experiments at 23 °C, 50% rh, and <0.15 ppm CH_2O for >7 days prior to measurement (11). Such precautions were followed in both the FSEM and environmental chamber experiments.

culated as the average of the intercepts determined from measurements of the nine wall panels.

Literature Cited

1. Girman, J. R.; Geisling, K. L.; Hodgson, A. T. "Sources and Concentrations of Formaldehyde in Indoor Environments," Lawrence Berkeley Laboratory Report LBL–14574, 1983.
2. Girman, J. R.; Allen, J. R.; Apte, M. G.; Martin, V. M.; Traynor, G. W. "Pollutant Emission Rates from Unvented Gas-Fired Space Heaters: A Laboratory Study," Lawrence Berkeley Laboratory Report LBL–14502, 1983.
3. Matthews, T. G.; Reed, T. J.; Tromberg, B. R.; Fung, K. W.; Thompson, C. V.; Simpson, J. O.; Hawthorne, A. R. "Modeling and Testing of Formaldehyde Emission Characteristics of Pressed-Wood Products," Report XVIII to the U.S. Consumer Product Safety Commission, 1985.
4. Calvert, J. G. "Formaldehyde and Other Aldehydes"; National Academy: Washington, D.C., 1981.
5. Macriss, R. A., personal communication, 1984.
6. White, P.; Porter, W., memo to Preuss, Peter, U.S. Consumer Product Safety Commission CPSC Kerosene Heater Briefing Package, 1983.
7. Department of Agriculture "The Tobacco Situation" by Horn, D.; Government Printing Office: Washington, 1978; Publication TS166.
8. Hawthorne, A. R.; Gammage, R. B.; Matthews, T. G.; Blackman, G. S.; Howell, T. C.; Allen, R. J. "An Evaluation of Formaldehyde Emission Potential from Urea–Formaldehyde Foam Insulation: Panel Measurements and Modeling," Oak Ridge National Laboratory Report ORNL–TM–7959, 1981.
9. Emery, J. A., American Plywood Association, personal communication, 1984.
10. Pickrell, J. W.; Mokler, B. V.; Griffin, L. C.; Hobbs, C. H. *Environ. Sci. Technol.* 1983, *17*, 753–57.
11. Matthews, T. G.; Hawthorne, A. R.; Daffron, C. R.; Corey, M. D.; Reed, T. J.; Schrimsher, J. M. *Anal. Chem.* 1984, *56*, 448–54.
12. U.S. Consumer Product Safety Commission "Determination of Formaldehyde Emission Levels from Ceiling Tiles and Fibrous Glass Insulation Products" by Matthews, T. G.; Westley, R. R.; Government Printing Office: Washington, 1983.
13. Wadden, R. A.; Scheff, P. A. "Indoor Air Pollution"; Wiley: New York, 1982.
14. Crank, J. "The Mathematics of Diffusion"; Clarendon: Oxford, 1979.
15. Hawthorne, A. R.; Gammage, R. B.; Dudney, C. S.; Womack, D. R.; Morris, S. A.; Westley, R. R.; Gupta, K. C. *Proc. Spec. Conf. Meas. Monit. Non-Criter. (Toxic) Contam. Air* 1983, 514–26.
16. Osborn, S. W.; Lee, L. A.; Heller, H. L.; Hillman, E. E.; Colburn, G.; Landau, P.; Thorme, E. J.; Krevitz, K., Franklin Research Institute Technical Report F–C5316–01, 1981.
17. Housing Industry Dynamics, Crafton, Md., 1980.
18. Traynor, G. W.; Anthon, D. W.; Hollowell, G. D. *Atmos. Environ.* 1982, *16* (*12*), 2979–87.
19. Moschandreas, D. J.; Gordon, S. M.; Eisenberg, W. D.; Relwani, S., presented at the APCA Spec. Conf. Monit. Non-Criter. (Toxic) Contam., Chicago, March 1983.
20. Traynor, G. W.; Allen, J. R.; Apte, M. G.; Girman, J. R.; Hollowell, C. D. *Environ. Sci. Technol.* 1983, *17* (6), 369–71.
21. Spengler, J. D. "Indoor Pollutants"; National Academy: Washington, D.C., 1983.
22. Weber, A.; Fischer, T.; Grandjean, E. *Int. Arch. Occup. Health* 1979, *43*, 183–93.
23. Matthews, T. G.; Reed, T. J.; Tromberg, B. J.; Fung, K. W.; Merchant, E. R.; Hawthorne, A. R. "Modeling and Testing of Formaldehyde Emission Charac-

teristics of Pressed-Wood Products," Reports XV, XVI, XVII to the U.S. Consumer Product Safety Commission, Oak Ridge National Laboratory Report ORNL–TM–9104, 1985, in press.

24. Matthews, T. G.; Reed, T. J.; Daffron, C. R.; Tromberg, B. J.; Gammage, R. B. "Modeling and Testing of Formaldehyde Emission Characteristics of Pressed-Wood Products," Reports XII, XIII, XIV to the U.S. Consumer Product Safety Commission, Oak Ridge National Laboratory Report ORNL–TM–9100, 1985, in press.

25. Dudney, C. S., personal communication.

RECEIVED for review September 28, 1984. ACCEPTED April 5, 1985.

11

Field Evaluations of Sampling and Analytical Methods for Formaldehyde

EUGENE R. KENNEDY, DAVID L. SMITH, and CHARLES L. GERACI, JR.

National Institute for Occupational Safety and Health, Cincinnati, OH 45226

A field evaluation to compare NIOSH method P&CAM 354 [2-(benzylamino)ethanol-coated sorbent] with NIOSH P&CAM 318 (oxidative sorbent), P&CAM 125 (impinger and chromotropic acid), and a 2,4-dinitrophenylhydrazine-coated silica gel sorbent tube method was conducted in three different environments: a formaldehyde production facility, a hospital dialysis unit, and a fiberglass insulation production facility. Results from these studies indicated that one of the methods, P&CAM 318, had sample stability problems and was not suitable for monitoring that required sample storage for any length of time. No apparent difference was found between the other methods in all of the studies, but the limitations of P&CAM 354 at levels less than 0.25 mg/m³ (0.2 ppm) were documented. This limitation was due mainly to the presence of a background level of formaldehyde on the sampling tube.

IN THE DEVELOPMENT OF SAMPLING AND ANALYTICAL METHODS for workplace contaminants, one of the most difficult and often overlooked tasks is field evaluation of the method. Previous work in our laboratories has shown that field testing of sampling and analytical methods has defined certain problem areas in various sampling and analytical methods. In a study performed by Hill (1) the instability of certain pesticides under field conditions was noted. Other field studies on polynuclear aromatic hydrocarbons have been used to determine which stage of a two-stage sampler contains the various compounds of interest. This type of information is not readily available from laboratory-generated samples unless extensive study of the field environment has been done. Variations in concentration due to drafts, currents in the workplace, and other variables and conditions under which the contaminant is formed and exists cannot be duplicated easily in the laboratory and may cause variation between laboratory and field evaluation studies. Also, the cost of duplicating field conditions in the laboratory may be prohibitive.

A major problem in performing a field evaluation is that the "true" concentration of the contaminant in the workplace cannot be determined. However, if a number of sampling and analytical methods are used for the measurement, an estimation of the actual concentration can be made. With the number of sampling and analytical methods for formaldehyde, this compound provided a good candidate for field evaluation.

Experimental

Sampling Sites.   Three different sites were used in these field evaluations. These were a formaldehyde production facility, a hospital dialysis unit, and a fiberglass insulation production facility. Each of these locations had very different workplace environments. The source of formaldehyde in each location was quite different. In the production facility, formalin (ca. 50%) and paraformaldehyde were the sources; in the dialysis unit, 5% and 37% formalin solutions were the only source; in the fiberglass production facility, potential sources of formaldehyde included free formaldehyde, formaldehyde adsorbed on particulate or fibers, and particulate and formaldehyde-generating particulate from the resin system chemically degrading in the sampling system.

Method Description.   Four methods were used in these studies. They were a published procedure by Beasley et al. (2) [2,4-dinitrophenylhydrazine (DNP) method], NIOSH P&CAM 354 (3, 4), NIOSH P&CAM 318 (5, 6), and NIOSH P&CAM 125 (7-9). The DNP method uses silica gel sorbent coated with a solution of hydrochloric acid, N,N-dimethylformamide, and 2,4-dinitrophenylhydrazine and packed into sampling tubes. The 2,4-dinitrophenylhydrazone of formaldehyde is formed on the coated sorbent, desorbed with acetonitrile, and determined by high-performance liquid chromatography.

NIOSH P&CAM 318 is based on the oxidation of formaldehyde to formate by a proprietary charcoal sorbent. The formate ion is desorbed with hydrogen peroxide solution and quantitated by ion chromatography. NIOSH P&CAM 354 uses a reagent-coated sorbent for collection of formaldehyde. This reagent [2-(benzylamino)ethanol] reacts with formaldehyde to form 3-benzyloxazolidine, a five-membered cyclic derivative. This compound is desorbed from the sorbent with isooctane and determined by capillary gas chromatography. NIOSH P&CAM 125 was adapted by NIOSH from the Intersociety Committee. It uses an impinger filled with 1% sodium bisulfite for collection of formaldehyde. This solution is reacted with chromotropic acid and sulfuric acid and turns purple in the presence of formaldehyde. The amount of formaldehyde in the resulting solution is determined spectrophotometrically at 580 nm.

Cellulose ester membrane prefilters were used on half of the samplers in the field evaluation in the fiberglass production facility. After sampling, the prefilters were placed in Nalgene cross-linked polyethylene (CPE) sample bottles with 10 mL of 1% sodium bisulfite solution. These samples were analyzed for formaldehyde by method P&CAM 125 described earlier.

All samples and filters were blank corrected. Blank values are noted in Tables I–III.

Apparatus.   Samples were collected with MDA model 808 Accuhaler sampling pumps with 50-cm^3/min orifices installed for the field evaluations in the formaldehyde production facility and hospital dialysis unit. In the fiberglass production facility a large vacuum pump (Gast MDL 0522) with critical orifices (0.05 L/min, made from sections of capillary tubing and Millipore critical orifices, 1.0 L/min) was used for sample collection. A device shown in Figure 1 was used to hold the samples during all of the field evaluations. This device was constructed from

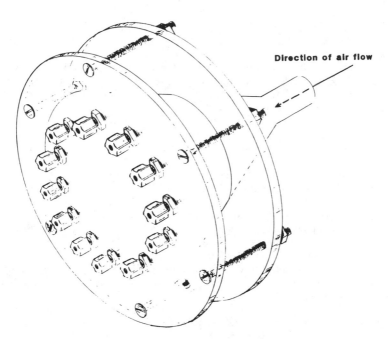

Figure 1. Device used to hold samples during all field evaluations.

0.5-in. (12.8-mm) plexiglas sheet stock, a 7-in. (17.8-cm) plastic laboratory funnel, and 0.25-in. (6.4-mm) polypropylene male connectors to hold the sampling tubes. Pumps or a vacuum source was connected to the tubes, and the air being sampled was drawn in through the funnel opening. This device provided a homogeneous atmosphere for the sampling. Laboratory tests with this device indicated relative standard deviations for sets of 12 samples of less than 10% could be obtained.

Sampling Protocol. Samples were collected 12 at a time with the device shown in Figure 1. Each run consisted of two sets of 12 samples collected over 4–6 h. Four samples of each of three methods (P&CAM 318, P&CAM 354, and 2,4-DNP) were used in each set for the first two evaluations. The position of samples in each device was randomized with at least one tube of each method in each quadrant of the sampler to eliminate any bias that might be due to the collection device. All samples were collected at 0.05 L/min to eliminate differences in flow between the methods as a variable in the study for the first two evaluations.

The protocol for the evaluation in the fiberglass production facility was slightly modified to study the effect that particulate material might have on form-aldehyde monitoring. For this study only two methods were used: P&CAM 125 and P&CAM 354. At this time these were the only two methods recommended by NIOSH for use on their industrial hygiene surveys because sample instability problems had been experienced with P&CAM 318. Samplers were configured both with and without prefilters to study the effect of potential formaldehyde-containing particulates on the sampling results. Again the sample holder device shown in Figure 1 was used. Cellulose ester membrane prefilters were used on half of the samplers in this field evaluation in the fiberglass production facility. After sampling, the prefilters were placed in Nalgene CPE sample bottles with 10 mL of 1% sodium bisulfite solution. These samples were analyzed for formaldehyde by method P&CAM 125 described earlier. In this instance samplers were run at their recommended flow rates for 4 h (P&CAM 125, 1.0 L/min; P&CAM 354, 0.05 L/min). Although this sampling rate and time exceeded the recommended sample volume of P&CAM 125, it allowed the collection of a greater amount of particulate material and a better chance to see contribution to overall formaldehyde levels from formaldehyde adsorbed on or hydrolyzed from this material. Prefilters were analyzed by the method described earlier, and their contribution to the overall sample was reported separately. Analysis of variance was used to study differences between the method configurations. Samples were collected over 2 days in two 4-h sets each day.

Several assumptions were made in the basic design of these studies. These included the following: differences in position in the sample holder are not statistically significant; a 10% coefficient of variation for each method is assumed; levels will be above the limit of detection; and no run-method interaction occurs. Previous evidence from NIOSH studies indicated no statistically significant differences between positions when this type of sample collection device was used. Two runs (48 total samples) were required to detect with 80% confidence a 10% true difference between any two methods with the significance level fixed at 0.05. Analysis of variance and Duncan's test were used to determine if means of the methods were statistically different (10).

Included as an additional experiment, a sample stability study was performed on the P&CAM 125 samples. Because the analytical method requires that only 4 mL of sample be used in the actual analysis, it was possible to preserve the sampling solutions for later analysis. This analysis was done as a check on sample stability and also as an indicator of formaldehyde generation from particulate. Samples were analyzed immediately upon receipt in the laboratory as well as 10 and 30 days following receipt.

Results and Discussion

Before the field evaluation of the methods was attempted, a laboratory test of the sample holder was undertaken. Six samples of both P&CAM 354 and P&CAM 318 were taken in a formaldehyde atmosphere generated by vaporization of formaldehyde in a heated injection block and diluted with clean air. Results from the six P&CAM 354 samples and three of the P&CAM 318 samples and their associated relative standard deviations (RSD) were 1.70 mg/m^3 (11%) and 1.68 mg/m^3 (11%), respectively. This experiment confirmed our assumption of an RSD of 10% for the methods under study. The second set of three samples from P&CAM 318 was retained for quality control samples for submission with the field samples to be collected. This second set of samples was submitted for analysis approximately 6 weeks after collection. The result from this set of samples was 0.55 mg/m^3 (RSD = 29%). These results were the first indication that a sample instability problem existed with method P&CAM 318.

Site 1.   The first field evaluation took place in a formaldehyde production facility. The methods used were the DNP method, P&CAM 354, and P&CAM 318. These methods were chosen because they were all sorbent tube methods and could be used at 0.05 L/min with no apparent problems. The results of the samples taken over 2 days are shown in Table I.

Table I. Formaldehyde Production Facility Field Data

Sample	DNP	P&CAM 354	P&CAM 318
Day 1			
Sample holder #1[a]	4.58 ± 1.55	4.07 ± 1.79	2.74 ± 5.6
Sample holder #2	4.17 ± 0.44	4.26 ± 0.66	2.07 ± 4.44[a]
Day 2			
Sample holder #1	1.34 ± 0.56	0.97 ± 0.05	0.62 ± 0.62[a]
Sample holder #2	1.33 ± 0.04	1.18 ± 0.07	0.79 ± 0.35

NOTE: Concentrations are the average of four samples except where noted and are expressed in parts per million with their 95% confidence limits. Averages of the blank values are as follows: 2,4-DNP, 3.1 μg/sample; P&CAM 318, 6.0 μg/sample; and P&CAM 354, 2.7 μg/sample.
[a]Values are the average of three samples.

Samples were collected for 6 h each day. With some of the samples the pumps stopped working before the end of the 6-h sampling period. Although the sampling pumps were fully charged before use, the conditions of high humidity and temperature (25–30 °C and 80–90% rh) may have adversely affected pump battery life. These low-volume samples contribute to the variability observed in the samples because the low-volume samples are included in the data table. In the statistical analysis, these low-volume samples were treated as outliers, and the DNP method and P&CAM 354 were found to give statistically equivalent results. With P&CAM 318, the nonoutlier results were low and significantly different from the other two methods. Also the RSD with this method was larger than the results obtained during the preliminary laboratory work with the sampler holder. These findings led to further study of sampling and storage parameters of P&CAM 318. The results of that study indicated sample instability when samples were stored for periods longer than 1 week (*11*).

Site 2.   The second field evaluation took place in a hospital dialysis unit. Formalin solutions (5% and 37%) were used to sterilize the peritoneal dialysis machines by pumping the solution into the machines and letting them sit for 2 h. The time required for the filling of two machines with formalin was approximately 1 h. Samples were collected over this 1-h period. Environmental conditions during this time were approximately 25 °C and 50% rh. Methods used were the DNP method, P&CAM 318, and P&CAM 354. At this time the instability problem of P&CAM 318 was defined, and the analysis request for these samples was cancelled. Results for the DNP method and P&CAM 354 are shown in Table II. A total of two sets of four replicates of each method were collected. In this study formaldehyde levels were quite low. These low levels were also confirmed by a CEA 555 continuous formaldehyde monitor. This monitor gave a time-weighted average concentration during the sampling period of 0.15 ppm. The levels measured by the DNP method and P&CAM 354 were only slightly larger than the blank amounts, which were 3.1 and 3.4 μg/sample,

Table II. Hospital Dialysis Unit Field Data

Sample	P&CAM 354	DNP
Sample holder 1	0.14 ± 0.03	0.28 ± 0.06[a]
Sample holder 2	0.11 ± 0.20	0.30 ± 0.1

NOTE: Concentrations are the average of four samples except where noted and are expressed in parts per million with their 95% confidence limits. Averages of the blank values are as follows: 2,4-DNP, 3.1 μg/sample; and P&CAM 354, 3.4 μg/sample.

[a]Values are the average of three samples.

respectively. This fact contributes to the high variability of each set of samples. Although a statistically significant difference between the two methods was observed, these differences are probably unimportant because the actual amounts of formaldehyde measured are at the bottom of the analytical range for each method.

Site 3.   The third field evaluation took place in a fiberglass insulation production facility. At this sampling site, fiberglass was blown, resin coated, and shaped, all within a single structure. The potential existed for free formaldehyde from the resin system as it was sprayed onto the fiberglass, for particulate-bound formaldehyde from its adsorption onto airborne materials, and for formaldehyde-generating particulate from the resin system chemically degrading in the sampling systems.

Formaldehyde can be trapped in dust as either an adsorbed species or in a chemically bound state (12). During sample workup, the formaldehyde can be released either by desorption or hydrolysis. To attempt to study this problem, a field evaluation was undertaken in a fiberglass insulation production facility. Samples were collected by using methods P&CAM 354 and P&CAM 125 with and without prefilters on each method, to give a total of four sampler combinations. The samplers were arranged in the sample holders in the four configurations in triplicate.

Samples were collected in two 4-h sample sets per day over 2 days. Environmental conditions were 25 − 30 °C and 80 − 90% rh. Results from the sample analysis are shown in Table III. Good agreement was found between impingers with and without prefilters, and the amount of formaldehyde found on the prefilters was minimal. The P&CAM 125 data demonstrate a rather consistent formaldehyde concentration ranging from 0.1 to 0.2 ppm. One significant observation was that glass fibers were found to have passed through the unfiltered P&CAM 125 samples and were collected on the prefilters before the critical orifices. Without the critical orifice prefilters, significant clogging and reduction of flow might have occurred. This problem was not observed with the prefiltered P&CAM 125 samples.

A red color and absorption maximum shift to 520 nm was noted with the desorbed impinger prefilter samples. Possible explanations include the extraction of organics from the filter media or the presence of the phenol in

Table III. Fiberglass Production Facility P&CAM 125 Data

Sample	Impinger with Prefilter	Impinger Alone
Day 1		
Sample holder 1	0.19 ± 0.02	0.20 ± 0.02
Sample holder 2	0.10 ± 0.01	0.11 ± 0.0
Day 2		
Sample holder 1	0.13 ± 0.02	0.12 ± 0.03
Sample holder 2	0.14 ± 0.01	0.13 ± 0.01

NOTE: Concentrations are the average of three samples and are expressed in parts per million with their 95% confidence limits. Averages of the blank values are as follows: prefilter, 0.8 µg/sample; and P&CAM 125, 0.7 µg/sample.

the resin system. Certain organic compounds and phenol, in particular, are negative interferences in P&CAM 125 (7). If phenol were present, then the true amount of particulate-bound formaldehyde determined from the prefilters would be different from the reported values.

The results for the P&CAM 354 tubes, both with and without prefilters, were below the limit of detection for the method. The results for the prefilters for the tubes were also below the limit of detection.

To study the stability of samples collected with P&CAM 125, impinger samples were transferred to Nalgene CPE bottles and analyzed immediately on receipt in the laboratory, 10 days after receipt, and 30 days after receipt. A second purpose for this study was to see if any formaldehyde-generating particulate had been captured by the impingers. If any of this particulate were present, then formaldehyde concentrations should have increased with time. As can be seen from the data in Table IV, the levels appeared to remain constant for at least 10 days. Statistical analysis of the results verified this observation. After storage for 30 days, significant differences were observed between day 1 and day 30 analyses. The stability of P&CAM 125 samples for 10 days is in conflict with the reported instability problems found with this method when evaluated under the joint OSHA–NIOSH Standards Completion Program (*13*).

Conclusions

The results from the first field evaluation in a formaldehyde production facility indicate that good correlation is found between the DNP method and P&CAM 354. Problems of sample instability were discovered with P&CAM 318 and have been defined in other work (*11*). The discovery of this problem with P&CAM 318 demonstrates once again the need for field evaluation of sampling and analysis methodology. In the second field evaluation in the hospital dialysis unit, the sensitivity and precision of the methods were found to be questionable at low levels. The last field evaluation once again confirmed the lack of sensitivity of P&CAM 354 at low formaldehyde levels.

Table IV. P&CAM 125 Storage Stability Study

	Recovery (μg/ sample)[a]		
Sample	1 day	10 days	30 days
301	49.7	49.9 (100.5)	48.6 (97.8)
302	48.6	48.1 (99.0)	46.4 (95.4)
303	42.0	41.0 (97.5)	35.5 (84.5)
304	30.3	30.6 (100.9)	29.7 (98.0)
305	21.2	20.6 (97.1)	16.2 (76.4)
306	29.8	30.2 (101.3)	24.2 (81.6)
307	41.5	40.6 (97.7)	39.2 (94.5)
308	29.9	29.1 (97.4)	24.4 (81.6)
309	31.6	30.2 (95.6)	26.1 (82.6)
310	29.7	29.1 (98.0)	30.2 (101.7)
311	27.3	25.9 (94.8)	24.0 (87.9)
312	29.4	28.7 (97.6)	27.9 (94.9)

[a]Percent recoveries enclosed in parentheses are relative to quantity of formaldehyde found on day 1.

This last study also demonstrated the ability of P&CAM 125 to measure formaldehyde in the 0.1- to 0.2-ppm range with a high degree of precision. Situations in which airborne particulate might be present, especially particulate that might potentially decay and release formaldehyde, warrant the use of a prefilter when P&CAM 125 impinger samples are collected. A second reason for using prefilters is that particulate can pass through the impinger and be trapped in the sample pump or vacuum source. The sample stability study done on P&CAM 125 samples collected during this evaluation indicated good sample stability for up to 10 days after sample collection when stored in Nalgene CPE bottles.

Acknowledgments

We acknowledge the assistance of Julia R. Okenfuss for the analysis of the 2,4-DNP-coated silica gel tubes, the UBTL Division, University of Utah Research Institute for the P&CAM 318 and P&CAM 354 analyses, Southern Research Institute for the P&CAM 125 and prefilter analyses, Larry Elliot for assistance in dialysis unit field evaluation, Pierre L. Belanger for assistance in the fiberglass production facility field evaluation, and Richard Hornung for statistical assistance in the preparation of the sampling protocol.

Mention of company names or products does not constitute endorsement by NIOSH.

Literature Cited

1. Hill, R. H., Jr.; Arnold, J. E. Arch. Environ. Contam. Toxicol. 1979, 8, 621–28.

2. Beasley, R. K.; Hoffmann, C. E.; Rueppel, M. L.; Worley, J. W. *Anal. Chem.* **1980**, *52*, 1110–14.
3. Kennedy, E. R.; Hill, R. H., Jr. *Anal. Chem.* **1982**, *54*, 1739–42.
4. "National Institute for Occupational Safety and Health Manual of Analytical Methods"; Taylor, D. G., Ed.; NIOSH: Cincinnati, 1981; DHHS (NIOSH) Publication No. 82–100, P&CAM 354.
5. "National Institute for Occupational Safety and Health Manual of Analytical Methods"; Taylor, D. G., Ed.; NIOSH: Cincinnati, 1980; DHHS (NIOSH) Publication No. 80–125, P&CAM 318.
6. Kim, W. S.; Geraci, C. L.; Kupel, R. *Am. Ind. Hyg. Assoc. J.* **1980**, *41*, 334–39.
7. "National Institute for Occupational Safety and Health Manual of Analytical Methods"; Taylor, D. G., Ed.; NIOSH: Cincinnati, 1977; DHHS (NIOSH) Publication No. 77–157–A, P&CAM 125.
8. Intersociety Committee on Methods of Air Sampling and Analysis "Methods of Air Sampling and Analysis," Method 111; American Public Health Association: Washington, D.C., 1972; pp. 194–98.
9. Intersociety Committee on Methods of Air Sampling and Analysis "Methods of Air Sampling and Analysis," 2d ed., Method 116; Katz, Morris, Ed.; American Public Health Association: Washington, D.C., 1977; pp. 303–7.
10. Miller, I.; Freund, J. E. "Probability and Statistics for Engineers"; Prentice-Hall: Englewood Cliffs, New Jersey, 1977; Chap. 12.
11. Smith, D. L.; Bolyard, M.; Kennedy, E. R. *Am. Ind. Hyg. Assoc. J.* **1983**, *44*, 97–99.
12. Cocalis, C. L.; Beaulieu, H. J., presented at the Am. Ind. Hyg. Conf., Houston, Tex., 1980.
13. National Institute for Occupational Safety and Health–Occupational Safety and Health Administration Standards Completion Program Failure Report for Formaldehyde, S327, NIOSH contract report, 1976.

RECEIVED for review September 28, 1984. ACCEPTED January 7, 1985.

12

Formaldehyde Measurements in Canadian Homes Using Passive Dosimeters

CLIFF J. SHIRTLIFFE[1], MADELEINE Z. ROUSSEAU[1], JUDITH C. YOUNG[1], JOHN F. SLIWINSKI[2], and P. GREIG SIM[1]

[1]Division of Building Research, National Research Council of Canada, Ottawa, Ontario, Canada K1A 0R6
[2]IEC Beak Consultants Ltd., 6870 Goreway Drive, Mississauga, Ontario, Canada L4V 1P1

Formaldehyde dosimeters have been tested in urea–formaldehyde foam insulated Canadian homes in which formaldehyde levels range from 0.02 to 0.2 ppm. Performance of commercial and experimental passive dosimeters was evaluated over 2- to 4- and 7-day exposures. Effects of transport, storage, humidity, and air velocity on blanks and overall performance were investigated. Modifications have improved reproducibility and blanks. Changes have involved design, handling, removal of contamination, and production control. The changes have resulted in the precision of dosimeters becoming almost equal to that of impingers. Devices under development are discussed. Design requirements include precision, accuracy, cost, use by untrained personnel, and blanks. Design requirements identified in the work for low-level dosimeters are discussed. Studies and the types of devices used in each are tabulated.

During 1980–84 several hundred thousand formaldehyde (HCHO) measurements were made in Canadian homes. Most of the measurements were in homes that contain urea–formaldehyde foam insulation (UFFI). During 1980 and early 1981, following a ban on UFFI sales in Canada, most measurements were made by technicians with the NIOSH P&CAM 125 chromotropic acid method and water-filled impingers (bubblers) to collect the formaldehyde over 1–2 h (1). In late 1980 a change was made to 3- and 4-h collection periods, sodium bisulfite solutions in the impingers, and improved analytical procedures (2). In September 1981 the Federal Government of Canada initiated a 10-week survey and study of 1978 UFFI

0065–2393/85/0210/0161$08.75/0

homes and 383 non-UFFI homes spread across Canada (3) (the "2300-home" survey). The collection and analysis were done by five commercial firms. The program started with the modified NIOSH chromotropic acid procedure and 3-h impinger collection, with 1% sodium bisulfite solution, after a 24-h preconditioning of the house. The latter part of the program included measurements with two 1-day and one 7-day type of passive formaldehyde dosimeters that sampled by permeance or diffusion. The results showed that the dosimeters were not subject to the wide variations found in the impinger measurements, and that they tended to give mean values approximately 25% higher. The differences in duplicate impinger readings, even with quality control, varied from 38% at the 0.1-ppm level, to 54% at the 0.05-ppm level, to 100% at the 0.015-ppm level (3).

Smaller studies followed in which the sodium bisulfite solution in impingers was replaced by molecular sieve to produce dry impingers. Studies were also made in which results from several types of long-term formaldehyde dosimeters were compared to repeated short-term measurements. Development of new formaldehyde dosimeters and improvement of existing dosimeters were initiated. The results of the measurements with passive dosimeters were promising even for the levels of 20–200 ppb by volume (0.02–0.20 ppm by volume) normally encountered in homes. The precision of dosimeter measurements was similar to that of active sampling techniques used in previous large studies (Figure 1). The tendency with the dosimeter was to indicate somewhat more formaldehyde than was present;

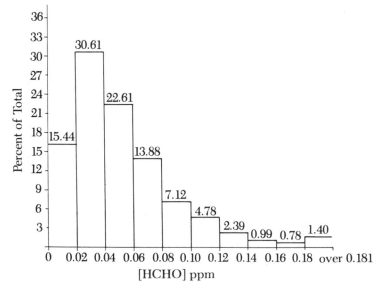

Figure 1. Distribution of HCHO levels in houses (national testing survey).

that is, additional formaldehyde was absorbed in shipment rather than that already sampled lost, and field blank readings were higher than those of laboratory blanks (3). The potential long-term exposure to gaseous formaldehyde could be measured in homes with the dosimeter over the basic 7-day family living cycle.

Short-term measurements were given lower priority because observations of diurnal and annual variations in UFFI homes indicated that formaldehyde concentrations in some homes responded rapidly to changes in the UFFI cavity environment and therefore to meteorological conditions. Peak-to-peak variation by factors of 2–4 and occasionally 10 had been measured (Table I). Peaks occurred in the evening (4) and during the autumn, and minimums occurred in late winter (Figure 2) when the formaldehyde concentration in the walls dropped by a factor of as much as 4 and occasionally 10. Diurnal and seasonal variations were generally larger in UFFI than in non-UFFI homes. Similar variations occurred in the wall cavities.

The Federal Government of Canada initiated an assistance program for homeowners in December 1981. The program involves several dosimeter measurements of formaldehyde in up to 80,000 UFFI homes during the period 1982–86. The deployment, collection, and analysis of dosimeters are to be done primarily by commercial laboratories and the Federal Government of Canada is to handle all collection of data. Houses entering the program are given a pretest in which dosimeters are sent to and returned from the homes by mail. Houses in the program have formaldehyde concentrations measured by a pair of 7-day dosimeters one to four additional times during the program, depending on their routing through the program and the results of previous tests. Thus, a relatively inexpensive but precise and accurate device to permit comparisons between results is needed.

As of February 1984, some 150,000 to 200,000 formaldehyde dosimeters had been used in the program and in related research. The majority of these were specially ordered, commercially manufactured units. Units that originally had a marginal performance have evolved into reasonably reliable devices. The evolution is continuing, along with development of improved new devices. Some devices are evolving into higher precision reference units and others into low-cost, somewhat less precise or accurate devices for use in larger studies.

Table I. Diurnal Variations in Formaldehyde Levels (ppm)

House Number	10:00 A.M.	2:00 P.M.
24	0.04	0.36
24	0.08	0.16
19	0.16	0.092
49	0.13	0.29

NOTE: All results are from Study 9 of Table IV.

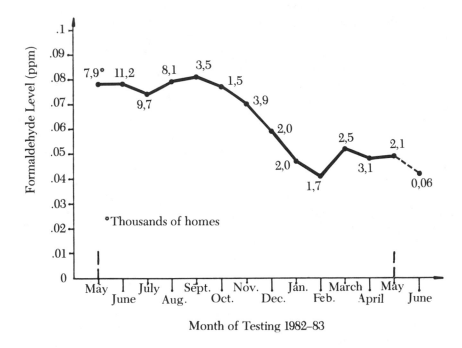

Figure 2. Average formaldehyde measurements in UFFI homes (seasonal variation Canadian UFFI–ICC program).

pollutants to much the same extent as the primary device. This technique has not been widely used. Field and laboratory blanks are of most use when

The UFFI program has provided a rare opportunity to examine details of many types of equipment and methods for measuring low-level air pollutants. The data collected from the homes in the UFFI program may be useful in defining the average and extremes in exposure of occupants to formaldehyde under different weather conditions and at different times of the year. The evolution phase has involved an extensive review of the fundamentals of what might be called indoor air dosimetry.

Many of the findings about the design, manufacture, use, and analysis of dosimeters and about the performance of other measurement devices are directly applicable to other studies or surveys of low-level pollutants.

An overview of the experience gained in Canada during the studies up to early 1984 will be presented in five sections.

Dosimeters for Measuring Air Pollutants

Definition of a Dosimeter.   Dosimeters are devices for collecting one or more substances in air; subsequent analysis is usually required to indicate quantities collected, which in turn allow calculation of exposures. There are two basic types: passive dosimeters for gaseous pollutants (less often dust or spores) and active dosimeters for particulate matter, aerosols, and gaseous pollutants. Passive dosimeters are also called personal samplers, and active dosimeters are often called sampling tubes or, when filled with liquid, impingers or bubblers.

Description of a Dosimeter System. Dosimeters collect the molecules or particles of a pollutant from a known volume of air over a given period of time. The components of a dosimeter system can be some or all of the following: the collection medium, or media, including desiccant layers or a liquid-coated matrix; the primary container with nipples to allow collected air to flow through openings or the pollutant gas to diffuse into the collection medium; caps or an outer-sealed container to prevent the device from sampling before the sampling period begins; screens, porous plugs, or membranes to hold the collection medium in place or to prevent spillage; screens, porous plugs, or membranes to filter out dust entering the device; screens, porous plugs, or membranes to guide or control the air flow or sampling at the input to the device; for an active dosimeter, a pump and flowmeter or flow controller; for a passive dosimeter, one or more columns of air of known dimensions to control the rate of diffusion and therefore the rate of sampling of the pollutant gas or gases from the surrounding air; components to suspend or support the device during deployment; seals for openings in the containers and packages for shipment; and labels for identifying the device and recording its exposure.

Fundamental Need for Field Blanks. Dosimeters become activated when they are produced, so they collect pollutant material from the surrounding environment by diffusion or mass transfer during production; from the surfaces of the components after production; from and through the components during storage and shipment; by diffusion and mass transfer, caused by pressure and temperature cycles, through leaks in the caps or outer container or other joints during storage and shipment; and during the eventual analysis.

In such cases, the measurement made with the exposed primary measuring device must always be compared with an identical secondary device that has the same occasions for collection as the primary measuring device, except during the sampling period. The secondary device must also have an identical aging period. The secondary device is called a field blank and differs from a laboratory blank. A laboratory blank is a testing device that has never left the laboratory; it has had a similar aging period but has not had identical occasions to collect the gas as the primary device or the field blank. Alternatively a field blank may have a period of exposure to the same atmospheres as the primary device but for a shorter duration. The collection medium of the field blank is exposed to water vapor and other the variations of the blanks are small compared to the changes from sampling and aging.

Some dosimeters are activated only after deployment and can only sample from that time until the device is analyzed. If the medium is analyzed immediately, blanks may not be required or may be of lesser importance.

The equation for conversion of the mass of gas collected in a dosimeter and the field blank to the equivalent parts per million by volume is given in Appendix 1.

Collection Medium in a Dosimeter.   The collection medium (or media) is the heart of the dosimeter. The forms of medium vary from simple to complex: a liquid that absorbs and stores the gas, gases, or dust by forming a simple solution; a solid medium such as a filter bed for dust; a desiccant for water vapor or adsorbent bed such as a molecular sieve for gases and vapors; a solid bed of dry chemicals or a film of viscous liquid chemicals, including water films, which chemically rather than physically trap the pollutant (reactions may form adducts or derivatives); and liquid solutions that contain chemicals.

For some dry collection media the presence of water vapor may be essential or may affect the collection rate or the recovery rate.

The collection medium may not be 100% effective; all the pollutant may not be collected from the volume of air sampled, or the collection medium may re-release pollutant to the air. Passive devices may not respond linearly to changes in concentration of the gas because of the establishment of a pollutant vapor pressure over the collection medium. Failure to achieve 100% collection may be partially compensated for in the calibration process; this compensation may be reliable provided it is under 10%. Chemical reactions may convert the gases to other chemicals that are not recovered in the analysis. Biological activity in the medium may consume the gases or break down dust collected. No simple way to compensate for such losses is known, so they must be minimized by the selection of a reliable medium.

Molecules of formaldehyde may be stored in their original form, dissolved in a liquid, stored as adducts as with sodium bisulfite solutions, reacted directly with chemicals to form derivatives as with triethanolamine or 2,4-dinitrophenylhydrazine, or reacted directly with chemicals to produce a colorimetric reaction in the solution or film as with chromotropic-sulfuric acid solution. The derivatization or reactions to produce an immediate colorimetric reaction are least subject to loss during storage.

Factors Affecting the Accuracy of Dosimeters for Gas Measurement.   The collection medium is not the only component that can affect the accuracy of the dosimeter. Dosimeters are such simple devices that their operation is often taken for granted; major sources of both random and fixed errors are often overlooked, especially errors involving the complete dosimeter package including shipping materials. Calibration techniques and approximations adequate for minimizing errors that arise in the measurement of occupational levels of gases may be inadequate for measurement of levels inside homes. When collecting a gas that has unpredictable and relatively complex behavior, such as formaldehyde, at levels approaching outdoor levels, large and highly significant errors can occur. Small changes in design, manufacture, deployment, and analysis can introduce unexpected random or fixed errors.

The accuracy of dosimeters can be affected by factors that influence the collection, storage, and release of gases from the collection medium. Some of the same factors and some additional ones will affect the sampling rate of the dosimeter, in particular the diffusion rate to the collection medium. The presence of dust, aerosols carrying the gases, and polymerized gas molecules can increase the sampling rate. Some factors that can affect the accuracy of dosimeters for gas measurement are listed in Table II along with the effect caused and a very approximate estimated error rating, which is given only as a prioritization parameter for error studies. In evaluating the performance of a dosimeter or in searching for causes of inaccuracies in results, the factors in Table II should not be overlooked. These factors cannot be precise because they depend on the design and deployment of the dosimeter, the gas being measured, and the environment of the sampling.

Formaldehyde is a very reactive gas, and this reactivity makes it difficult to establish reliable sampling rates. Many environmental factors affect the results. Some of the factors are discussed in later sections.

The errors involved in the colorimetric analysis of the eluant from exposed dosimeters are similar to the errors for normal colorimetric analysis of formaldehyde collected in water-filled impingers. The primary differences are the small quantities of water used to elute the gas from the collection medium and the need to remove particulate matter from the solution. The elution process must be checked for its effectiveness at recovering the correct amount of formaldehyde. Discussion of the analytical procedures has been for the most part excluded from the following discussion on dosimeter development, although considerable work has been done on the chromotropic, pararosaniline, and 2,4-dinitrophenylhydrazine methods to improve their performance and to adapt the methods for use in homes.

Adaptation of Commercially Available Dosimeters for Use in Homes

Three commercial formaldehyde dosimeters were evaluated as they became available and were modified in cooperation with the manufacturers to improve their performance for measurements in homes. Each of these dosimeters is discussed in detail in this section. A comparison of the various devices is given in Table III.

Pro-Tek C–60 Du Pont Dosimeter. BACKGROUND. The Pro-Tek

Table II. Some Factors That Affect Accuracy of Dosimeters

Factor	Type of Medium Affected[a]	Effect	Estimated Error Rating[b]
High absolute humidity	dry	capacity reduction	1
Low absolute humidity	wet	evaporation	3
Large humidity variations	dry	loss from media	3
Large temperature changes	dry	loss from media	2
	both	sampling rate changes	4
Barometric pressure variations (±10%)	both	diffusion rate changes	4
	dry	vapor pressure on medium changes	3
Diffusion of other gases to surface	both	small changes in sampling rate	4
Adsorption of gas on diffusion tube	both	nonconstant sampling	3
High face velocity (turbulence)	both	increased sampling rate	3
	both	decreased sampling rate	1
Low face velocity	both	decreased sampling rate	1–2
Swelling of permeance membrane (dimethyl silicone)	both	decreased sampling rate	1–3
Dust collection at mouth of device (on screen)	both	increased sampling rate at low concentrations	1–2
Sampling of aerosols	both	increased sampling rate by adsorbed gas	?
Sampling of gas polymers	both	changed sampling rate	?
Development of vapor pressure over medium (back desorption)	dry	collection not true integration	1–2
	both	loss of collected gas	2–3
	both	reduced sampling rate	2–3
	both	calibration impossible	2–3
	wet	similar to dry, but effect generally less	1–2

[a] "Both" indicates both wet and dry media.
[b] The estimated error ratings are defined as follows: 1, 50–100%; 2, 25–50%; 3, 10–25%; 4, 0–10%; and ?, impossible to assess from data so far.

C–60 formaldehyde dosimeter [Du Pont (6)], composed of a single blister unit, was selected for initial studies as it was the only personal monitor specifically developed for measurement of formaldehyde that was commercially available in 1981. A review of the data indicated that the Du Pont formaldehyde dosimeter had considerable potential for low-level measurements and 7-day exposures. It was designed for occupational levels and had a capacity of 2–50 ppm · h. Du Pont had done extensive laboratory and some field evaluation of the device. Most of the data were for exposures of 8 h and concentrations of approximately 0.5–10 ppm. The mean blank

Table III. Formaldehyde Dosimeters Studied

| Device | Material[a] | | | | | Sampling Rate (cm^3/min) | Collection Medium | Colorimetric Method Used | | Special Considerations for Analysis | Recommended Exposure Duration (days) |
	Reservoir	Diffuser	Cap	Inner Bag	Outer Bag			Basic Method[b]	Modification in Analysis		
Du Pont Pro-Tek	serlin plastic	PE	PE	PE	foil	2.27	sodium bisulfite solution	CTA	Du Pont–Beak	precise make-up of evaporation	5–7
AQRI	glass	glass	PE	PE[c]		3.95	sodium bisulfite on glass fiber	CTA	AQRI–Beak	centrifuging	4–7
3M	PE	PE	PE	PE	PE[d]	65.9	sodium bisulfite on cellulose	CTA	3M–Beak	filtration	1–7[e]
Concord	PE	PE	PE	PE		4.16	mol. sieve 13X ($1/16$ in. pellets)	PRA	Concord	filtration	7
NAE–NRCC	glass	glass	PE	PE		4.03	mol. sieve 13X (beads)	CTA	NAE–Beak	filtration	3–4

[a] PE is polyethylene.
[b] CTA is chromotropic acid; PRA is pararosaniline.
[c] Unsealed.
[d] Semisealed.
[e] Insufficient documentation to establish well.

absorbance and variation were quality controlled so that accuracies of ± 25% could be achieved within that range at a 95% confidence level.

VERSION I: INITIAL DESIGN: A SINGLE BLISTER UNIT. The device consists of a diffusion strip that contains 34 holes approximately 0.6 mm in diameter and 6 mm long (the face velocity effect is eliminated by using these multiple small-diameter holes and low sampling rate per hole [length-to-diameter ratio is 10]); a diffusion membrane between the diffusion strip and the adsorbent medium to prevent leakage and to act as a separator for dust; a wet adsorbent medium, 2.4 mL of a 1% sodium bisulfite solution held in the diffusion blister; a plastic shell that covers and protects the plastic blister against breakage and UV light and supports the blister while it is being exposed; an aluminum pouch and two plastic closures that act as sealing materials for the device after exposure (no cap for the diffusion strip was used); some labels (stickers) for user identification, lot identification, and recording data; and instructions for analysis by colorimetry and Du Pont-modified NIOSH P&CAM 125 chromotropic acid method.

Field Studies. In 1981, the Federal Government of Canada conducted the 2300-home survey (Study 2, Table IV, and Ref. 3). Three-hour impinger tests were run on the first day of a 7-day measurement. Pro-Tek C–60 monitors were used in approximately 900 homes as a first trial for indoor air quality use. The dosimeters were deployed by technicians, exposed for 7 days, then mailed by homeowners to a laboratory in Toronto. Although the devices did meet manufacturer's specifications, neither the blank values nor the confidence level was adequate for studies in homes. Various problems had to be resolved to improve the performance of the Pro-Tek C–60 dosimeter for the monitoring of low levels of formaldehyde. These problems were as follows:

Problem 1: High field blank readings with large random error varying from lot to lot.

Cause: Contamination of the monitor inside the pouch.

Large self-adhesive labels had been placed on the outside of the plastic shell of unexposed badges; these labels were inside the pouches during the journey back to the analytical laboratory (2–20 weeks) and caused contamination of the sampling blister unit. Adhesive used for the sticker and ink used on the label, especially from felt marker pens used inadvertently, emitted various gases and contaminated the solution in the blister (Table V).

Problem 2: High and variable laboratory blank readings.

Cause: Inadequate control over the cleanliness of the production line.

A contributing factor to the contamination of the pouch was an inadequate control over the cleanliness of the production line and of the storage of components before assembly. Further laboratory experiments showed

Table IV. Formaldehyde Measurement Studies

Study	Number of Buildings	Objectives	Date	Formaldehyde Measurement Technique: Number of Devices per Building per Trial				
				Du Pont	AQRI	3M	NAE	Active Device
1. National survey, pilot run	100	Monitor levels in UFFI homes where health problems identified.	fall 1981	3				2 Impinger 1 Impinger(O)[a]
2. National survey[b]	2300	Monitor levels in homes across Canada; 1978 UFFI homes and 383 non-UFFI controls.	fall 1981	3[c]	2			2 AORST[d] 1 AORST(O)[a]
3. New Brunswick survey	83	Repeat of homes from Study 2 because of QA–QC problems.	fall 1981	2	2			2 AORST 1 AORST(O)[a]
4. University of Western Ontario	28	Monitor seasonal variations in levels in UFFI and non-UFFI homes. Each home monitored up to five times.	1982–84	3[e]	3[e]	3[f]	2[g]	1 AORST 1 Impinger
5. Lavoie study	2	Monitor levels and investigate effects of varying exposure duration.	Aug. 1982	20	20			2 or 3 AORST
6. Building 88 study	1	Comparison of performance of monitors.	1982–83	80	80	20	10	1 Impinger 1 AORST 1 DNPH[h]
7. SHQ study	1	Monitor elevated levels and compare devices.	1983–84	5[i]	5[j]			1 Impinger
8. AQR study	2	Compare performances of two lots of AQR devices with Du Pont at various formaldehyde levels.	mid-1983	45	45			1 Impinger
9. St. John's study[k]	50	Determine levels and compare techniques in UFFI and non-UFFI homes.	summer–fall 1981	2 or 3				MIRAN[l] 1 Impinger

Continued on next page.

Table IV (continued)

Study	Number of Buildings	Objectives	Date	Formaldehyde Measurement Technique: Number of Devices per Building per Trial				Active Device
				Du Pont	AQRI	3M	NAE	
10. Newfoundland study								
a.		Determine levels in homes and schools and compare techniques.	summer–fall 1981	33	45		33	
b.		Compare 3M version II to AQRI version II and AORST.	summer–fall 1981		2–3	2–5		1 or 2 AORST
11. Du Pont study	5	Evaluation of Du Pont double blister device.	winter 1982–83	5–8	2			
12. MacLaren study								
a.	12	Measure levels in non-UFFI homes (19 wk).	winter 1981–82	4[m]				1 Impinger
b.	10	Measure levels in some air-conditioned homes (10 wk).	summer 1982	4[m]				2 Impinger[n]
13. PRL study								
a.	50	Measure relationship between levels and airtightness in non-UFFI homes.	1982	2				
b.	46	Measure levels in non-UFFI homes with low air-change rates.	Jan. 1983	2				

14. Canadian study[o]	200	Compare performances of experimental and commercial devices.	summer 1983	3	2	8	
15. UFFI Homeowner Assistance Program[p]							
a. Screening pretest	40,000	Establish levels in UFFI homes before any remedial measures.	1982–86		2		15–30 Detector tubes
b. Premeasure (full scale)	13,000	Establish levels before major remedial measures.	1982–86		2		15–30 Detector tubes
Premeasure (medium scale)	14,000	Establish levels before lesser remedial measures.			2		15–30 Detector tubes
c. Postmeasure	17,000	Confirm adequacy of separation between occupants and cavities.	1982–86		2		5 Detector
16. Schools							
a.	18	Establish levels in UFFI and new non-UFFI schools.	1981–82	15–35	0–15	0–12	Impingers[q] AORST[q] DNPH[q]
b.	2	Compare levels in schools to exposure of pupils. Six students per school.	1982	3	3[r]		

[a] (O) indicates reading taken outdoors. [b] The 2300-home survey; surveys 1 and 2 together are sometimes referred to as the 2400-home survey. [c] In 450 homes only, 2 inside and 1 outside the home. [d] AORST denotes active Oak Ridge sampling tube. [e] Sets of 3 devices, deployed on 5 or more separate occasions. [f] Sets of 3 devices, deployed on 2 or more separate occasions. [g] Single time deployment of a pair of devices. [h] DNPH denotes 2,4-dinitrophenylhydrazine sampling tube. [i] Sets of 5 devices, deployed on 9 separate occasions. [j] Sets of 5 devices, deployed on 13 separate occasions. [k] Program of Health and Welfare Canada. [l] MIRAN denotes continuous IR monitor. [m] Two devices deployed inside the home and 2 deployed outside. [n] One impinger inside the home and one outside. [o] Four Concord devices and 8 Concord wall cavity devices were also used per building per trial. [p] Ongoing program; total program registration as of March 1984 was 60,000. [q] Varying numbers of devices were used in different homes. [r] Each student wore 3 consecutive badges for times of 2, 2, and 1 day.

that released gases increased the contamination of the chemical medium in the diffusion blister (Table V).

Problem 3: Leakage of chemical adsorber solution from the blister.
Cause: Failure of the diffusion membrane over the diffusion strip, evaporation of the blister solution, or inadequate control in filling the blister with the chemical solution.

Spillage of the chemical solution was observed in the foil pouch upon receipt in the laboratory. Some blisters were empty with only a slight solid deposit left. Diffusion membranes were broken, and the solution spilled out and sometimes evaporated; this result was assumed to be induced by squeezing of the device during transportation. The device was relatively fragile. Padded envelopes had been selected because their use permitted faster and far less expensive shipping than the use of small boxes. In many cases, the volume of the chemical collection medium that remained in the blister was small; adjustments had to be made during the analysis of the dosimeter, and thus a source of error was introduced.

Problem 4: Improper usage by homeowner or technician.
Cause: Complicated deployment procedures.

The dosimeter was complicated for use by homeowners and even trained technicians. Pouches were received that were improperly sealed, labels were either not completed at all or not properly completed. Opening, sealing, and labeling operations were necessary prior to exposure. The design of the plastic closures used to seal the foil pouch made it difficult to seal and even more difficult to reopen after exposure and reseal once the

Table V. Formaldehyde Contamination Inside Pro-Tek Version I Pouch: Test Results

Components Tested for HCHO Emission					Conditions		Equivalent ppb for 7-day Exposure	
D.B.U.	Shell	Pouch	Collector	Label	Heating 55 °C (h)	Cooling 5 °C (h)	Diffusion Blister	Sealed Blister
					69	27	27	7
•	•	•			69	17	25	9
					69	27	19	9
•	•				69	27	16	7
					64	23	17	4
			•	•	64	23	32	9
					64	23	121	9
•	•	•		•	64	23	129	6
					64	23	39	7
•	•	•	•	•	64	23	37	5

NOTE: The components tested for HCHO emission are defined as follows: D.B.U., double blister unit with diffusion strip; shell, white plastic shell to hold blister unit; pouch, tedlar-coated aluminum foil pouch to hold badge; collector, dilute sodium bisulfite solution; and label, large: 1 in. × 2 in., hand written, and small: 1/4 in. × 3/4 in., machine numbered.

dosimeter was placed inside. The overall difficulty in using the device generated various field-related problems.

DESIGN MODIFICATIONS FOR VERSION II.   Arrangements were made for the following changes to be made in a single lot of Pro-Tek C–60 dosimeters: (1) A sealed blister was added to the blister unit above the diffusion blister. The chemical adsorbent solution in the sealed blister was isolated from any inside-the-pouch contamination. Its analysis could be used to evaluate the effects on the chemical solution of aging and cycling of temperature during transportation. The aim of this change was reduction of the high mean blank value. (2) The instructions for its use were modified so that only one small factory-printed identification sticker was put into the aluminum pouch. This procedure eliminated unknown inks and labels from the pouch. The production line was cleaned prior to production of the modified devices. The aim of this change was reduction of contamination.

VERSION II: THE DOUBLE BLISTER UNIT.   *Field Studies.*   In early 1982 a field study in 83 homes was carried out in New Brunswick (3). It compared the Pro-Tek C–60 double blister with the Air Quality Research (AQR) PF–1 version I (to be described later) and an active sampling technique based on a molecular sieve collection medium.

Also, several hundred devices were used for surveys in schools insulated with UFFI (Table IV). The problems and observations recorded during these studies are as follows:

Problem: Field blank values much higher than laboratory blank values.
Cause: Solution aging and pouch contamination.

Analysis of blisters from unopened pouches showed that the sealed blister had a much lower background formaldehyde level than the diffusion blister and could not act as a representative field blank (Table VI, Version II). This result indicated that a major contamination source in the pouch had not yet been eliminated. The white plastic shell that held the device in place and provided a clip with which to suspend it was one possible source of contamination: formaldehyde sticking to the surface of the pouch could be released in high levels that could be absorbed by the chemical solution during extended transportation periods.

The effects of the storage conditions and handling of the diffusion blister were unknown. Laboratory simulation of temperature cycling during transportation and storage was necessary to determine the magnitude of the effect.

The following conclusions were reached as a result of the simulation: The absorbance of the sodium bisulfite solution in the sealed blisters of 25 dosimeters aged at 50 °C increased according to the following equation, where t denotes the time of exposure (in hours), and r is the correlation coefficient:

absorbance $= 0.0078 + 0.000177\,t$ $r = 0.93$
equivalent parts per billion for 7-day exposure $= 3.2 + 0.074\,t$

The absorbance of the sodium bisulfite solution in the diffusion blister that sampled the air in the coated foil pouch increased according to the following equation:

absorbance $= 0.030 + 0.00365\,t$ $r = 0.83$
equivalent parts per billion for 7-day exposure $= 12 + 1.5\,t$

The difference in the second term gives an approximate measure of the contribution of the pouch at 50 °C and equals $0.0019t$ or $0.79t$ ppb. The difference in the first term gives an approximate measure of the collection by diffusion from the pouch prior to the aging test and equals 0.022 absorbance units or 9 ppb.

Deviation of 24 of the 25 data points from the equation was less than 0.012 absorbance units (5 ppb). The data indicate a positive correlation with both solution aging and pouch contamination.

Table VI. Blank Analyses for Various Versions of Pro-Tek C–60 Dosimeters

Dosimeter	Quantity of Blisters Analyzed	Geometric Mean (ppb)	Standard Deviation (ppb)	Range of Readings (ppb)
Version I				
Nine production lots	425	10–43	2–9	6–67
Version II				
Diffusion blister				
Lot 16[a]	73	29	24	14–55
Lot 19	20	9	2	7–16
Lot 14–2	21	14	3	10–23
Lot 15–8	14	17	3	13–33
Sealed blister				
Lot 16[a]	75	6	2	3–13
Lot 19	20	6	1	3–8
Lot 14–2	21	5	2	3–11
Lot 15–2	14	9	2	6–12
Version III (Lot 12)				
Diffusion blister	88	12	6	7–39
Diffusion blister	176	16	7	7–57
Sealed blister	42	7	2	4–14
Sealed blister	58	8	6	3–41
Sealed blister	48	7	2	4–16

NOTE: Measured absorbances have been converted to equivalent parts per billion for 7-day exposure.
[a] New Brunswick survey.

Various small experiments were also carried out to isolate possible contamination sources in the pouch. Table V gives results for formaldehyde collected from different combinations of the dosimeter components when exposed for 64 or 69 h at 50 °C, expressed in equivalent parts per billion for a 7-day exposure. The labels inside the pouch are the major source of formaldehyde, and the plastic shell and even the inside liner of the foil pouch are significant sources of uncontrolled contamination.

DESIGN MODIFICATIONS FOR VERSION III. A new batch of devices was produced with modifications arrived at in discussions with the Du Pont production department. These were as follows: (1) Another blister unit having both sealed and diffusion blisters was added to the pouch. One of the diffusion blister units was to remain in the pouch from the production stage to the laboratory analysis and was meant to monitor the formaldehyde level in the pouch. The formaldehyde concentration was to be subtracted from that present in the diffusion blister that was exposed to the room air. The other two sealed blisters were to provide information on storage conditions and could be used for evaluating the reproducibility of the analysis combined with constancy of production. The pouch containing the extra blister unit was to be resealed and suspended in close proximity to the exposed blister unit. (2) The white plastic shells were eliminated to reduce contamination. Blister units would be less protected against physical abuse and UV light but would have better protection during shipping.

VERSION III: TWIN DOUBLE BLISTER UNITS WITHOUT PLASTIC SHELL. *Field Studies.* Field comparisons with other commercial dosimeters were conducted in schools and homes to evaluate reproducibility of results with the dosimeters and to determine the adequacy of the field blanks. The following observations were made:

Problem: Field blank readings not sufficiently low and reproducible.
Cause: Diffusion blister sampling during storage and transportation.

The sources of contamination were reduced to a minimum but there was no mechanism to reduce the diffusion blister sampling during storage and transportation in case of accidental puncture of the pouch or contamination during the long period of storage before deployment or after sampling. These could all increase the amount of formaldehyde absorbed by the blanks (Table VI, Version III).

DESIGN MODIFICATIONS FOR VERSION IV. A large batch of dosimeters was purchased with the following changes: numbers were embossed on each blister unit to avoid use of labels inside the pouch; a cap was placed on the diffusion strip during production, and it was to be removed before exposure and replaced afterwards; individual blister units were sealed into polyethylene bags during production; and white plastic shells were sealed in separate aluminum pouches. Shells were only to be used when excessive exposure to light occurred during the measurement after exposure.

After exposure, the two blister units were to be placed in one of the polyethylene bags and sealed inside the pouch with the same closures as for the pouch.

VERSION IV: VERSION III WITH DIFFUSION STRIP CAPS AND SEALED BAGS. *Field Studies.* Evaluation of Version IV has not been completed. One general observation is possible.

The ease of use of the device has decreased with every added complication in the design; simplicity has been sacrificed for an overall improvement in performance. The cost has also increased. The device has become the high-reliability reference dosimeter in field surveys conducted by trained personnel.

AQRI PF–1 Monitor. BACKGROUND. A second commercial dosimeter was selected in 1982 for use in the UFFI Homeowner Assistance Program. The device was originally designed for monitoring of indoor air quality levels over 7-day periods. It also underwent evolutionary changes. To reduce the number of changes that might be required, a visit was made to the plant and extensive discussions were held on production techniques and quality control. The changes that were subsequently required relate to the selection of formaldehyde-free components, details of production, and improvement of analytical procedures, rather than to changes in form.

VERSION I: INITIAL DESIGN. The AQR PF–1 monitor (Air Quality Research International [AQRI]) consists of a glass vial that provides the diffusion column, sized to reduce face velocity effects for indoor conditions; a dry chemical collection medium consisting of a glass fiber filter pad coated with sodium bisulfite and situated at the bottom of the vial; a plastic snug-fit cap that is placed on the closed end of the vial during sampling; on the outside of the bottle, a label that identifies the production lot number; attached to the vial, a ribbon that suspends the inverted device during sampling; and directions for the analysis that is to be initiated in the vial. The collection medium eluent is centrifuged and analyzed by colorimetry by using an AQR-modified NIOSH P&CAM 125 chromotropic acid method.

Field Studies. A certain number of devices were analyzed as laboratory blanks as part of the Quality Assurance–Quality Control (QA–QC) program for the UFFI Homeowner Assistance Program. The problems found and observations made in early 1982 are as follows:

Problem: High contamination of the device after use of a few lots.
Cause: Change in components.

After producing several lots that had good performance, the manufacturer changed to a more precise type of vial and new cords to suspend the device. These vials had a slightly smaller diameter, and, therefore, the original caps were not as tight. In addition, the new cords emitted formaldehyde at levels of several parts per million and escaped checks made on the original cord. The formaldehyde entered the vials during shipment and storage and produced a serious contamination that was not detected during the manufacturer's QA–QC procedures.

DESIGN MODIFICATIONS FOR VERSION II.   Caps and precision vials were carefully sized. Low formaldehyde emission ribbons were obtained and checked for formaldehyde emission. The manufacturers improved their QA–QC program. The blank values became acceptably low and more reproducible within lots.

Field Studies.   In 1982–83, the Lavoie study and the AQRI study (Table IV) were conducted. The first aimed at comparing Pro-Tek C–60 Version II with AQR Version I for various exposure times ranging from 3 to 10 days. The second study included UFFI houses and particle board buildings in which formaldehyde levels varied from 0.04 to 0.8 ppm. In this study, two AQR Version II dosimeters of different lots and one Du Pont Version I were exposed for a 7-day period. The problems observed were as follows:

Problem: Sampling rate not constant.
Cause: Not determined.

The sampling rate decreased after approximately 5 days at formaldehyde levels of approximately 0.1 ppm. At higher concentrations, between 0.15 and 0.4 ppm, the recovery rate of the device compared to the Pro-Tek C–60 Version I was approximately 50%. The sampling rate was different between different production lots.

The cause of the problem could not be positively identified, so a review of available laboratory data on the device was initiated.

In the meantime, the manufacturer recommended that the duration of exposure be reduced to 4 or 5 days; this modification was considered unacceptable to the Canadian UFFI Homeowner Assistance Program.

DESIGN MODIFICATION FOR VERSION III.   The manufacturer increased the capacity of the device for formaldehyde collection by increasing the amount of bisulfite on the coated glass filter pad. Also the manufacturer modified the analytical procedure to increase its sensitivity and the recovery capability of the dosimeter.

Problem: Calibration factor uncertainty.
Cause: Not determined.

The measurements with the AQRI device were still lower than those of the Du Pont device. Further investigation of possible causes was performed, and the calibration data was reviewed.

DESIGN MODIFICATION FOR VERSION IV.   The manufacturer increased the calibration factor by 9% after a review of the data and calculation procedure. Where measurements have been made in the same houses before and after the change in calibration factors, an adjustment must be made to one of the readings.

VERSION V: THE FUTURE.   Some questions are not completely resolved, and these may require additional modifications to the device. First, the effects of environmental factors such as air moisture content on the recovery and the stability of the sodium formaldehyde bisulfite adduct are not well understood, especially in regard to their effect on absorption under conditions of varying concentration. Also, accuracy in measuring the formaldehyde level when other gases are present needs to be documented. Controlled calibration chamber exposures do not simulate exposure to other gases, dust, and cycling of environmental factors. Only carefully monitored field studies or more complex simulations can improve the knowledge of real-life performance of the monitors. Finally, an indicator is needed to show clearly if the cap has been removed.

Use of 3M Formaldehyde Monitor 3750 in Canada and Subsequent Modifications.   BACKGROUND.   The 3M formaldehyde monitor 3750 was originally designed as a personal monitor for occupational areas where relatively high formaldehyde levels are present. Initial devices were supplied with data of extensive laboratory evaluations and some field evaluations.

The objective of the study in Canada was to determine if the device could provide adequate field performance over 168 h in order to be part of the Canadian UFFI Homeowner Assistance Program.

VERSION I: INITIAL DESIGN.   The 3M formaldehyde monitor 3750 consists of a plastic enclosure measuring 12 × 33 mm (length/diameter = 2.8); dry chemical absorption media consisting of a sodium bisulfite impregnated cellulose disc at the bottom of a vial; a microporous plastic screen at the mouth of the vial to control turbulence effects and to prevent dust entry, held in place with a snap-on ring; a clip for attaching to a cord or ribbon; a snap-on cap that replaces the snap-on ring after exposure and during initial analysis; and instructions for analysis by a 3M-modified P&CAM 125 chromotropic acid method with the elution in the badge but analysis completed outside the monitor.

Field Studies.   A small-scale field study done in 1982 consisted of deploying the Pro-Tek C-60, AQR PF-1, 3M, and experimental dosimeters side by side in a room where the formaldehyde level was known to be approximately 0.2 ppm. The 3M monitors were exposed for various times from 6 to 72 h. This procedure was to determine the period of constant sampling rate. At that time, the recommended exposure duration for the 3M monitor was 1–2 days. The test was repeated in the same room while a small fan circulated the air.

Problem: Actual sampling rate lower than that measured in laboratory.
Cause: Probably face velocity effect.

A face velocity effect could partly explain the low sampling rate. The sampling rate increased while the fan was working (Figure 3).

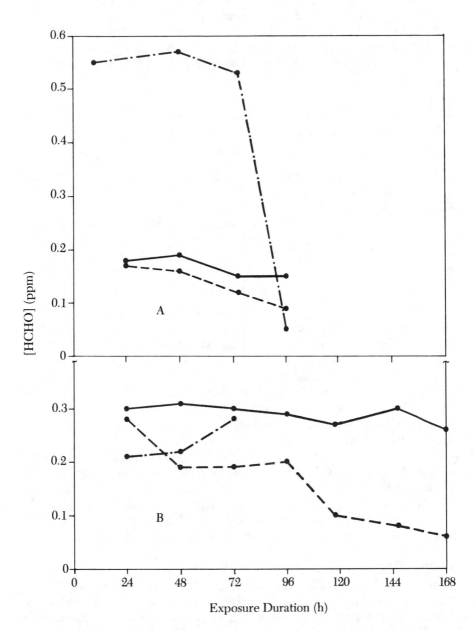

Figure 3. *Effect of air velocity on collection rate. Key: A, static air (no fan);
B, fan running (air circulating);* ———, *Pro-Tek C-60;* ---, *AQR PF-1;
and* —·—·, *3M.*

Data supplied by the manufacturer state that the 3M monitor sampled within 25% accuracy with air velocities greater than 0.1 m/s. Between 0.05 and 0.1 m/s, the sampling rate decreases by 25%. Between 0.04 and 0.05 m/s, the rate decreases by an additional 25% of its original value. The face velocity measured in the room rather crudely was the order of 0.1 m/s and could explain the low result.

Problem: Deactivation after about one day of exposure.
Cause: Probably insufficient collection medium for the sampling rate.

The 3M device failed to collect formaldehyde after 1-day exposure and appeared to lose formaldehyde if exposed for a longer period.

Problem: Complicated procedure for sealing after exposure of the do-
         simeter.
Cause: Poor design.

The snap ring and protective screen must be removed after exposure, and a complex snap cap must be installed. Dexterity is required to place the snap cap on the dosimeter, and many users were unable to accomplish the installation without training.

The identification label at the back of the device is too small.

DESIGN MODIFICATIONS FOR VERSION II.   The manufacturer tackled the problem of increasing the sampling period and reducing the sampling rate by adding a perforated plastic disc between the chemical medium and the screen to provide face velocity and dust protection.

Field Study.   A small number of the 3M devices were deployed for 168 h at the same time as a small number of the C–60 Version III (Table IV, Study 6). The results were promising but were obscured by contamination of the absorbent discs at the plant; this contamination gave higher and more variable blank readings.

FUTURE.   The manufacturer was requested to provide further laboratory evaluation and QA–QC data for the new device before further field testing could be undertaken.

A change in the application would require a complete redesign of the device; this redesign was not contemplated by the manufacturer. Use of the 3M was discontinued except for short-term exposures and research projects.

Evaluation of Existing Experimental Devices

Active Oak Ridge National Laboratories Dosimeter (Sampling Tube). The experimental active sampling tube system designed by Oak Ridge National Laboratories (ORNL) was first used as designed (7). It was then modified, and extensive work was undertaken to check its performance and to obtain low and reproducible laboratory blank readings. The original design specified standard polyethylene dryer tubes filled with molecular sieve 13X and held in place with metal screens. Either 30 or 60 L of air was drawn through the devices over a 15- or 30-min period with a sonic orifice and a timer-controlled pump. Either a specially modified pararosaniline or the chromotropic acid methods were used for analysis (2). These samplers displayed various advantages over the liquid impinger method, including increased sample retention after sampling, ease of use and transportation, and improved reproducibility. The modified system was especially suitable for large studies and for obtaining reliable values in relatively stable conditions against which to compare the results from the passive dosimeters. Problems were encountered with obtaining consistently low blanks when several commercial laboratories were asked to apply a standard procedure. Rather than instituting the special training required, we eliminated the device from the larger studies. The study did indicate that the same technology might be applicable to passive devices and provided valuable information for their development.

Passive Oak Ridge National Laboratories Dosimeters (Permeance Membrane). Two experimental passive devices were tested during one 1981 field survey of approximately 80 homes (Table IV, Study 3). Both were based on the ORNL water-filled passive dosimeter design (8). Dimensions were approximately 35 mm in diameter by 15 mm high. A 25-μm membrane of dimethyl silicone was placed over the mouth of the polyethylene devices; the adsorbent was water in one case and molecular sieve 13X in the other. The devices were exposed for 24 h, analyzed by the pararosaniline method, and the results were compared with those of the Du Pont dosimeter and liquid impinger values. Performance of both was erratic; good agreement was found for a minority of the devices. Concern arose over the use of the thin membrane and its sensitivity to relative humidity. The permeance of the membrane appeared to change by a significant percentage when the average humidity to which the membrane was exposed changed from 25% to 75%.

Development of New Formaldehyde Dosimeters. The availability and adaptability of formaldehyde dosimeters for use in homes was reviewed in 1981. The only dosimeter commercially available to monitor HCHO was the Du Pont Pro-Tek C–60 Version I, designed for monitoring at occupational levels of HCHO. The development and eventual production of a dosimeter in Canada appeared to be the only course that would result in a device uniquely suited to the long-term needs of the Canadian UFFI Homeowner Assistance Program. Achievement of the strict quality control in manufacturing and shipment that was required seemed improbable unless the device could be made under close scrutiny.

The objective of the program was to develop a dosimeter that was inexpensive and accurate and gave a reproducible measurement of formaldehyde. Furthermore, any device that was developed would have to be suited to the measurement of gases other than formaldehyde that were emitted by UFFI or UFFI-contaminated materials. The gases were not identified at that point, as only preliminary work on UFFI offgases had been performed. The simplest scenario was that the same dosimeter be used for all gases. Various techniques for producing and analyzing dosimeters were considered. Investigations of effectiveness of different absorbents, constructions, materials, sampling durations, analytical methods, etc. were initiated.

Two laboratories were enlisted to undertake the development of the dosimeter. One was charged with designing a dosimeter that used gas chromatograph analysis and the other was charged with designing a dosimeter that used colorimetric analysis. Each used its specific area of expertise and experience. The solution was approached from two directions, and at the same time the evolution and use of the commercially available dosimeters described earlier was continuing. Experience gained in that activity was applied to the development of the new devices.

Development of National Aeronautics Establishment–National Research Council of Canada Dosimeter. The developmental work on one dosimeter (9) was undertaken at the National Research Council of Canada (NRCC) by the National Aeronautics Establishment (NAE) under the direction of L. Elias, who has experience in the measurement of trace quantities of gases in both the lab and the field using active dry sampling tubes with gas chromatographic analysis. An attempt was made to adapt these techniques to the measurement of UFFI offgases. The first phase was to adapt the device for formaldehyde, then to adapt it to other aldehydes, and eventually to other offgases. The development of the unit for formaldehyde is nearing completion and has been used in comparative studies. The design phase of the NAE–NRCC device included the following studies and selection of components:

ADSORBENT SELECTION. Adsorbents that were examined include Alumina, Carbosieve B, Tenax GC, Poropak T, N, and R, Chromosorb 104 and 105, OV–17 on platinum mesh, charcoal, molecular sieve 13X, and silica gel. Trapping efficiencies and recovery were investigated. Because of long experience and good initial laboratory results, silica gel was selected. Silica gel proved to have poor long-term formaldehyde storage, so a change was made to molecular sieve 13X.

ANALYTICAL TECHNIQUES. Thermal desorption into a gas chromatograph was originally selected as the most efficient method to sample a variety of gases. Analyses for formaldehyde were made with both flame and photoionization detectors, although photoionization detectors were found to offer little advantage for formaldehyde measurements with the types of gas chromatographs available.

Difficulties with precision of formaldehyde determinations, the greater sensitivity required, and the lack of any reference method that could deal with the small quantities collected made it necessary to change to a larger amount of collection medium in a larger device so comparisons could be made to colorimetric analysis. The better reproducibility and accuracy of the colorimetric analysis for formaldehyde and increasing experience in commercial laboratories with colorimetric analysis dictated a change to such a technique. The chromotropic acid method of colorimetric formaldehyde analysis with water elution, followed by filtration before color development, was selected.

DIMENSIONS OF DOSIMETERS. As gas chromatography was the method initially selected, the first versions were slightly modified gas chromatographic adsorption tubes. A supported 2–3-cm length of granular adsorbent was placed in a very narrow (6.3 mm o.d.) piece of 7.5-cm long Pyrex tubing.

Although this design looked promising for high concentrations of formaldehyde typical of occupational exposure, not enough formaldehyde was collected for proper analysis of the lower levels found in UFFI dwellings. The design of the device was changed to increase the sampling rate. A glass specimen bottle approximately 2.3 cm o.d. × 5.6 cm long, having a screw cap and approximately 1-cm long neck, was employed.

The desire for greater reliability of formaldehyde measurements resulted in the choice of an even larger glass bottle approximately 2.3 cm o.d. × 8.5 cm long. A very simple screw top design was chosen for low cost. A bed 0.2 cm deep of molecular sieve was measured into the bottom of the tube, and a diffusion length of approximately 8 cm was left. The ratio of diameter to length was then approximately 1.9.

The NAE–NRCC dosimeter was first tested and calibrated in the laboratory, then was compared initially to various versions of the Du Pont and AQR dosimeters in a number of small studies (Table IV, Study 14). A number of problems were found with the performance of the initial version.

Problem: Collection capacity at humidities above 70% was low.
Cause: Tendency of the molecular sieve to collect water vapor.

The trapping and retention of formaldehyde on molecular sieve decreased with time at higher humidities. At levels of relative humidity below 50% the sampling rate did not change significantly with exposure durations of up to 120 h. However, at relative humidity levels of 70% and higher, decreases in sampling rate of approximately 6%, 12%, and 35% were found after exposure durations of approximately 1, 2, and 5 days, respectively. All experiments were performed at room temperature (22 °C).

Modification 1. The bed depth was increased to 0.5 cm but still did not provide adequate capacity at high relative humidities. Preconditioning the molecular sieve by exposure to a humid formaldehyde-free atmosphere did not ameliorate the situation. Cycling humidity levels could cause some of the trapped formaldehyde to be released.

Modification 2. The sampling duration was reduced from 7 to 3 or 4 days to reduce humidity-induced effects. This modification improved the accuracy.

Problem: Cap loosening.
Cause: Poor design.

The screw-on caps presented problems in transportation and home-owner handling. Some caps loosened during transportation and resulted in leakage of air and occasional spilling of molecular sieve.

Modification 3. Further experimentation with different types of snap-on and more positively locking caps is continuing.

Problem: Different deployment from other dosimeters.
Cause: Different design parameters.

The NAE–NRCC devices were originally designed to be placed up-right on top of furniture. Such placement could permit spillage of sieve and contamination from formaldehyde sources (i.e., shelving and ashtrays), which decreased the choice of placement.

Modification 4. For comparison with other devices, a narrow ribbon was attached to allow the devices to be hung from the ceiling.

FUTURE WORK. More investigations are underway to evaluate the dosimeter's performance. Field and laboratory trials are planned to test the modified units. Use for other aldehydes and other gases is being evaluated.

Development of the Concord–NRCC Dosimeter. Before the UFFI Homeowner Assistance Program was launched, Concord Scientific Corporation had performed work for the Ontario Government on dosimeter monitoring of SO_2 at levels found in homes. For Health and Welfare Canada they had also conducted studies on SO_2 and NOx dosimetry at indoor levels.

In 1982, as the UFFI Homeowner Assistance Program got underway, Concord Scientific Corporation was asked to assist in the development of a formaldehyde dosimeter under a contract with the National Research Council of Canada. Various requirements were established including accurate, reproducible formaldehyde measurements at low levels in UFFI houses; selection of formaldehyde-free materials and machine-tooled, tightly fitting components; and applicability of the device to monitor other UFFI offgases. Two 7-day monitors have been developed—one for room air and the other for wall cavities. The design of the Concord–NRCC dosimeter included a number of comparative studies and a selection of components.

ADSORBENT SELECTION. The adsorbent selected was molecular sieve 13X. The selection was based on data showing long-term stability of the adsorbed formaldehyde after sampling. The use of the sieve allowed a number of analytical methods to be used when the tube was adapted to measurement of other gases.

Desiccant beds added above the sieve beds in the modified active Oak Ridge sampling tube (AORST) to increase the capability of the device to sample at high humidities gave results that were not reproducible. The desiccant removed formaldehyde from the air along with the water vapor.

ANALYTICAL TECHNIQUES. The modified pararosaniline method of analysis was selected for formaldehyde measurement (2) because of higher sensitivity obtained in initial studies in which comparisons were made with chromotropic acid analysis. The pararosaniline method was modified beyond Miksch's modification and optimized for the configuration of the dosimeter (e.g., volume of water used in the solution, etc).

DIMENSIONS OF THE DOSIMETER. Various prototypes were constructed with sampling tubes and bed diameters of 8 and 15 mm and lengths of 100 to 300 mm. The ratio of diameter to length was determined for each. In addition, water uptake tests were performed and were followed by actual field measurements of formaldehyde where results were compared with those from other devices.

Sampling rates were determined for the prototype devices, and a diffusion length of 8 cm was chosen as optimum for the device.

SELECTION OF MATERIALS. For reasons of cost, ease of manufacture, reproducibility of dimensions, and tightness of fit, the components of the devices were specially machined. Polyethylene was selected because it fulfilled the preceding requirements. Stringent tests and controls have been established to ensure absolute cleanliness of the components before assembly.

Field Studies. Table IV lists field studies in which early and more advanced prototypes have been used. Field studies were undertaken only after extensive laboratory evaluation. The field studies yielded the following observations:

Problem: Variable laboratory blanks.
Cause: Not yet determined.

Although initial studies yielded a procedure that was adequate for prolonged laboratory studies, the results of a major field study indicate that the blank values are variable and tend to increase with time.

Further investigation is planned on blanks, including improvements in sieve cleaning procedures (e.g., higher temperature baking, drawing a vacuum while baking, and cycling of temperature). The effects of varying humidity and temperature cycles on the collection and retention of formaldehyde will also be examined in laboratory calibration chambers.

Current Status of Low-Level Formaldehyde Dosimeters

Comparisons of various dosimeters have been made under field conditions because of the difficulty in simulating the conditions adequately in calibration chambers. Although considerable data have been generated from these comparisons, much of this data no longer represents the state of the art with low-level formaldehyde dosimeters. Sufficient data has been obtained on the later versions to justify conclusions on certain performance factors: reproducibility of pairs of devices in the field; stability of the devices before and after exposure, including an aging factor; ability to integrate formaldehyde concentration over a 7-day exposure period, which includes an element of accuracy; and loss of formaldehyde during elution and analysis.

Du Pont Pro-Tek Version I in 40 houses with two to four dosimeters in a set per house and levels of 2–215 ppb gave one difference of 49 ppb and the rest below 22 ppb. Of 40 comparisons, 30 were within 10 ppb. For 39 houses, the standard deviation was 6.7 ppb (*11*).

Pro-Tek Version I was used in 12 non-UFFI houses over 19 weeks and compared to 19 1-h modified NIOSH P&CAM 125 impinger measurements analyzed in 4-cm path length cuvettes (*12*). The differences between average values in the houses measured by the impinger and dosimeter methods ranged from − 5 to + 9 ppb at the 17–58-ppb level. The averages for the 12 houses over the 19 weeks were 36 ppb for impinger and 35 ppb for dosimeter. Standard deviations of the impinger results were on the average 2 ppb higher than those of the dosimeter (13 vs. 11 ppb), but differences between impinger and dosimeter readings for individual houses were − 10 to + 14 ppb, when compared for the 19-week period.

Integration of formaldehyde concentrations by multiple low-flow-rate sodium bisulfite impingers with 6-, 8-, or 12-h exposures, compared to Pro-Tek C–60 Version I, showed the dosimeter higher by approximately 30 ppb at the 120-ppb level (*11*).

Reproducibility of measurements with the Pro-Tek C–60 Version III is illustrated by data from 200 pairs exposed in 200 homes in which the levels ranged from 10 to 200 ppb. Eighty-five percent of the pairs agreed to within 10 ppb, and 95 % agreed to within 20 ppb. Another set of measurements made in four houses with levels of 40–60 ppb, using four to eight dosimeters at the same location, gave standard deviations of the readings of 10, 8, 5, and 8 ppb (*11*).

Comparisons of C–60 Version III and AQRI PF–1 Version II in six houses with four to eight C–60 dosimeters and PF–1s gave differences of 0, − 12, 11, − 4, − 6, and − 6 ppb, for an average of − 3 ppb at the 35–60-ppb level. Comparisons of large numbers of C–60 Version III to PF–1 Version II in schools in 1982, at levels of 9–125 ppb, gave differences of − 17 ppb to + 48 ppb, a mean difference of 10 ppb, and a standard deviation of 14 ppb (*11*). Similar comparisons in one house of four C–60 Version II with four PF–1 Version II gave a mean difference of − 18 ppb and a standard deviation of 8.6 ppb at the 110-ppb level. At the 272-ppb level in a mobile home, the mean difference was 53 ppb, and the standard deviation was 15 ppb.

The standard deviation in the difference between pairs of PF–1 Version II used in 1982 in 59 homes was 10 ppb over the 20–190-ppb range (*11*).

The shelf life of the devices can be illustrated by simultaneous exposure and analysis of dosimeters of different ages, exposed at levels of 75 and 137 ppb. The results from two pairs of each of eight lots of PF–1 Version II, aged 3–13 months, show that lots older than 12 months are unusable and lots aged 10–11 months are inconsistent. The reproducibility of a lot aged 3 months was 5 ppb for four pairs of dosimeters.

Recovery of formaldehyde in later versions of both the Pro-Tek C–60 and the PF–1 at the 20–200-ppb levels is between 90% and 96% for exposures of 7 days, compared to spiking of the dosimeters with refluxed paraformaldehyde solution. Recovery is adequate in both devices, though the elution must be done carefully and according to manufacturers' instructions in the PF–1 device to achieve these results.

Results from the 200-home study using 310 Pro-Tek C–60 Version III and 155 PF–1 Version II gave the following equation:

$$(PF–1) = 0.002 + 1.3 \ (C–60) \ (ppm)$$

The square of the correlation coefficient was 0.61.

The standard deviation for 66 pairs of 3M Version I (1982) was 6.7 ppb at the 24–181-ppb level (*11*). Differences between 17 sets of 3M Version II and PF–1 Version II (1983) showed the 3M to be 9.7 ppb higher on average and the standard deviation to be 13.5 ppb at the 20–70-ppb level (*11*).

These errors may be compared to those of the 3-h impinger method using the modified chromotropic acid method, where uncertainties were ± 66 ppb at the 200-ppb level, ± 27 ppb at the 50-ppb level, and ± 18 ppb at the 20-ppb level and showed a strong bias toward the negative error (*3*). The dosimeters are generally more precise.

Conclusions

Dosimeters can integrate 7-day concentrations of formaldehyde in air at least as well as, and often better than, low-flow-rate or multiple sample impinger methods. Formaldehyde dosimeters have evolved to the state where different models will provide adequate accuracy and operational characteristics for either field surveys or scientific studies in homes in the 30–200-ppb range. The bulk of the houses studied in Canada fall between these limits. Dosimeters constitute a system that includes all components used in transport and handling. Use of dosimeters at low levels requires constant attention to every detail of manufacture, transport, deployment, retrieval, and analysis. Use of field blanks is imperative to obtain reliable results for low-level formaldehyde measurements.

An indicator for displaying the life of dosimeters is highly desirable, or gaseous lab spikes can be used for checks.

Storage of formaldehyde on the collection medium is not adequately understood, so there is a need to continue to investigate and improve the design of most commercial devices.

Evaluation of dosimeters is expensive and should only be undertaken when it can be ensured that batch-to-batch variations are small.

A 95% confidence level is adequate for surveys, but not for studies on individual homes, especially in measuring changes in levels within houses.

Seasonal and diurnal variations in formaldehyde levels in homes make measurements of formaldehyde concentration and comparisons difficult. Even when compared to the mean equilibrium formaldehyde concentration of the source, short-term meteorological variations can mask changes and introduce large uncertainties.

Houses containing UFFI have diurnal and hourly variations that are different and potentially more extreme than many non-UFFI houses. The use of long-term passive dosimeters is necessary to integrate these variations. There is little chance to reduce these variations to an adequate level by a simple protocol for house preparation.

In the Canadian climate, levels of formaldehyde in all houses (those containing UFFI and those without) approach similar values in January and February. As the absolute humidity values inside and outside the houses decrease dramatically, the equilibrium formaldehyde concentrations of the sources decrease and the sources have little capacity to emit formaldehyde.

The studies have yielded a certain initial understanding of the variation of formaldehyde in houses, which may be a useful starting point for further studies. UFFI houses may be categorized into two classes: responders and nonresponders. Responders are houses in which the formaldehyde concentrations in the UFFI-filled cavities and the living space air change dramatically and proportionately from summer to winter. Nonresponders are houses in which the formaldehyde concentrations in the cavities and in the living space vary little with time. The mixing of results of measurements from responders with those from nonresponders may well mask differences between UFFI houses and non-UFFI houses.

Appendix I. Calculation of Concentration from Measurement: Gases

The reading from analysis of the dosimeter is in mass of collected or sampled gas minus the mass of gas collected in the field blank. The dosimeter does not measure concentrations directly. The following steps and approximations are involved in reducing a measurement to a concentration:

 1. The volume of air sampled per unit time must be known.
  The precision of the measurement of volume sampled can be

difficult to establish. If the device is passive and relies on diffusion to collect the gas, the equivalent volume of air sampled per unit time can be calculated theoretically from the geometry, but it must be checked by calibration under several conditions. If the device is an active system, the flow rate must be controlled constant so it is known, but it must be corrected to standard temperature and pressure.

2. The density of the measured gas at the same reference temperature and barometric pressure as the air from which it is sampled must be used. This requirement is somewhat difficult when the temperature and pressure have changed dramatically during the sampling period.

3. The duration of the exposure in hours must be known. This information requires clear records and is complicated by bookkeeping errors, such as converting 1½ h to 150 min rather than to 90 min.

These considerations accepted, the equation for the time-weighted average exposure [TWA (ppb by volume)] is the following:

$$\text{TWA} = \text{weight recovered}/(\text{sampling rate} \times \text{density} \times \text{exposure time})$$

$$= \frac{1000 \ \mu g}{\text{L/h} \cdot \mu g/\mu L \cdot h}$$

$$= \mu g \cdot \frac{1000}{\mu g/h \cdot L/\mu L} \cdot 1 \ h$$

$$= \mu g \times \text{conversion factor} \times \text{time}^{-1}$$

Thus, the conversion factor is $1000[\mu g/\text{ppm} \cdot h]^{-1}$. Caution is necessary because the reciprocal of this conversion factor is sometimes used by manufacturers or investigators.

Acknowledgments

The authors would like to acknowledge the technical contributions of the following people without whose cooperation this work would not have been possible: Paris Georghiou, Geortec Ltd.; Robert Caton and Claude Davis, Concord Scientific Corporation; Richard Young, IEC Beak Consultants Inc.; Bert Kring, E. I. du Pont de Nemours and Co.; Robert Miksch, Air Quality Research Inc.; E. Yolanda Stankiewicz, 3M Canada Inc.; Kristen Geisling–Sobotka, Lawrence Berkeley Laboratories; Thomas Matthews, Oak Ridge National Laboratories; Mark Richardson, UFFI Information Coordination Center, Consumer and Corporate Affairs Canada; David T. Williams, Health and Welfare Canada; Lorne Elias and Irena

Short, National Aeronautics Establishment; and George Tamura and Robert Dumont, Division of Building Research. This paper is a contribution of the Division of Building Research, National Research Council of Canada.

Literature Cited

1. "National Institute of Occupational Safety and Health Manual of Analytical Methods," 2d ed.; Taylor, D. G., Ed.; NIOSH: Cincinnati, 1977; Vol. 1, 125–29.
2. Miksch, R.; Anthon, D. W.; Fanning, L. Z.; Hollowell, C. D.; Rezvan, K.; Glauville, J. *Anal. Chem.* 1981, 53, 2118.
3. Urea–Formaldehyde Foam Insulation Information and Coordination Center (UFFI-ICC), Consumer and Corporate Affairs Canada, Final Report of the National Testing Survey, Ottawa, September 1983.
4. Georghiou, P. E., personal communication.
5. Georghiou, P. E., personal communication.
6. Kring, E. V.; Thorley, G. D.; Dessenberger, C.; Lautenberg, W. J.; Ansul, G. D. *Am. Ind. Hyg. Assoc. J.* 1982, 43, 786.
7. Matthews, T. G.; Howell, T. C. *Anal. Chem.* 1982, 54, 1495.
8. Matthews, T. G.; Hawthorne, A. R.; Howell, T. C.; Metcalfe, C. E.; Gammage, R. B. *Environ. Int.* 1982, 8, 143–51.
9. Elias, L.; Krzymien, M. K.; Short, I. "Development of a Versatile, Low-Cost Passive Air Sampler for UFFI Homes"; National Research Council of Canada: Ottawa, March 1983; National Aeronautics Establishment, Laboratory Technical Report LTR–VA–66.
10. Georghiou, P. E.; Snow, D. A. "Formaldehyde Studies: Part I, Evaluation of the New Du Pont 'Double Blister' Formaldehyde Dosimeter; Part II, Field Evaluations and Comparisons of Commercially Available Formaldehyde Passive Monitors; Part III, Formaldehyde Levels in Wall Cavities of UFFI-Containing Homes"; Research Contract Reports, National Research Council of Canada, 1983.
11. Manley, P. J.; Helmeste, R. H.; Tamura, G. T., presented at the 5th A.I.C. Conf., Reno, Nev., October 1984.

RECEIVED for review October 25, 1984. ACCEPTED March 22, 1985.

Generation of Standard Gaseous Formaldehyde in Test Atmospheres

MAT H. HO

Department of Chemistry, University of Alabama at Birmingham, Birmingham, AL 35294

Generation of standard formaldehyde in test atmospheres is essential in studies of toxicological effects, development and testing of analytical methods, as well as evaluation and calibration of air monitors, analytical instruments, and personal sampling devices. As the need for measurement of formaldehyde at lower concentrations increases, the preparation of known concentrations of this compound in standard gaseous mixtures becomes more critical. Because formaldehyde is highly reactive and has a tendency to polymerize, a suitable generation system is required. In this chapter, several dynamic methods for generating formaldehyde standards are reported. The principle, instrumentation, and performance characteristics of each system are described. Several factors that affect the accuracy and precision are also discussed.

Preparation of accurate, reproducible, and controllable standard concentrations of gaseous or vapor pollutants at part-per-million (ppm) or part-per-billion (ppb) levels is essential in environmental and occupational health sciences. Studies of the toxicological effects on humans, animals, and vegetables; development and testing of analytical methods; and evaluation and calibration of air monitors, analytical instruments, and personal sampling devices all require standard concentrations. Effective application of any analytical method or instrument is heavily dependent on the calibration to verify its accuracy and precision. In recent years, the development of several highly sensitive instruments has further complicated the calibration problem. The improvement in the detection limit of many analytical methods and the demand to measure lower concentrations of toxic gases have also increased the need to prepare accurate standards at low levels for calibration.

The procedures for the preparation of gaseous mixtures may give less accurate results than those for liquid mixtures. However, many methods have been developed for the generation of gas standards with very good

0065–2393/85/0210/0193$06.25/0

accuracy (1–4). In general, these methods can be divided into two categories: static methods and dynamic methods. In static methods, a known amount of pure gas or vapor is added into a known volume of diluent (usually nitrogen or purified air) in closed devices such as Teflon bags, stainless steel cylinders, or glass vessels in which they are mixed and contained until use. Static methods are simple and inexpensive; however, they suffer from a number of disadvantages such as losses due to absorption and condensation on the walls of the container. Only limited volumes can be prepared, leaks can occur, and pressure changes can exert an effect. In addition, the errors associated with the introduction of small volumes of a component to be measured into a dilution gas give poor accuracy. Consequently, static methods cannot be used to prepare standards of very low concentrations. In dynamic methods, pure gas or vapor of a known generation rate is continuously introduced into a known flow rate of diluent in a flowing system. Although more elaborate and relatively more expensive equipment is involved, dynamic systems offer several advantages compared to the static methods. Because standards are generated continuously, losses due to absorption onto the walls of the system are negligible after equilibrium. Large volumes of gas standards can be generated and concentrations can be varied to provide a wide dynamic range. Flexibility in concentration range, volume, and flow rate of the generated standards is especially important for calibration and evaluation of analytical instruments or sampling devices.

Formaldehyde is one of the widely used chemicals both in industrial and nonindustrial environments (5, 6). The health effects on humans and recent evidences of carcinogenicity associated with formaldehyde exposure in animals have created great concern for controlling and monitoring of this chemical (7, 8). Several methods have been developed for the determination of formaldehyde in air (9–14). In general, these methods can be divided into two categories: those involving direct monitoring and those involving the collection of formaldehyde in suitable media followed by analysis with appropriate techniques. The validation and calibration of all of these methods require the generation of accurately known concentrations of formaldehyde vapor in the standard test atmospheres that simulate the conditions (i.e., concentration range, humidity, and interferences) similar to those found in the field. As the need for measurement of formaldehyde at lower concentrations and with greater sensitivity increases, generation of known concentrations of this pollutant in standard gas mixtures becomes more critical.

The preparation of standard concentrations of nonreactive gases and vapors is relatively straightforward; however, for formaldehyde it is considerably more difficult. Formaldehyde is a highly reactive compound that is stabilized either in aqueous solution or as its solid polymers (paraformaldehyde and trioxane). In the gas phase, this compound is very unstable and tends to polymerize. Gaseous formaldehyde standards for calibration must fulfill a number of practical requirements. They must be stable, accurate, reproducible, available in sufficient amounts, and preferably simple to prepare. In addition, gaseous formaldehyde standards should be prepared by using measurements of fundamental quantities such as mass, temperature, or pressure, and all sources of errors should be known and precisely defined.

Because formaldehyde vapor is unstable, can combine with many substances, and has a tendency to polymerize, a suitable dynamic method is required for the preparation of this compound at the part-per-million and part-per-billion levels in air. One way of preparing formaldehyde vapor is to bubble nitrogen or purified air through an impinger containing formalin solution, that is, a solution of approximately 37% formaldehyde in water and 10–15% methanol to prevent polymerization. Another way is to inject dilute formalin solution into a heated airstream (*15–17*). The major drawback of these methods is that large amounts of water, methanol, and other substances will be generated in the gas mixture. Standard formaldehyde free of methanol and water in the test atmosphere can be generated by thermal depolymerization of paraformaldehyde, a long chain polymer, or trioxane, a trimer. In this chapter, several dynamic systems using permeation and diffusion methods for generating known concentrations of formaldehyde in simulated standard atmospheres will be described.

Permeation Methods

Principle of Permeation Methods.   Permeation methods for the generation of standard gases were first described by O'Keefe and Ortman (*18*). These methods are based on the mixing of a small but known volume of gases or vapors passing through a membrane of a permeation device with a known volume of a diluent gas. Since the development of these methods, the use of permeation devices as primary standards in dynamically generated gaseous mixtures has been studied in great depth, and many applications have been reported (*1–4, 19*).

Permeation tubes are made of TFE Teflon (polytetrafluoroethylene) or other plastic materials. Solid, liquid, or liquified gas of the material to be used in preparing the calibration gas is introduced into the tube, and both ends of the tube are sealed with a Teflon plug or glass beads. Following an initial induction period, the vapor continuously permeates through the walls of the tube at a constant rate if the tube is held at a constant temperature. The permeation of the component gas is produced by a combination of diffusion through the microporous structure of the membrane and the solubility in the membrane. If the permeation tube is immersed in a flowing stream of purging gas with known flow rate, the permeated component is then mixed with the purging gas that passes across the tube to provide a standard mixture.

At equilibrium, the permeation rate of a gas through a membrane can be expressed by the following equation (18, 20–22):

$$r = DS(P_1 - P_2)(A/d) \tag{1}$$

where r is the permeation rate, D is the diffusion coefficient, S is the solubility constant, P_1 and P_2 are the partial pressures of the permeant gas on the two sides of the membrane, A is the membrane area, and d is the membrane thickness. The permeation coefficient, B, of a membrane for a particular gas can be expressed by the Arrhenius equation:

$$B = DS = B_o \exp(-E_p/RT) \tag{2}$$

where E_p is the permeation activation energy, R is the gas constant, and T is the absolute temperature of the membrane. Substituting Equation 2 into Equation 1 yields the following:

$$r = B_o(P_1 - P_2)(A/d)\exp(-E_p/RT) \tag{3}$$

This equation shows that the permeation rate is proportional to the area, inversely proportional to the thickness, and dependent on the material of the membrane. Thus, a desired range of permeation rates of a particular gas can be obtained by choosing the appropriate membrane material, area, and thickness. For high permeation rates, a large area and thin membrane can be used. In general, tubings are widely employed in the preparation of standard gases because they have good surface area to internal volume ratios, are naturally rugged, and require a minimum of sealing surface. In this case, the membrane area can be varied by changing the length of the tube. In addition, Equation 3 shows that the permeation rate varies logarithmically with the inverse operating temperature ($1/T$). Many authors have reported that the permeation rate varies 10% for every 1 °C change in the operating temperature (3, 4, 20). Thus, control of the temperature of the permeation tube to within ±0.1 °C is necessary to maintain 1% accuracy in the permeation rate. The permeation rate of a given gas through a given membrane can be increased by increasing the operating temperature. If a constant temperature is maintained, the permeation rate will remain essentially constant for several months or years until nearly all of the material has permeated through the wall of the tube.

Generation of Standard Formaldehyde Using Permeation Methods. PERMEATION TUBES. Standard concentrations of gaseous formaldehyde in the test atmosphere can be generated by thermal depolymerization of paraformaldehyde or α-polyoxymethylene in a permeation device. Several permeation systems using these polymers have been described (23–27). In general, these systems consist of three components: a permeation source, flows of purging and diluent gas, and a sampling manifold.

Permeation tubes can be prepared by sealing paraformaldehyde or α-polyoxymethylene in a TFE Teflon tube with glass beads or Teflon plugs (23–25). Figure 1A shows the cross-sectional diagrams of such a device. TFE Teflon is the construction material of choice because it is chemically inert. However, other plastic materials can also be used. The tubes are sealed tightly to prevent formaldehyde vapor from leaking and thereby disrupting the normally steady permeation rate. The permeation tube is placed inside a chamber, and the temperature of the system is controlled to ±0.1 °C with a thermostated oven or a thermostated oil bath. Figure 1B shows some typical chambers used in our laboratory for the permeation method. A gastight chamber is made of Pyrex glass and sealed with a Teflon stopper with the help of a rubber "O" ring. A glass midget impinger or an air sampling impinger can be used as a chamber. The permeation tube can also be inserted in the central tube of a glass condenser maintained at constant temperature by a thermostated circulating water supply. Figure 1C shows the dynamic generation system using a permeation tube. The purging and dilution gas are thermally conditioned by passing through a copper or stainless steel coil, a few meters long, immersed in the constant temperature oven or oil bath. The flow rate of gas is monitored by a flow meter and is held constant with a differential flow controller. Under controlled temperature, paraformaldehyde or α-polyoxymethylene depolymerizes to release formaldehyde vapor which then permeates through the wall of the permeation tube. A flow of purging gas is passed over the tube and mixed with the permeated formaldehyde to provide a standard gas mixture with known concentration. The mixture can be further diluted by mixing with a stream of dilution gas. The flow rate of the diluent should also be controlled precisely. Permeation tubes for formaldehyde are now commercially available from many vendors.

A gravimetric method has found most widespread use for the calibration of the permeation tubes. The permeation rate is determined by measuring the weight loss of the tube over a given period of time while the tube is held at a constant temperature in a flowing stream. After an initial induction period required for obtaining a constant rate, the tube is weighed and then immediately placed into the chamber. The tube is reweighed after a certain time interval. Because the loss of mass of the tube is equal to the mass of the permeating formaldehyde, the permeation rate is calculated as follows:

$$r = W/t \tag{4}$$

where r is the permeation rate (ng/min), W is the weight loss (ng), and t is the time between weighings (min). For weighing, the tube is removed from the chamber, sealed in a closed vessel, cooled to room temperature, and then weighed to the nearest 0.1 or 0.01 mg. The weighing should be carried

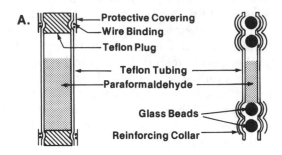

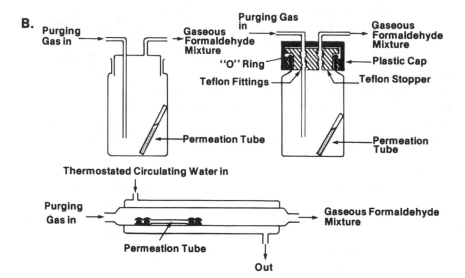

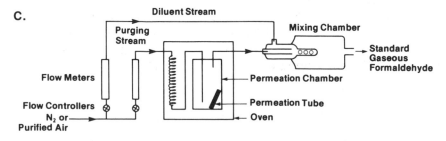

Figure 1. Schematic diagrams of A, permeation tubes; B, permeation chambers; and C, apparatus for formaldehyde generation using the permeation method.

out with extra care because fingerprints or additional dirt on the outside of the tube can seriously affect the accuracy of the results.

Van de Wiel et al. (28) showed that the major source of error in dynamic generation using permeation sources is in the gravimetric determination of the permeation rate. The accurate generation of a known concentration of formaldehyde in the standard mixture is highly dependent on the accurate measurement of the permeation rate. Because of a relatively high dependence of the permeation rate on temperature, the temperature should be maintained constant to ±0.1 °C during the calibration. Usually, we assume that paraformaldehyde or α-polyoxymethylene is composed of 100% formaldehyde. If the polymer is composed of less than 100%, this factor should be included in the calculation of the permeation rate provided that impurities permeate at the same rate. For example, Meadows and Rusch (29) reported that paraformaldehyde used in their study was composed of 96.3% formaldehyde by the sulfite–HCl assay, and the factor 0.963 was applied to the weight loss to prolonged time ratio to obtain the permeation rate. For best results, the tube must be used at the calibration temperature.

Permeation tubes are relatively durable, and no special attention is required to ensure the calibrated rate is maintained throughout their lifetime. Usually, the permeation rate will remain essentially constant until nearly all of the material has permeated through the wall of the tube if the tube is stored under the normal conditions of temperature, pressure, and humidity. High temperature, high humidity, and contamination of the permeation membrane during handling or storage should be avoided. Operating the permeation tube under high humidity is not a good practice. If high humidity in the standard mixture is desired, water vapor should be added downstream from the tube with the dilution stream.

PERMEATION CELLS. Another permeation system for the dynamic generation of formaldehyde using paraformaldehyde was described by Muller and Schurath (26). A schematic representation of the system is shown in Figure 2. A gastight permeation cell was made of Pyrex glass or stainless steel. Paraformaldehyde (approximately 10 g) was loosely packed into the cell with quartz wool. A Teflon permeation tube (3.2 mm o.d., 2 mm i.d.) with an active length up to 266 cm was placed inside the cell and sealed tightly into its top with Swagelok fittings. One end of the permeation tube was connected to the purging gas flow and the other was connected to the dilution chamber. The temperature of the cell was controlled to ±0.1 °C with a thermostated oven or oil bath. Formaldehyde vapor, which was generated from the depolymerization of paraformaldehyde inside the cell, permeated through the wall of the Teflon tube and then was mixed with a controlled flow of purging gas inside the tube to generate standard mixtures.

The stability of the permeation rates was investigated, and the results

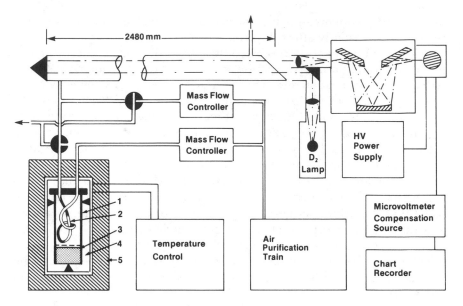

Figure 2. Schematic diagram of permeation system for formaldehyde generation: 1, permeation chamber; 2, permeation tube; 3, paraformaldehyde and quartz wool; 4, air space between permeation cell and thermostated brass container; and 5, stonewool packing. (Reproduced from Ref. 26. Copyright 1983 American Chemical Society.)

are shown in Figure 3 (26). The formaldehyde concentrations in the effluent gas were measured at intervals over a period of 9 days by using the UV absorption spectroscopic method under the conditions shown in the legend of Figure 3. The formaldehyde concentrations, and therefore the permeation rate, decreased during the first 100 h of operation and then became more stable. After an initial period of approximately 5 days, the stability of formaldehyde concentration was better than $\pm 2\%$. This value includes the possible errors in the measuring method of formaldehyde and the fluctuations of the flow rate. The long-term stability was also checked by storing the permeation cell at room temperature for 118 days, and the formaldehyde was measured under the same conditions as described in Figure 3. The results yield 97% of the average permeation rate previously observed (26). The concentration of formaldehyde in the standard mixture was inversely proportional to the total flow rate of dilution gas, and this relationship was confirmed over a large range of flow rate. Muller and Schurath (26) also reported that the plot of the logarithms of the permeation rate versus $1/T$ gave an excellent straight line. A linear regression through the data points yielded the following equation:

$$\ln r = (59.186 \pm 0.920) - [(10107 \pm 152)/T] \qquad (5)$$

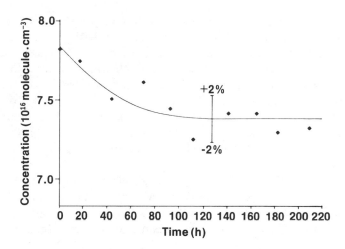

Figure 3. Stability of permeation rates. Conditions: permeation cell is filled with 10 g of paraformaldehyde and thermostated at 121.8 °C; permeation tube: 3.2 mm o.d., 2 mm i.d., 266 cm long; and flow rate of diluent air: 5 L/h. (Reproduced from Ref. 26. Copyright 1983 American Chemical Society.)

where r is the permeation rate of formaldehyde (molecule/cm/s) and T is the absolute temperature (K). At moderate temperatures, formaldehyde vapor may decompose into CO and H_2, particularly on metal surfaces. A Pyrex glass permeation cell, which contained a 90-cm Teflon permeation tube and generated 170 ppm of formaldehyde at 90 °C, produced 10 ppb of CO and 20 ppb of H_2 in the standard gas mixture, whereas a stainless steel cell under the same conditions produced 30 ppb of CO and 50 ppb of H_2. Although the thermal decompositions of formaldehyde in both cases are quite low and were not studied extensively, glass as construction material for the cell and operation at low temperature are preferable if possible.

A new design of the permeation cell using a silicone membrane and α-polyoxymethylene was described by Godin et al. (27). This design allows the cell to operate at relatively low temperatures and therefore minimizes the thermal decomposition of formaldehyde and the formation of degradation products of the polymer other than formaldehyde. Figure 4A shows the diagram of the cell. The cell is made of glass and consists of two chambers separated by a dimethylsilicone membrane of 2-mm thickness. Two Teflon or nylon nets were used to support the membrane and to avoid its deformation. Polyoxymethylene was placed inside chamber A. Temperature of the cell was controlled to ±0.1 °C with a thermostated oven or oil bath. Under the controlled conditions and after an initial induction period, formaldehyde vapor, which was generated by the thermal depolymerization of polyoxymethylene, permeated across the membrane into chamber B at a constant rate. An accurately known flow of purging gas, which was preheated to the same temperature as the permeation cell, was passed through chamber B and mixed with formaldehyde to produce a standard gas mixture. The mixture could be further diluted to produce lower concentrations as shown in Figure 4B. Depending on the desired concentration range, several chambers with different diameters, for example, 15, 25, and 50 mm, could be used to construct the permeation cell. For a permeation cell having a 12-mm diameter membrane and operating at 110 °C, a permeation rate of 1.1 mg/h and a relative standard deviation of 3.4% were observed (27).

Permeation cells can be calibrated gravimetrically as described for permeation tubes. Permeation cells are, however, heavier than the permeation tubes, and the weight loss of the cells should be large enough to give accurate results. The permeation rates can also be determined by measur-

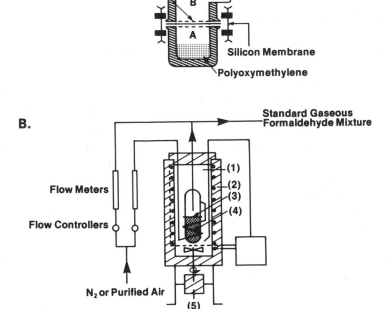

Figure 4. Schematic diagram of A, permeation cell; and B, apparatus for generation of formaldehyde using permeation cell: 1, temperature-controlled oven; 2, insulator; 3, silicone membrane; 4, polyoxymethylene; and 5, fan. (Reproduced with permission from Ref. 27. Copyright 1978 Marcel Dekker.)

ing the amount of formaldehyde generated in the effluent stream using a UV-absorption spectroscopic method at 285 nm (*26*) or by collecting the effluent formaldehyde in a suitable medium and subsequently analyzing the solution using the chromotropic acid method (*13*). An absolute pressure method has also been developed for the calibration of permeation devices (*30*). For accurate results, several factors and conditions such as those described for the calibration of the permeation tubes must be considered in the calibration of the permeation cells.

By measuring the weight loss of the permeation cell and determining the amount of formaldehyde in the standard mixture, Godin et al. (*27*) found that for the same period of time the weight loss of polyoxymethylene is higher than the amount of formaldehyde generated, particularly at high temperatures. This result may be caused by several factors such as the formation of degradation products other than formaldehyde, the thermal decomposition of formaldehyde, and the impurities of the polymer. The differences between the weight loss of polyoxymethylene and the amount of formaldehyde liberated are 5% and 50% at 100 °C and 130 °C, respectively. Although no extensive study has been reported, operation of these permeation cells at a temperature lower than 100 °C seems necessary. If high permeation rates are desired, one should use a large membrane area rather than operate at high temperature. One other conclusion that can be drawn from the work of Godin et al. (*27*) is that the calibration of the permeation cell by the gravimetric method is unreliable at high temperature.

Diffusion Methods

Principle of Diffusion Methods.   The use of diffusion cells in the preparation of gaseous standards was first introduced by Fortuin (*31*) and McKelvey and Hoelscher (*32*). The theoretical background of the method was described by Altshuller and coworkers (*33, 34*). Various designs of the diffusion systems and their applications for a wide range of volatile organic liquids were reported by many investigators (*3, 4*). Next to the permeation devices, the diffusion methods are currently the most widely used methods in the generation of standard gaseous mixtures.

In the diffusion methods, the vapors of a component to be prepared are evaporated from a liquid reservoir and then diffused through a capillary tube into the stream of purging gas to produce a known concentration. If the temperature and tube geometry remain constant, the vapors will diffuse through the tube at a constant rate. By assuming that the concentration of the generated vapor in the upper part of the diffusion tube (mixing chamber) is nearly zero and that the lower part (reservoir) is saturated, the diffusion rate is given by the following equation:

$$r = (DMPA/RTL) \ln [P/(P - P_v)] \tag{6}$$

where r is the diffusion rate (g/s), D is the diffusion coefficient (cm²/s) at pressure P and temperature T, M is the molecular weight of the vapor (g/mol), P is the pressure of the diffusion cell at the open end of the capillary tube (atm), A is the cross sectional area of the diffusion path (cm²), R is the gas constant (cm³ atm/mol K), T is the absolute temperature of the diffusion cell (K), L is the length of the diffusion path (cm), and P_v is the partial pressure of the diffusion vapor (atm) at temperature T.

The diffusion rate of a particular vapor depends on the operating temperature, on the pressure, and on the geometric dimensions of the diffusion path. Generally, the diffusion rate will change approximately 5% with a change in temperature of 1 °C. To obtain a standard mixture with an accuracy of 1%, the temperature should be controlled to ±0.2 °C. Another parameter to adjust the diffusion rate is the dimensions of the tube. High diffusion rates can be obtained by increasing the diameter or decreasing the length of the diffusion path as shown in Equation 6.

Generation of Standard Formaldehyde Using Diffusion Methods. DIFFUSION CELL WITH PARAFORMALDEHYDE. Figure 5 shows the diagrams of some typical designs of the diffusion cells that have been reported in the literature (19, 32, 33). Among these, diffusion tubes are the most widely used devices in the generation of standard gas mixtures. Diffusion tubes are constructed from Pyrex glass and consist of a reservoir and a long-neck capillary tube. The dimensions of the tube, that is, diameter and length, are different for each desired concentration range. The reservoir is partially filled with paraformaldehyde (usually 4–6 g), and the cell is kept at constant temperature. Formaldehyde vapor in the reservoir, which was generated by the depolymerization of paraformaldehyde, diffuses through the tube into the mixing chamber where it is mixed with the purging gas stream passing over the open end of the tube to provide a standard mixture (29, 35). Figure 6A shows two designs of the gastight chambers, which were developed in our laboratory, that have been used to house the diffusion

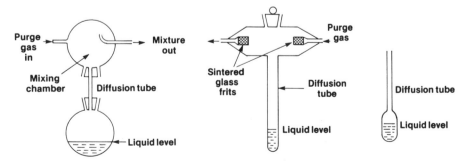

Figure 5. Typical designs of diffusion devices. (Reproduced from Ref. 32 and 33. Copyright 1957 American Chemical Society.)

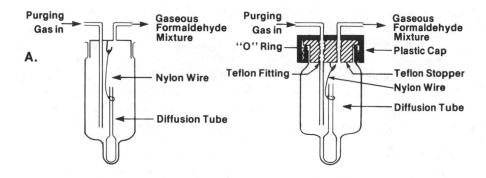

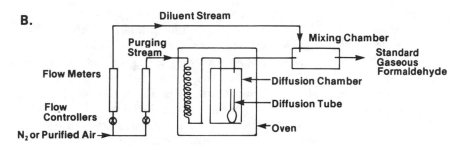

Figure 6. Apparatus for the generation of formaldehyde using diffusion devices: A, diffusion chambers; and B, generation system.

tubes. The chambers are made of glass, and the diffusion tube can be easily removed for calibration with the help of a nylon wire attached to the neck of the tube. Figure 6B shows the diagram of a typical generation system used in the diffusion method. It is similar to the permeation system. Because of a relatively high dependence of the diffusion rate on temperature, the temperature is controlled to ±0.1 °C with a thermostated oven or oil bath. The flow rates of purging and diluent streams are held constant with flow controllers and monitored with flow meters. During the filling procedure, the polymer is introduced into the reservoir without leaving any particles clinging to the inner bore of the diffusion path. During handling and storage, care should be also taken to ensure that the polymer does not get into the diffusion path.

Because the diffusion coefficients are available from the literature, the diffusion rate can be calculated from Equation 6. In practice, however, the diffusion rate is determined gravimetrically. Although theory is very useful in predicting the diffusion rates and in getting tube dimensions near the size of interest, the weight-loss method is more accurate and reliable for calibration. Because the diffusion bore geometry may vary slightly throughout its path length and the measurement of the diffusion path length and area may introduce errors, the diffusion rate obtained from Equation 6 is not accurate.

The techniques and procedures for the calibration of diffusion tubes are similar to those described earlier for permeation devices. The tube is held at constant temperature, and a small flow of purging gas is passed across the tube to remove any of the diffused material. Following thermal equilibrium, the tube is removed from the chamber, sealed with aluminum foil, cooled to room temperature, and then weighed to the nearest 0.1 mg. Immediately after weighing, the diffusion tube is put back into the thermostated chamber. After a certain period of time the tube is removed, and the entire procedure is repeated again. To minimize the errors due to the weighings, the weight loss of the diffusion tube between the time intervals should be at least 50 mg. The diffusion rate is determined from the weight loss with Equation 4. The cooling period is not counted as diffusion time; only the elapsed time inside the chamber is used for this calculation.

Once the diffusion tube has been calibrated, the concentration of formaldehyde can be varied by changing the operating temperature, dimension (bore or length) of the diffusion path, or flow rate of the diluent gas. Changing the temperature or dimension of the tube should be avoided because it may introduce errors if the tube is not recalibrated. For a given tube, the simplest way is to vary the flow rate of diluent gas. The flow rate of purging gas across the open end of the diffusion tube should be kept below 1 L/min because a high flow rate creates turbulences and, thus, reduces the effective diffusion path length. High flow rates may also alter the temperature of the chamber. Low concentrations of formaldehyde can be obtained by diluting the mixture further downstream from the tube with another stream of diluent gas.

DIFFUSION CELL WITH TRIOXANE.   Trioxane can be depolymerized to produce formaldehyde. Schnizer et al. (36) reported the use of trioxane vapor in the generation of percent concentrations of formaldehyde in air. Nitrogen or purified air was bubbled through molten trioxane, and the vapor of this material was swept over a catalyst where it was depolymerized to formaldehyde. With this system, conversion yields higher than 98% were not attainable because of the repolymerization of formaldehyde at high concentrations. Recently, this technique was modified by Geisling and Miksch (37) for generating part-per-million and part-per-billion levels of formaldehyde. Instead of bubbling nitrogen, a diffusion tube was used to generate trioxane vapor.

Figure 7 shows the apparatus for the generation of formaldehyde using the diffusion cell and catalyzed depolymerization of trioxane. The diffusion cell, which consists of a diffusion path (7.9 cm long and 4.9 mm i.d.) and a reservoir, is made of Pyrex glass. Molten trioxane was introduced into the reservoir and then allowed to solidify at room temperature before the cell was placed inside the thermostated chamber. The temperature of the

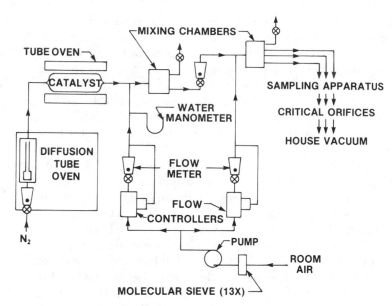

Figure 7. Schematic diagram of formaldehyde generation system using tri-oxane and diffusion tube. (Reproduced from Ref. 37. Copyright 1982 American Chemical Society.)

diffusion system was controlled to ±0.05 °C with an oven. Trioxane vapor from the reservoir diffused through the diffusion tube into the chamber and then was mixed with a flowing stream of nitrogen or purified air that was preheated to the same temperature as the diffusion cell. The desired concentration of trioxane vapor was obtained by controlling the temperature of the diffusion cell, the dimensions of the diffusion path, and the flow rate of diluent gas. The second part of the system is the converting chamber which contains a catalyst. A quartz tubing (7.5 cm long and 1 cm i.d.) was packed with approximately 18 g of 40-mesh carborundum, which had been coated to saturation with 85% phosphoric acid and held at 160 °C in a furnace. The trioxane vapor was passed through the catalyst bed and converted to formaldehyde with essentially 100% efficiency. Because formaldehyde tends to repolymerize at high concentrations, the mixture stream emerging from the catalyst was immediately diluted with diluent gas. Before and after use, the catalyst bed should be heated and purged with N_2 to clean and prevent the accumulation of formaldehyde.

The diffusion rate of trioxane was determined gravimetrically as described earlier. For the diffusion cell designed by Geisling and Miksch (37), as described earlier, the maintenance of the cell at 35.2 ± 0.05 °C produced a constant diffusion rate of 740 ± 10 μg/h. The conversion efficiency of trioxane to formaldehyde in the catalyst chamber was investigated. The gas mixture downstream from the catalyst was collected by bubbling the air through two distilled water bubblers that were connected in series and maintained at 2 °C. The bubbler solutions were analyzed for formaldehyde and formaldehyde plus trioxane by using the pararosaniline method (14) and the chromotropic acid method (13), respectively. No detectable trioxane was found in the solutions. A gas sample collected from the effluent of the catalyst was analyzed by mass spectrometry, and no trioxane or other anomalous compounds in amounts exceeding 2% of the total formaldehyde were detected.

Formaldehyde concentrations in the range from 0.05 to 2 ppm can be generated dynamically with the system just described. Higher or lower concentrations can also be prepared by modifying the geometry of the diffusion path or adjusting the temperature and flow rate of the system. Muller and Schurath (26) tested this technique and found that some of the formaldehyde generated may decompose to produce CO and H_2 under conditions suitable for the depolymerization of trioxane catalyzed by carborundum at 160 °C. The thermal decomposition of 90 ppm of formaldehyde in the standard mixture produces more than 180 ppb of CO and 40 ppb of H_2.

DIFFUSION CELL COMBINED WITH PYROLYSIS.   Another interesting dynamic system for the generation of formaldehyde as well as other reactive aldehydes over wide concentration ranges (from parts per million to parts per billion) was developed by Tsang and Walker (38). This system is based on the use of the diffusion cell to generate a mixture of large "parent" molecules in an inert diluent followed by passage of this mixture through a hot tube in which it undergoes pyrolytic decomposition to produce formaldehyde. In his previous study, Tsang (39) showed that a suitable parent compound can be pyrolyzed in a gold tube reactor held at high temperature to produce the reactive gas of interest. This interesting concept was applied to the generation of formaldehyde, acetaldehyde, and acrolein vapors using 3-methyl-3-buten-1-ol, 4-penten-2-ol, and 5-methyl-1,5-hexadien-3-ol, respectively. The reactions of the pyrolytic decomposition are the following:

$$CH_2 = C(CH_3)CH_2CH_2OH \longrightarrow HCHO + i\text{-}C_4H_8 \tag{7}$$

$$CH_2 = CHCH_2CHOHCH_3 \longrightarrow CH_3CHO + C_3H_6 \tag{8}$$

$$CH_2 = CHCHOHCH_2C(CH_3) = CH_2 \longrightarrow CH_2 = CHCHO + i\text{-}C_4H_8 \tag{9}$$

The success of this method is critically dependent upon the choice of the parent compound.

The apparatus for the generation of formaldehyde using this technique consisted of two thermostated compartments as shown in Figure 8. The temperature of the first compartment, which contained diffusion and

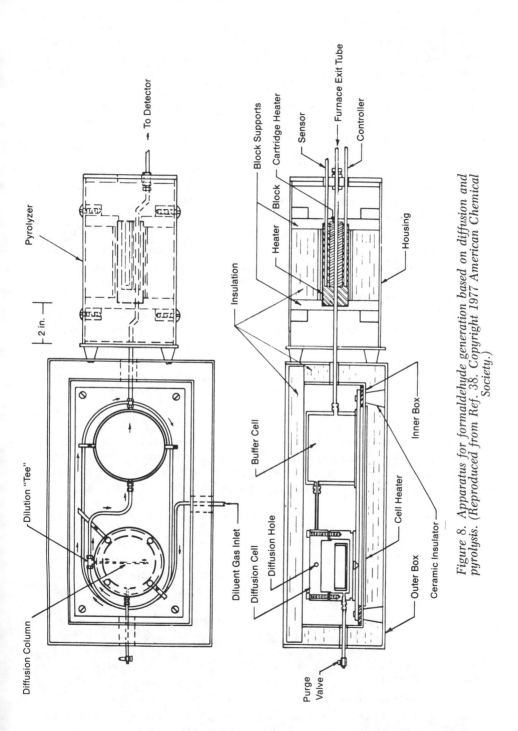

Figure 8. Apparatus for formaldehyde generation based on diffusion and pyrolysis. (Reproduced from Ref. 38. Copyright 1977 American Chemical Society.)

buffer cells, was controlled to ± 0.1 °C. The diffusion cell was used to generate known concentrations of 3-methyl-3-buten-1-ol vapor. The length of the diffusion path was kept at 6.3 cm, and its diameter was varied up to 0.63 cm depending on the desired range of concentrations. The reservoir of the cell was filled with approximately 15 mL of 3-methyl-3-buten-1-ol. A flow of helium or other nonreactive gases such as nitrogen or argon, which was preheated in the chamber, was passed over the diffusion cell and mixed with 3-methyl-3-buten-1-ol vapor to provide a "parent" gas mixture. The diffusion rate of the cell can be calibrated gravimetrically as described earlier, and the concentration of the parent component in the gas mixture can be calculated. The buffer cell was used to minimize the pressure surges that may arise from the switching of valves or the connection of fittings. After passing over the buffer cell, the stream of the generated gas mixture entered the second compartment containing the 1/8-in. o.d. gold tubing in which the pyrolysis occurred. The temperature of this compartment can vary from 0 to 700 °C and could be held to within ± 12 °C of the desired temperature. The entire system was operated at approximately 0.5 atm above ambient pressure. The flow rate and pressure of the system were controlled by needle valves and regulators. Gas chromatography with a flame ionization detector was used to analyze and identify the products in the generated gaseous standards. Formaldehyde was hydrogenated to methane before passage into the detector.

The optimum conditions, operational characteristics, and performance of the system were reported in detail by Tsang and Walker (38). At low flow rates of carrier gas (~ 30 mL/min), complete decomposition in the gold tube ($>97\%$ conversion) was obtained at a temperature of 600 °C. For the higher flow rates (up to 200 mL/min), however, a minimum temperature of 650 °C was required. The relationship between formaldehyde output and the temperature of the diffusion cell (1/8 in. i.d. $\times$ 0.63 cm long diffusion path) fits the following equation:

$$\log [\text{HCHO}] \ (\mu g/min) = [(-2704 \pm 30)/T] + (9.09 \pm 0.09) \qquad (10)$$

The standard deviation over the entire range was 4%, and the long-term stability of the formaldehyde output was $\pm 2\%$. The proper behavior of the apparatus was also confirmed over a large range of flow rates and cell pressures. The long-term stability of the system was investigated over several hundred hours. The results showed that the concentration output was quite stable, and the reproducibility of the generated gas was very good not only for formaldehyde but also for acetaldehyde and acrolein.

Calculation of Formaldehyde Concentration in the Standard Gaseous Mixture

Concentration of formaldehyde in the standard mixture can be calculated from the permeation or diffusion rates and the total flow rate of diluent gas. The equation for this calculation is given as follows:

$$C = [f/(f + F)] \times 10^6 \tag{11}$$

where C is the concentration of formaldehyde (ppm, v/v), f is the permeation or diffusion rate of formaldehyde across the membrane (mL/min), and F is the total flow rate of the air passing through the permeation tube chamber and the diluent gas (mL/min).

Because F is normally very much larger than f, this equation can be reduced to

$$C = (f/F) \times 10^6 \tag{12}$$

Usually, the permeation or diffusion rate, f, is determined gravimetrically as described earlier and is given in nanograms per minute at a specified temperature and pressure rather than in milliliters per minute. Consequently, f can be expressed as follows:

$$f = (22.4/M) \, (T/273) \, (760/P) \, r \times 10^{-6} \tag{13}$$

where M is the molecular weight of formaldehyde (g), T is the absolute temperature (K) at which permeation is occurring, P is the pressure (mm Hg) at which permeation is measured, and r is the permeation or diffusion rate (ng/min) at T and P.

The temperature and pressure at which the flow rate of diluent gas, F, is measured must be in agreement with the temperature and pressure used in the determination of the permeation or diffusion rate. If the flow rate is measured at a temperature and pressure that are significantly different from the temperature T and pressure P used in Equation 13, it should be corrected to the conditions at which f was determined:

$$F = (T/T_A) \, (P_A/P) \, F_A \tag{14}$$

where F_A is the total flow rate of diluent gas (mL/min) measured at T_A and P_A. Substituting Equations 13 and 14 into Equation 12 yields

$$C = (22.4/M) \, (T_A/273) \, (760/P_A) \, (r/F_A) \tag{15}$$

Usually, the flow rate is measured at ambient temperature (25 °C) and atmospheric pressure (760 mm Hg). In this case Equation 15 can be reduced to

$$C = (24.45/M) \, (r/F_A) \tag{16}$$

Equations 15 and 16 show that the concentration of formaldehyde is cut in half if the flow rate of diluent gas is doubled. This result is extremely useful for dynamically generating formaldehyde standards over a wide range of concentrations with the same permeation or diffusion device thermostated at the same temperature.

Concentrations of trioxane in the Geisling and Miksch system (37) and of 3-methyl-3-buten-1-ol in the Tsang and Walker method (38) can be calculated from Equations 15 or 16. In these cases r is the diffusion rate and M is the molecular weight of the respective compound.

General Requirements for the Use of Permeation or Diffusion Devices for Formaldehyde Generation

To obtain an accurate concentration of formaldehyde in the standard mixture, two major requirements must be fulfilled. First, the permeation or diffusion rate must be known to within 1%, and second, the flow rates of the purging and diluent gas must be measured to within 1% accuracy. The precision of the concentrations of the resulting gaseous mixtures is directly related to the precision of the permeation or diffusion rate and diluent flow rate; both of these parameters must be maintained constant throughout the preparation of the standard.

Because the permeation rate and diffusion rate vary 10% and 5%, respectively, for every 1 °C change in the operating temperature, as discussed earlier, it is necessary to control the temperature to better than ±0.1 °C. This requirement is for long-term stability. For most applications, a thermostated water or oil bath can be used. Several controlled-temperature ovens for analytical applications are capable of controlling to within ±0.02 °C, thus the ±0.1 °C requirement could easily be obtained. High-quality, fixed-set-point, mercury-in-glass thermostats are suitable for this purpose. The initial induction period, usually from a few days to several weeks for permeation tubes, is required only at the time of the filling of the tube. For daily use, a time period of several hours is needed for a tube to come to thermal equilibrium so that a constant permeation or diffusion rate is produced. The time required to reach thermal equilibrium depends on the temperature difference between operating and room temperature and is an inverse exponential function of the volume of the permeation or diffusion chamber. Usually, low-volume chambers are desired for rapid start-up and for changing concentration.

The flow rates of the purging and diluent gas should be calibrated by using the soap-bubbling method. Flow meters are usually not adequate as the primary flow-measuring devices. However, they can be used as a flow indicator and for returning to a previously calibrated set point. The stability of the flow rate can be maintained with a differential flow controller. Usually, the purging gas is preheated to the same temperature as in the permeation or diffusion chamber, and its flow rate is low to preserve the thermal stability of the chamber. In view of the critical temperature-control requirement, it is preferable to vary the concentration by changing the flow rate of diluent gas.

Materials used for construction of the mixing chamber sampling manifold and tubing lines must be inert and have no interaction with formaldehyde. Teflon and glass are the most suitable materials for this purpose. Dry nitrogen or purified air can be used as purging or diluent gas. To provide purified air, dust particles and oil droplets are removed from the compressed air with particle filters (pore size <0.2 m) and conventional oil traps. The air is then purified with molecular sieves or silica gel to remove water and activated charcoal to remove organic materials.

Literature Cited

1. Nelson, G. O. "Controlled Test Atmospheres, Principles and Techniques"; Ann Arbor Science: Michigan, 1971.
2. "Calibration in Air Monitoring"; American Society for Testing and Materials: Philadelphia, 1976; STP598.
3. Barratt, R. S. *Analyst* 1981, *106*, 817.
4. Namiesnik, J. *J. Chromatogr.* 1984, *300*, 79.
5. Borzelleca, J. F. "Formaldehyde—An Assessment of Its Health Effects"; NAS: Washington, D.C., 1980.
6. Consumer Product Safety Commission *Fed. Regist.* 1980, *45*, 34031.
7. "Formaldehyde: Evidence of Carcinogenicity," NIOSH Current Intelligence Bulletin No. 34, April 15, 1981.
8. OSHA *Occup. Hazards*, 1980, *January*, 21.
9. Papa, L. J.; Turner, L. P. *J. Chromatogr. Sci.* 1972, *10*, 744.
10. Beasley, R. K.; Hoffman, C. E.; Rueppel, M. L.; Worley, J. W. *Anal. Chem.* 1980, *52*, 1110.
11. Kim, W. S.; Geraci, C. L., Jr.; Kupel R. E. "Sampling and Analysis of Formaldehyde in the Industrial Atmosphere"; NIOSH: Cincinnati, 1978.
12. Septon, J. C.; Ku, J. C. *Am. Ind. Hyg. Assoc. J.* 1982, *43*, 845.
13. "Methods of Air Sampling and Analysis," 2d ed.; American Public Health Association: Washington, D.C., 1977; pp. 303-7.
14. Miksch, R. R.; Anthon, D. W.; Fanning, L. Z.; Hollowell, C. D.; Revzan, K.; Glanville, J. *Anal. Chem.* 1981, *53*, 2118.
15. Cares, J. W. *Am. Ind. Hyg. Assoc. J.* 1968, *29*, 405.
16. Kim, W. S.; Geraci, C. L., Jr.; Kupel, R. E. *Am. Ind. Hyg. Assoc. J.* 1980, *41*, 334.
17. Matthews, T. G.; Howell, T. C. *Anal. Chem.* 1982, *54*, 1495.
18. O'Keffe, A. E.; Ortman, G. C. *Anal. Chem.* 1966, *38*, 766.
19. "Calibration Standards"; Analytical Instrument Development: Avondale, Pa., 1978.
20. Lucero, D. P. *Anal. Chem.* 1971, *43*, 1744.
21. Crank, J.; Park, G. S. In "Diffusion in Polymers"; Crank, J.; Park, G. S., Eds.; Academic: New York, 1968; p. 1.
22. Ibusuki T.; Toyokawa, F.; Imagami, K. *Bull. Chem. Soc. Jpn.* 1979, *52*, 2105.
23. Ho, M. H., unpublished data.
24. Beasley, R. K.; Hoffman, C. E.; Rueppel, M. L.; Worley, J. W. *Anal. Chem.* 1980, *52*, 1110.
25. Bisgaard, P.; Molhave, L.; Rietz, B.; Wilhardt, P. *Am. Ind. Hyg. Assoc. J.* 1984, *45*, 425.
26. Muller, R. E.; Schurath, U. *Anal. Chem.* 1983, *55*, 1440.
27. Godin, J.; Bouley, G.; Boudene, C. *Anal. Lett.* 1978, *A11*, 319.
28. Van de Wiel, H. J.; Uiterwijk, J. W.; Regts, T. A. In "Studies in Environmental Science: Air Pollution Reference Methods and Systems"; Schneider, T.; de Koning, H. W.; Brasser, L. J., Eds.; Elsevier: Amsterdam, 1978, Vol. 2.
29. Meadows, G. W.; Rusch, G. M. *Am. Ind. Hyg. Assoc. J.* 1983, *44*, 71.

30. Dietz, R. N.; Smith, J. D. In "Calibration in Air Monitoring"; American Society for Testing and Materials: Philadelphia, 1976; STP598, pp. 164–79.
31. Fortuin, J. M. H. *Anal. Chim. Acta* **1956**, *15*, 521.
32. McKelvey, J. M.; Hoelscher, H. E. *Anal. Chem.* **1957**, *29*, 123.
33. Altshuller, A. P.; Cohen, I. R. *Anal. Chem.* **1960**, *32*, 802.
34. Altshuller, A. P.; Clemons, C. A. *Anal. Chem.* **1962**, *34*, 466.
35. Kulle, T. J.; Cooper, G. P. *Arch. Environ. Health* **1975**, *30*, 237.
36. Schnizer, A. W.; Fisher, G. J.; MacLean, A. F. *J. Am. Chem. Soc.* **1953**, *75*, 4347.
37. Geisling, K. L.; Miksch, R. P. *Anal. Chem.* **1982**, *54*, 140.
38. Tsang, W.; Walker, J. A. *Anal. Chem.* **1977**, *49*, 13.
39. Tsang, W. *J. Res. Natl. Bur. Stand. Sect. A* **1974**, *78*, 157.

RECEIVED for review January 31, 1985. ACCEPTED March 8, 1985.

TOXICOLOGY

Toxicology of Formaldehyde

JOHN W. GOODE

Gulf Life Sciences Center, Gulf Oil Corporation, Pittsburgh, PA 15230

Concern about the toxicity of formaldehyde has increased since the report by the Chemical Industry Institute of Toxicology (Third CIIT Conference on Toxicology, November 1980, Raleigh, North Carolina) that carcinomas were found in animals after exposures to formaldehyde. The present toxicology of formaldehyde is reviewed along with a description of the current toxicological testing methodologies. The most significant properties of the compound are its potential to cause irritation and nasal tumors. Considerations for interpreting the results are discussed.

OUR PROGRESS SEEMS TO BE THROUGH CHEMISTRY, as an advertisement implies, and this progress has meant an increase in exposure of our population to chemicals. Each year it becomes more evident that we must know if these chemical entities present a hazard to human health or our environment; in fact, we have laws that require this information. Because some of the effects of these substances can take years to develop and immediate removal of the product from the marketplace would still leave a generation of developing cases, the use of animals as indicators of the toxicities of these substances is prudent. We assume that results of these animal studies can be extrapolated to humans. There are many arguments about extrapolation of extremely high doses and findings of animal tests to humans and whether or not it is moral to use animals for this purpose. None of the arguments have been resolved to everyone's satisfaction, but they do stimulate continuing investigations for better methodologies.

Increased interest in the hazards to ourselves and our environment has caused toxicology to gain rapid recognition during the last two decades. The Society of Toxicology, formed in 1961, is rather young when compared to other scientific societies. This fact may give some indication of the public's awareness of its environmental problems. My purpose in writing this chapter is derived from this awareness; formaldehyde is one of the more important chemicals in the production of thousands of industrial and commercial products and one with which a large portion of our population has contact.

A better title for this chapter might be "Toxicology Methodologies and

0065–2393/85/0210/0217$06.00/0

Formaldehyde" because it is my intent to outline, in general, the methods used in testing along with results reported for formaldehyde by using approximate methodologies. This chapter is not intended to be an exhaustive review of all studies conducted on formaldehyde but rather to give typical data from a number of common toxicology testing areas.

The purposes of toxicity studies are the following: to identify if a substance causes harm; to identify the route and dose that cause harm; to identify what organ(s) or processes may be harmed; to determine the extent and course of the injury; and to identify "safe levels" of the substance.

Acute Mortality Testing

Investigation of a substance is usually begun by conducting the short-term or acute mortality study. Death is the end point in these studies. Most of these studies consist of a single exposure to the substance of a few animals via various routes followed by a 14-day holding and observation period. The holding period allows us to determine latent responses to the substance and whether or not reversibility of the response is likely to occur. Body weight gains and losses, as well as gross pathology, are determined in all cases. Histopathology is performed if findings are indicated during the gross pathology.

Compounds that are undergoing acute study can be compared by use of some guidelines. At our laboratories, Table I is used for categorizing toxicity of a substance. This table is derived from the various classification schemes used by such agencies as the Department of Transportation and the Consumer Product Safety Commission.

The acute tests will determine if the substance can cause lethality within physiological dose limits and if a median lethal dose (LD_{50}) can be determined for comparison with other substances. The median lethal dose, regardless of the route of exposure, is a statistically derived single dose of a substance that can be expected to cause death in 50% of the animals. It is customary to dose at least 10 animals (5 male, 5 female) for each level tested. "Limit studies" are conducted first at our laboratory. A group of animals is given the "limit dose" as specified in Table I. If no mortality is observed, further dosing is considered unnecessary. This approach reduces the number of animals used in testing.

The acute study may also determine the route of exposure to the substance that presents the greatest hazard and the possible existence of a dose–response relationship to help set dose levels for longer term toxicity studies. From gross observation of the animals during the in-life phase of the test and the observation of organs during the gross pathology, some initial indication of mode of action of the substance may be determined.

According to Table I, when formaldehyde (formalin) is tested orally, it is slightly toxic to rats as indicated by its reported LD_{50} value of approximately 800 mg/kg of body weight (Table II).

Table I. Acute Toxicity Classification Scheme

Rating Scale	Single Dose Oral LD$_{50}$ 14-day Observation	Single Dose Dermal LD$_{50}$ 14-day Observation	4-h Inhalation LC$_{50}$ 14-day Observation		Primary Eye Irritation (Draize Scoring System)	Primary Skin Irritation (Draize Scoring System)
			Particulate	Vapor		
0	practically nontoxic LD$_{50}$ > 5.0 g/kg	practically nontoxic LD$_{50}$ > 2.0 g/kg	practically nontoxic LC$_{50}$ > 5.0 mg/L	practically nontoxic LC$_{50}$ > 10,000 ppm	nonirritating max. mean irr. score ≤ 1.0	nonirritating primary irr. score ≤ 0.1
1	slightly toxic LD$_{50}$ ≤ 5.0, but > 0.5 g/kg	slightly toxic LD$_{50}$ ≤ 2.0, but > 1.0 g/kg	slightly toxic LC$_{50}$ ≤ 5.0, but > 2.0 mg/L	slightly toxic LC$_{50}$ ≤ 10,000, but > 1000 ppm	slightly irritating mean irr. score ≤ 6.0 at 24 h, no corneal or iridal irritation, and irritation completely reversible by Day 7.	slightly irritating primary irr. score > 0.1, but ≤ 2.0
2	moderately toxic LD$_{50}$ ≤ 0.5, but > 0.05 g/kg	moderately toxic LD$_{50}$ ≤ 1.0, but > 0.2 g/kg	moderately toxic LC$_{50}$ ≤ 2.0, but > 0.2 mg/L	moderately toxic LC$_{50}$ ≤ 1000, but > 100 ppm	moderately irritating: mean irr. score > 6.0 at 24 h, but max. mean irr. score ≤ 30.0, and all corneal or iridal findings completely reversible by Day 14.	mod. irritating primary irr. score > 2.0, but < 5.0
3	highly toxic LD$_{50}$ ≤ 0.05 g/kg	highly toxic LD$_{50}$ ≤ 0.2 g/kg	highly toxic LC$_{50}$ ≤ 0.2 mg/L	highly toxic LC$_{50}$ ≤ 100 ppm	severely irritating max. mean irr. score > 30.0, or any corneal or iridal findings persisting at Day 14.	severely irritating primary irr. score ≥ 5.0
					corrosive irreversible damage	corrosive irreversible damage

For substances for which a common route of exposure might be through dermal contact, the determination of a dermal LD_{50} is desirable. The rabbit is commonly used in this type of study and is chosen because of its size, skin permeability, and established data base. In this test, enough hair is clipped from the back of the animal to allow application of the substance to 10% of the body surface. This area is then covered with a porous dressing to hold the substance in contact with the body surface for 24 h. The substance is removed following this period, and observations are recorded on the animals for 14 days. A dermal LD_{50} value of 270 mg/kg for formaldehyde is reported in Table II. According to Table I, we would classify formaldehyde as moderately toxic via the dermal route.

If the substance might be inhaled as a gas, vapor, dust, or aerosol, then an acute inhalation study should be conducted, and the LC_{50} or lethal airborne concentration for 50% of the animals is determined for the substance. Again, the rat is the animal of choice, and each dose group is exposed in a few-hundred-liter chamber for a period of hours. Four hours is the most common duration used.

In addition to the parameters measured in other short-term tests, we must be concerned with chamber airflows and hourly concentrations of the chamber atmosphere. In this test, particle size of the airborne substance is important because we are interested only in those particles that may be inhaled; for humans, particles smaller than 10–15 μm are considered to be inhaled. When administered via inhalation, formaldehyde is moderately toxic as indicated by its reported 4-h LC_{50} of 482 ppm in the rat (Tables I and II). A 30-min LC_{50} of 820 ppm in the rat has also been reported for formaldehyde.

Table II. Acute Toxicity Potential of Formaldehyde

Effect	Species	Dose	Response
Oral	rat	550–800 mg/kg	LD_{50} (1–2)
Dermal	rabbit	270 mg/kg	LD_{50} (3)
Inhalation	rat	820 ppm × 30 min	LC_{50} (4)
	rat	482 ppm × 4 h	LC_{50} (5)
	mouse	414 ppm × 4 h	LC_{50} (5)
Eye irritation	rabbit	0.5 mL	8–10 severe (6)
	rabbit	40–70 ppm × 10 days	no adverse findings (7)
	guinea pig	40–70 ppm × 10 days	no adverse findings (7)
Skin irritation	rabbit	0.1 → 20%	mild → moderate (8)
	guinea pig	0.1 → 20%	mild → moderate (9)
Sensitization	guinea pig	1% (open application)	negative (10)
		3% (open application)	positive (10)
		1% (intradermal)	positive (9)
		airborne	none reported (11)

Acute Irritation Testing

The tests discussed so far all use death as their end point. Other short-term effects of substances that we are concerned with on a daily basis do occur. One of these effects is irritation, either of the eyes or skin. The use of six adult rabbits is common to both tests. For the dermal irritation test, the hair is removed from the back of the animal so that 0.5 mL of a liquid or 5 mg of a solid can be placed on the site and covered with a gauze patch. The substance is usually kept in contact with the skin for 4 h after which the residual is removed.

The observation period is at least 72 h during which the skin is scored at 30 min, 1 h, and each 24-h period thereafter. The scoring system allows a 0–4 reading for erythema (redness) and a 0–4 reading for edema (swelling) at each observation interval. Formaldehyde results would classify it as a moderate skin irritant.

Eye irritation testing is somewhat similar. Here, approximately 0.1 mL of a substance is placed in the conjunctival sac of one eye (the other eye serves as a control) and is left in contact for up to 24 h. The eyes are scored at similar intervals as those for the skin. The cornea is scored for density or opacity (0–4); the iris is scored for reaction to light, swelling, and congestion (0–2); and the conjunctivae are scored for redness (0–3) and swelling (0–4). The total for all scores may be 110, and this total falls into a classification from our table. Formalin is a severe eye irritant under this scheme.

Application of 0.5 mL of formalin to rabbit eyes caused edema of the cornea, conjunctiva, and iris, and was given a Grade 8 on a scale of 10. Exposure of rabbits and guinea pigs to airborne formaldehyde at 40–70 ppm for 10 days caused lacrimation but no corneal injury.

In addition to irritation tests, skin sensitization studies identify possible hazards of repeated exposure to a substance. Following initial exposures to the test substance, usually lasting 2 weeks, the animals are rested for a week and then a challenge dose is given to determine if a hypersensitive state has been induced. The guinea pig is the animal of choice, and skin reactions following the challenge dose are compared with those of the initial exposures. The results for formaldehyde are presented in Table II and demonstrate that it is a sensitizer but not via the airborne route.

Irritation and odor thresholds for formaldehyde are important and should be discussed. The odor of formaldehyde and its irritant property serve as built-in safety factors. The threshold for odor recognition of formaldehyde for individuals has been reported to be between 0.1 and 1 ppm (12). This threshold is a safety advantage for the compound because people can detect and leave an area contaminated by the substance. The sense of smell becomes fatigued quickly, but formaldehyde also causes irritant effects to the eyes, nose, and throat at concentrations between 1 and 5 ppm. These effects were shown in a study using the decrease in respiratory rate of mice as an index of irritation (13). The data showed that an effect occurs at

0.5 ppm for formaldehyde in the air. The study also demonstrated that tolerance to the irritant effects of formaldehyde did not occur.

Another nonroutine short-term test has demonstrated a dose–response relationship for inhibition of the mucociliary defense apparatus in the frog. No impairment of function at 1.37 ppm or below occurred, and impairment of this apparatus may account for the relationship found for the induction of nasal squamous cell carcinomas in rats and formaldehyde gas (14).

The new OSHA Hazard Communication Guidelines would classify formaldehyde as hazardous from the results of these acute tests with the exception of the oral and possibly skin irritation tests. The acute responses to this substance should not be overshadowed by its long-term effects.

In many cases, acute toxicity data will be sufficient to satisfy safety needs. But when exposure of a large population for an extended period is likely, repeated-dose tests will have to be conducted to ensure safety.

Subchronic Testing

After initial toxicity information is obtained from the acute testing, intermediate or subchronic tests may be indicated. Subchronic tests are generally 30–90 days in duration, and more animals are used to determine the repeated-dose effects of the substance. These tests do not usually allow assessment of carcinogenicity or those effects that have a long development period; but definitive pathology and clinical effects are obtained with these studies, and "no effect" levels, as well as levels for lifetime studies, can be selected following completion.

The subchronic or intermediate repeated-dose effects of formaldehyde are presented in Table III. No surprises are found here. Low doses up to 4 ppm caused no adverse effects after 90 days of exposure. The responses of the higher doses are most likely due to the irritant properties of formaldehyde that probably decrease appetite. The beginning of the nasal problem is evident in these results also.

Reproduction and Teratology Testing

The teratogenicity study is designed to determine the potential of the test substance to induce abnormalities in the developing fetus. The substance is dosed during the period of major organogenesis of the animal, that is, gestation days 6–15 for the rat. The test for reproduction determines the effects of the substance on fertility, gestation, and offspring development.

A reproduction study usually covers more than one generation of the animal and involves those that have been exposed to the test substance from conception to the time of bearing offspring, plus a study of the offspring. These studies last approximately 33 weeks in rats.

Table IV lists results of some of the teratogenic and reproductive effects of formaldehyde. These results indicate that formaldehyde does not appear to pose any reproductive or teratological hazard.

Table III. Subchronic Toxicity Potential of Formaldehyde

Species, Route, and Exposure	Dose (ppm)	Observations
Rat—Nose Only Inhalation		
6 h/day, 5 days/wk for 4 wk	3	no adverse findings (15)
6 h/day, 5 days/wk for 4 wk	16–100	antibody inhibition (15)
Rat—Inhalation		
22 h/day for 90 days	1.6	no adverse findings (16)
22 h/day for 45 days	4.5	decrease weight gain (16)
22 h/day for 60 days	8	decrease liver weight (16) eye irritation
22 h/day for 90 days	3.8	lung inflammation (17)
6 h/day, 5 days/wk for 13 wk	4	no adverse effect (18)
6 h/day, 5 days/wk for 13 wk	12	nasal erosion (18)
6 h/day, 5 days/wk for 2 wk	39	nasal ulceration (18)
Mouse—Inhalation		
6 h/day, 5 days/wk for 13 wk	12.7	decreased body weight (18)
1 h/day, 3 days/wk for 35 wk	82	squamous cell metaplasia (19)
1 h/day, 3 days/wk for 11 wk	161	death (19)

Table IV. Teratology and Reproduction Potential of Formaldehyde

Species	Dose	Observations
Dog	375 ppm/day	no abnormalities in two generations (20)
Rat	4.1 ppm/4 h/day × 19 days	no fetal abnormalities (21)
Rat (male)	0.1 ppm in water 0.4 ppm × 180 days	no adverse findings on reproduction function (22)
Mice	148 mg/kg/day	no fetal abnormality (1 death/ 35 dams) (23)
	185 mg/kg/day	no fetal abnormality (23 deaths/ 34 dams) (23)
Rat	1600 ppm HMT[a]	no reproductive effects (20)

[a]HMT is hexamethylenetetramine.

Mutagenic Testing

In the last decade, a new group of predictive oncogenic and mutagenic tests has grown in popularity for screening substances to determine whether or not further animal testing is necessary. These in vitro tests decrease the number of animals used in testing, and they are therefore more economical for screening purposes. Another advantage of these tests is that large populations of cells are exposed to the substance, and the tests take only a few weeks to complete compared to a few years for the animal tests. Disadvantages include the dissimilarities in translocation barriers and the absence of metabolic and other defense mechanisms found in the whole animal. The poor quantitative agreement between these tests and the animal tests makes it unlikely that they will be accepted as definitive indicators of carcinogenicity potential of a substance in the immediate future. However, their use as an indicator in combination with an animal study will increase as more of the tests are validated.

Table V lists results obtained when formaldehyde is tested with these methods. These results demonstrate an unconvincing but possible mutagenic potential for formaldehyde and the need for further animal testing. It appears from these results and those reported in Table IV that the lethality of formaldehyde may be a more important risk than its mutagenicity potential.

Chronic Testing

The purpose of chronic animal testing is to determine the long-term or lifetime effects of a substance. It is useful to draw out those effects that have a prolonged latency period occurring from reduced exposure levels. These studies give the most explicit information on hematological, biochemical, behavioral, and pathological potential of a substance. Because of their costs, these studies are generally conducted on compounds that have the potential for large population or environmental exposures. A minimum of 100 rodents per dose level is suggested, and more than one species is desirable. The number of parameters is more extensive than in other animal studies. Table VI outlines the results and major findings of long-term studies with formaldehyde.

Interpretation of these results should include the following considerations: species variation in the response occurs; the tumor produced is rare; the response is reproducible; the rat is an obligatory nose breather with a complex nasal structure; and a threshold of 1 ppm is suggested.

A mechanism of action is not suggested from the results. The mechanism may be associated with the irritant potential of the compound or may

Table V. Mutagenic Potential of Formaldehyde

Assay	Response
Ames	negative (24)
Ames (modified)	positive (25)
Chinese hamster ovary	negative (26)
Sister chromatid exchange	positive (27)
Transformation	negative (28)
Transformation (modified)	positive (28)
Drosophila, oral	positive (29)
Drosophila, vapor	negative (29)
Dominant lethal study	negative (30)

Table VI. Chronic Toxicity Potential of Formaldehyde

Species, Route, and Exposure	Dose (ppm)	Observations
Rat—Inhalation		
6 h/day, 5 days/wk for 2 yr	2	squamous metaplasia (31)
6 h/day, 5 days/wk for 2 yr	6–15	squamous cell carcinoma (31)
588 days→lifetime	14	squamous cell carcinoma (32)
588 days→lifetime	14 + HCl	squamous cell carcinoma (32)
22 h/day, 7 days/wk for 26 wk	1	no adverse findings (33)
22 h/day, 7 days/wk for 26 wk	3	squamous metaplasia (33)
Rat—Oral (H_2O) 2 yr	1–5 HMT[a]	no adverse findings (34)
Mouse—Oral (H_2O)	1–5 HMT[a]	no adverse findings (34)
Mouse—Inhalation		
6 h/day, 5 days/wk for 27 mo	2–6	no adverse findings (31)
6 h/day, 5 days/wk for 27 mo	14	squamous cell carcinoma (31)
1 h/day, 3 days/wk for 35 wk +		squamous cell metaplasia
29 wk at 125 ppm	41.5 + 125	(no tumors) (19)
Hamster—Inhalation		
5 h/day, 5 days/wk for 18 mo	10	hyperplasia (35)
5 h/day, 5 days/wk for 18 mo	50	squamous metaplasia (no tumors) (35)
22 h/day, 7 days/wk for 26 wk	1–3	no adverse findings (33)
Monkey—Inhalation		
22 h/day, 7 days/wk for 26 wk	1	no adverse findings (33)
22 h/day, 7 days/wk for 26 wk	3	metaplasia in nasal turbinates (33)

[a]HMT is hexamethylenetetramine.

be a genetic effect as demonstrated by the possible mutagenic potential of the compound. Other studies are necessary to determine formaldehyde's mechanism of action.

Conclusion

Limited resources are available for the assessment of substances. The laboratory capacity does not exist to conduct all the tests mentioned on every chemical. The desire for absolute protection from risks must be balanced with the desire to reduce regulations and costs. The costs in Table VII are based upon the particular protocols used in our laboratory but serve to indicate test cost magnitude as well as cost comparisons between the various tests.

All substances may be hazardous at a certain dose level under certain exposure conditions. Identifying these toxic effects with animal studies is currently the best way for assessing the risk to the human population. Currently, we regulate substances without requiring evidence of harm to humans. This course appears to be the most prudent as we find ourselves mov-

Table VII. Cost Estimates for Toxicity Tests

Test	Amount (dollars)
Acute (short-term)	
Oral LD_{50}	1500
Dermal LD_{50}	3500
Eye irritation	750
Dermal irritation	750
Dermal sensitization	3500
Inhalation (LC_{50})	7000
Genetic	
Gene mutation	1000
Chromosome aberration (in vitro)	2500
DNA damage–repair	4000
Cell transformation	4000
Subchronic (mid-term)	
90-day Inhalation	100,000
90-day Dermal	75,000
Chronic (long-term)	
Carcinogenic (dermal)	350,000
Carcinogenic (inhalation)	800,000
Reproduction	
Teratology	35,000
Reproduction	120,000

ing away from the "Delaney Clause mentality," where products were removed immediately from the market upon finding that they could cause cancer in any animal at any dose level, into a risk analysis of benefits.

During the last two decades we have seen the evaluation of risk and benefits move more and more to the social level. This change is partly because we have no Delaney rule for chemical substances; no formula for risk assessment satisfies the scientific community and the community at large.

The failure to establish a socially acceptable risk evaluation process hampers our technological process. It leaves committees rendering an array of decisions between the economic consequences of zero-level exposure to known carcinogens or an increase in hazard from higher levels of exposure. The prudent course is to continue to regulate chemicals on the basis of data obtained with the techniques outlined in this chapter and to improve not only these techniques but also the risk evaluation process.

Literature Cited

1. Tsuchiya, K.; Hayashi, Y.; Onodera, M.; Hasegawa, T. Keio J. Med. 1975, 24, 19–37.
2. Smyth, H. F., Jr.; Seaton, J.; Fischer, L. J. Ind. Hyg. Toxicol. 1941, 23, 259–68.
3. Lewis, R. J., Sr.; Tatken, R. L. "Registry of Toxic Effects of Chemical Substances"; National Institute for Occupational Safety and Health: Cincinnati, Ohio, 1980; Vol. 1, p. 695.

4. Skog, E. *Acta Pharamacol.* **1950**, *6*, 299–318.
5. Nagorny, P. A.; Sukakova, Z. A.; Schablenko, S. M. *Gig. Tr. Prof. Zabol.* **1979**, *1*, 27–30.
6. Carpenter, C. P.; Smyth, H. F., Jr. *Am. J. Ophthalmol.* **1946**, *29*, 1363–72.
7. Grant, W. M. "Toxicology of the Eye," 2d ed.; Thomas: Springfield, Ill., 1974; pp. 502–6.
8. Committee on Aldehydes "Formaldehyde and Other Aldehydes"; National Academy: Washington, D.C., 1981; p. 179.
9. Colburn, C. W. In "Formaldehyde—An Assessment of Its Health Effects"; NAS: Washington, D.C., 1980; p. 6.
10. U.S. Consumer Product Safety Commission "Reliable Animal Test for Predicting Human Skin Sensitizers" by Maibach H.; Government Printing Office: Washington, 1978.
11. Committee on Aldehydes "Formaldehyde and Other Aldehydes"; National Academy: Washington, D.C., 1981: p. 180.
12. Leonardos, G.; Kendall, D.; Barnard, N. *J. Air Pollut. Control Assoc.* **1969**, *19*, 91–95.
13. Kane, L. E.; Alarie Y. *Am. Ind. Hyg. Assoc. J.* **1977**, *38*, 509–22.
14. Morgan, K. T.; Patterson, D. L.; Gross. E. A. *Fund. Appl. Toxicol.* **1984**, *4*, 58–68.
15. American Industrial Hygiene Association *Industrial Hygiene News Report*, Chicago, Ill., March 1983.
16. Dubreuil, A.; Bouley, G.; Godin, J.; Boudene, C. *J. Eur. Toxicol.* **1976**, *9*, 245–50.
17. Coon, R. A.; Jones, R. A.; Jenkins, L. J., Jr.; Siegel, J. *Toxicol. Appl. Pharmacol.* **1970**, *16*, 646–55.
18. Committee on Aldehydes "Formaldehyde and Other Aldehydes"; National Academy: Washington, D.C., 1981; p. 180.
19. Horton, A. W.; Tye, R.; Stemmer, K. L. *J. Natl. Cancer Inst.* **1963**, *30*, 31–43.
20. Hurni, H.; Ohder, H. *Food Cosmet. Toxicol.* **1973**, *11*, 459–62.
21. Sheveleva, G. A. *Toksikol. Nov. Prom. Khim. Veschestv* **1971**, *12*, 78–86.
22. Guseva, V. A. *Gig. Sanit.* **1972**, *37*, 102–3.
23. Marks, T. A.; Worthy, W. C.; Staples, R. E. *Teratology* **1980**, *22*, 51–58.
24. Committee on Aldehydes "Formaldehyde and Other Aldehydes"; National Academy: Washington, D.C., 1981; p. 182.
25. Sasaki, Y; Edno, R. *Mutat. Res.* **1978**, *54*, 251–52.
26. Hsie, A. W.; O'Neill, J. P.; San Sebastian, J. R.; Couch, D. B.; Fuscoe, J. C.; Sun, W. N. C.; Brimer, P. A.; Machanoff, R.; Riddle, J. C.; Forbes, N. L.; Hsie, M. H. *Fed. Proc. Fed. Am. Soc. Exp. Biol.* **1978**, *37*, 1984.
27. Obe, G.; Beek, B. *Drug Alcohol Depend.* **1979**, *4*, 91–94.
28. Boreiko, C. J.; Ragan, D. L. "Proceedings of the Third Annual CIIT Conference: Formaldehyde Toxicity," 1982, p. 63.
29. Auerbach, C.; Moutschen-Dahmen, M.; Moutschen, J. *Mutat. Res.* **1977**, *39*, 317–61.
30. Epstein, S. S.; Arnold, E.; Andrea, J.; Bass, W.; Bishop, Y. *Toxicol. Appl. Pharmacol.* **1972**, *23*, 288–325.
31. Kerns, W. D.; Donofrio, D. J.; Pavkov, K. L. "Proceedings of the Third CIIT Annual Conference: Formaldehyde Toxicity," 1982, p. 111.
32. Albert, R.; Sellakumar, A.; Laskin, S.; Kuschner, M.; Nelson, N.; Snyder, C. *J. Natl. Cancer Inst.* **1982**, *68*, 597–603.
33. Rusch, G. M.; Bolte, H. F.; Rinchart, W. E. "Proceedings of the Third Annual CIIT Conference: Formaldehyde Toxicity," 1982, p. 98.
34. Della Porta, G.; Colneghi, M. I.; Parmiani, G. *Food Cosmet. Toxicol.* **1968**, *6*, 707–15.
35. Colburn, C. W. In "Formaldehyde—An Assessment of Its Health Effects"; NAS: Washington, D.C., 1980; p. 6.

RECEIVED for review October 29, 1984. ACCEPTED March 12, 1985.

History and Status of Formaldehyde in the Cosmetics Industry

HERMAN E. JASS

29 Platz Drive, Skillman, NJ 08558

Although formaldehyde has been found to be a mucosal and dermal irritant and contact sensitizer, regulatory and industry experts have concluded that the use of formaldehyde as a preservative is safe to the majority of consumers. A review of evidence indicates that formaldehyde is not carcinogenic when applied to the skin at normal concentrations, but evidence is insufficient to conclude that HCHO in sprayed cosmetics is safe. By regulation, formaldehyde, when used as a nail hardener, must be adequately labeled and limited to concentrations of 5%. Although formaldehyde is still the sixth most popular cosmetic preservative, used primarily in shampoos and bubble baths, its use appears to be declining. Equipment sanitation applications of HCHO are also decreasing.

Uses of Formaldehyde

Formaldehyde has been used historically in the cosmetics industry in three principal areas: preservation of cosmetic products and raw materials against microbial contamination, certain cosmetic treatments such as hardening of fingernails, and plant and equipment sanitation.

By far the major use of formaldehyde is as an antimicrobial preservative. Certain categories of cosmetic and toiletry products that contain high concentrations of surfactants, such as shampoos and bubble baths, have traditionally required formaldehyde to preserve them against contamination by Gram-negative microorganisms, particularly members of the *Pseudomonas* family (*1*). According to the U.S. Food and Drug Administration (FDA) computerized information file (*2*), formaldehyde was the sixth most frequently used cosmetic preservative in 1982. As shown in Table I, formaldehyde was listed as being used in 734 cosmetic products. This information is obtained from cosmetic formulation data submitted to FDA by companies participating in the FDA voluntary registration program (*3*). Because this regulatory program is voluntary, some formulations are not registered. Therefore, the frequency of use reported here is somewhat understated.

0065–2393/85/0210/0229$06.00/0
© 1985 American Chemical Society

Table I. Frequency of Preservative Use in Cosmetic
Formulas, 1982

Preservative	Frequency of Use[a]
Methylparaben	7148
Propylparaben	6274
Imidazolidinylurea	1820
Quaternium-15	1079
Butylparaben	739
Formaldehyde solution	734
BHA	618
2-Bromo-2-nitropropane-1,3-diol	546
BHT	475
Sorbic acid	361

[a] The number of cosmetic products in which each preservative is used.

Table II. Product Formulation Data for Cosmetics Containing Formaldehyde

Product Category	Total No. of Products	No. Containing Formaldehyde	Percent
Shampoos (noncoloring)	909	316	34.8
Bubble baths	475	109	22.9
Hair conditioners	478	95	19.9
Wave sets and hair rinses (noncoloring)	338	69	20.4
Other bath preparations	132	24	18.2
Other hair preparations (noncoloring)	177	13	7.3
Bath oils, tablets, and salts	237	10	4.2
Other products in categories listing HCHO	4999	78	1.56

According to the data obtained through the voluntary registration regulation (4), approximately 35% of marketed cosmetic shampoos, 23% of bubble bath formulations, and 20% each of hair rinses and hair conditioners contain HCHO as an antimicrobial preservative (Table II). The usual HCHO concentration in these products is approximately 0.10% or less, and probably does not exceed 0.20% in most commercial formulations (5). Concentrates that are to be diluted prior to application may have a higher concentration, usually up to approximately 1.0%. As shown in Table II, HCHO is used somewhat in other cosmetics, also. The concentrations shown in all of these tabulations of FDA data are usually expressed as 40% formalin solutions, the form used in the cosmetic industry. The frequency of use in these other products is low, and, in some cases, may merely represent obsolete information.

Another cosmetic use of formaldehyde is as a nail hardener. Fingernails treated with HCHO solutions become harder and more resistant to breakage. For this application, the HCHO concentration is significantly higher, up to approximately 5%. However, because of safety considerations, FDA has placed restrictions on the use and labeling of such products (6). As a result, such products represent only a very small segment of the cosmetic market.

Some drug products that are related to cosmetic applications use formaldehyde. For example, abnormally excessive perspiration, known as hyperhydrosis, may be controlled with topical applications of HCHO (7). However, this use is limited by law to a prescription-only basis.

Formaldehyde has also been used as a tooth desensitizer dentifrice for persons whose teeth are abnormally sensitive to touch and temperature variations. This type of product was reviewed by an FDA advisory panel (8), which determined that insufficient evidence was available to determine the effectiveness of this agent in over-the-counter drug use. If additional data that would be considered by FDA sufficient to establish the drug's effectiveness are not submitted by the dentifrice's sponsors by the time a final monograph for this category of drug is published by the agency, it will become illegal. Meanwhile, it cannot be determined if any products of this type, in which HCHO is present at 1.4%, are still being marketed.

Formaldehyde has had a long history of use by the cosmetic industry for plant and equipment sanitation. Sanitizing agents are not used on a routine basis but as a procedure to reduce or remove contamination by resistant microorganisms that can pose a serious preservation problem in finished cosmetics. Emergence of strains of microorganisms resistant to normal concentrations of germicidal agents has been a frequent occurrence in cosmetic plants. This resistance is due in part to the use of many natural and earth-derived ingredients, such as natural pigments and gums, that are difficult to render free of microorganisms. Also, many raw materials contain high concentrations of ingredients, such as ethoxylated surfactants, which form complexes with many preservatives including the parabens, the most commonly used agents. This complexing of preservatives reduces their effectiveness. Many cosmetic ingredients, like many food ingredients, actually provide a good nutritional medium for microorganisms. These factors, coupled with the continual use of processing equipment, have a tendency to produce dominant populations of mutant bacterial strains that are very difficult to eradicate once they are established in manufacturing facilities. Formaldehyde, in aqueous solution at concentrations of 0.1–0.5%, has proved to be extremely efficacious for ridding the plant and processing equipment of these resistant bacteria.

However, in recent years, a number of factors have combined to make outcropping of troublesome resistant bacterial strains much less frequent.

These factors include the introduction of better plant sanitation features such as sanitary piping, regular hot water or steam cleaning, and generally improved housekeeping. In addition, raw materials can now be purified by suppliers with methods that drastically reduce their natural microorganism content; such methods have had a major impact in the fight against bacterial contamination. As a result, the need for heroic measures to combat plant infections has lessened recently. Furthermore, because of occupational safety considerations, formaldehyde, once the primary agent for this use, has been largely supplanted by other agents such as hypochlorite or hydrogen peroxide solutions. However, HCHO is still employed when particularly difficult situations arise. One such situation involves water–deionizer beds. Cosmetic manufacturing requires huge amounts of deionized water, and although maintenance of deionizer bacterial standards has improved greatly over the years, an occasional contamination problem may still occur. Because of its nonionic character, HCHO is particularly suited for this type of use. However, current use concentrations of HCHO in sanitation tend to be lower than in the past and range from 0.05% to 0.10%. In all uses of formaldehyde by the cosmetic industry, the material typically used is formalin, an aqueous 40% solution of formaldehyde.

Safety of Formaldehyde

The main manifestations of toxicity by formaldehyde topical application to humans or animals have been irritation and sensitization. Thus, dermatitis due to occupational exposure to formaldehyde solutions is a well-recognized problem (9, 10). Eye irritation can occur due to formaldehyde vapor (10) or aqueous solutions of formaldehyde (11). However, minimal irritation, at most, was observed when cosmetic products containing 0.074% and 0.0925% formaldehyde were instilled into rabbit eyes (12). Most cosmetic applications of HCHO are in shampoos and similar compositions in which high surfactant content may produce much higher levels of eye irritation than would be expected from formaldehyde at preservative concentrations.

Similarly, the skin irritation levels resulting from normal use concentrations of HCHO or even exaggerated levels were found to be moderately irritating, at most, in animal tests (10, 13). The North American Contact Dermatitis Group (14) reported a 5% incidence of skin sensitization among 2374 patients exposed to 2% formaldehyde in aqueous solution.

Data submitted by cosmetic firms in cosmetic product experience reports to FDA under the voluntary cosmetics regulatory program for the period 1979–82 are shown in Table III (15). Shampoos, the leading product category employing formaldehyde as a preservative, had a reported incidence of 2.26 consumer experiences per million units sold. Bubble bath products, the next most frequent use of HCHO, had an incidence of 1.14 per million units sold. By contrast, product categories using formaldehyde

Table III. Selected Values of Cosmetic Experience Reports Submitted to the FDA Under the
Voluntary Cosmetics Regulatory Program, 1979–82

Product Category	Estimated Units Distributed in Millions	Total Number of Experiences	Number of Experiences per Million
Bath soaps	4157.64	1187	0.29
Shampoos	1070.29	2415	2.26
Colognes and toilet waters	850.11	377	0.44
Other personal cleanliness products	646.27	2170	3.36
Lipstick	625.35	587	0.94
Face, body, and hand care preparations	590.51	751	1.27
Hair sprays	501.01	339	0.68
Hair dyes and colors	273.20	3477	12.73
Make-up foundations	145.81	662	4.54
Face powders	116.51	162	1.39
Bubble baths	135.75	155	1.14

infrequently had incidences as follows (per million units sold): hair dyes, 12.73; make-up foundations, 4.54; lipsticks, 0.94; and face powders, 1.39. These data suggest that the inclusion of formaldehyde in cosmetics does not, per se, create a public health problem.

In an extensive review of the literature and of safety information submitted by individual companies, the Expert Panel of the Cosmetic Ingredient Review[1] (CIR) (4) concluded that formaldehyde "... is an irritant at low concentration, especially to the eyes and the respiratory tract in all people. It induces hypersensitivity, but not as often as might be expected considering the frequency and extent of exposure. Perhaps the single, most important attribute common to these toxic effects of formaldehyde is that they are all concentration–time dependent." The panel then concluded that formaldehyde in cosmetic products is safe to the great majority of consumers. However, because of skin sensitivity of some individuals to this agent, CIR recommended that formaldehyde be formulated in cosmetics at its minimal effective concentration, not to exceed 0.2% as free formaldehyde.

The attitude of FDA is reflected in a statement by Heinz Eiermann, director of the Division of Cosmetics Technology of FDA (5): "When used as a preservative, the concentration of HCHO is usually too low and the exposure too short to induce sensitization. However, it is then quite capable of eliciting a skin reaction in some HCHO-sensitive consumers." After citing the data of the International Contact Dermatitis Group, which ranked

[1] The Expert Panel of the Cosmetic Ingredient Review is an independent group of scientific experts funded by an industry association, the Cosmetic, Toiletry, and Fragrance Association, to review and determine the safety of cosmetic ingredients.

formaldehyde among the 10 most prominent contact sensitizers, Eiermann then stated: "However, we must remember that while a relatively large number of consumers may react to HCHO when patch tested, few of them are expected to present the same reaction under actual use conditions, and millions can use a HCHO-preserved cosmetic without showing any sign of harm."

However, nail hardeners, which employ a much higher concentration of formaldehye, present a greater concern to safety evaluators. The policy of FDA (6) is that the agency does not object to the use of formaldehyde as an active ingredient in nail hardeners provided that the product contains no more than 5% HCHO, provides the user with nail shields to restrict application to nail tip, furnishes adequate directions for safe use, and carries a label that warns of danger of misuse and potential for allergic reactions. The CIR Expert Panel reviewed submitted comments relating to the use of formaldehyde at a concentration of 4.5% in nail hardeners but concluded that the submitted evidence was inadequate to ensure that formaldehyde could be safely used above 0.2%.

Since the results of the Chemical Industry Institute of Toxicology (CIIT) inhalation study in rats and mice were made public, the significance of the findings to the use of cosmetics containing formaldehyde has been analyzed. However, although expressing concern, both FDA and CIR concluded that the current evidence does not indicate a risk of carcinogenicity for cosmetic use of HCHO. The conclusions include the fact that all tumors in the CIIT study occurred in the nasal cavity and not systemically and that cancers in the CIIT study occurred in only one of the two species exposed and none at the 2-ppm level. Also, the level of exposure to HCHO vapor from cosmetic use would be significantly less than the lowest dose to which the rodents in the CIIT study were exposed. However, conclusions reached at the Consensus Workshop on Formaldehyde in October 1983 (17) include the fact that epidemiological studies involving embalmers, medical personnel, and industrial workers exposed to HCHO have shown an excess of brain cancer and leukemia relative to the general population. They further stated that no dose–response information is currently available for quantitative risk assessment of these findings.

Taking all of these facts into consideration, the CIR panel stated, after a review of the evidence, that it could not be concluded that formaldehyde is safe in cosmetic products intended to be aerosolized, that is, sprayed. However, this reservation was based not on positive evidence of hazard, but on a lack of submitted data in support of aerosolized cosmetics containing formaldehyde. This lack, in turn, is due to the rarity in the marketplace of spray cosmetics that contain formaldehyde.

The use of formaldehyde for plant and equipment sanitation has also declined dramatically. As a result of the findings of CIIT, the National Institute of Occupational Safety and Health has recommended that formaldehyde in work environments be handled as a potential human carcinogen. This recommendation has been an important factor in leading companies to seek alternative sanitizing agents and systems in plants.

Other Factors Influencing Cosmetic Use of Formaldehyde

Federal regulations require the listing of ingredients on cosmetic package labels (*16*). Cosmetic marketers, concerned about the publicity over the CIIT study results, as well as the controversy over the possible hazards of formaldehyde from the use of urea–formaldehyde foam insulation, have become uneasy over the listing of formaldehyde as a cosmetic ingredient. The two major claims for cosmetics in marketing and advertising are that cosmetics enhance beauty and that they are safe to use. The presence on a product label of an ingredient such as formaldehyde, which has become associated in consumers' minds with potential hazards and other unpleasant aspects, is contradictory to these claims of benefit and innocuity. That the implied hazard is cancer is even more worrisome. This situation has led to an attempt by many marketers to replace formaldehyde with other preservatives and thus avoid listing this troublesome ingredient. However, when the product is one with a high concentration of surfactant, such as a shampoo, the discovery of a suitably effective alternative system is not an easy matter. Ironically, formulators have often turned to preservatives that are formaldehyde donors. That is, their preservative effectiveness is due to the formation of formaldehyde in situ at low levels by hydrolytic action. Such HCHO-donor preservatives include quaternium-15 [*cis*-1-(3-chloroallyl)-3,5,7-triaza-1-azoniaadamantane chloride], 2-bromo-2-nitropropane-1,3-diol, and DMDM hydantoin (dimethylol dimethyl hydantoin). Use of these preservatives in some cases provides the benefits of HCHO without the need to list formaldehyde as an ingredient on the label.

Newer preservatives with even greater effectiveness against Gramnegative organisms have recently been introduced and may further decrease the use of HCHO. However, adoption of new preservatives has traditionally been a slow, gradual process because of the desire of industrial microbiologists and safety specialists to acquire a history of safe use before incorporating a biologically active material into a valuable product. Such history includes the record of gradually increasing use of the product over an extended period of time with no apparent health problems by competitive companies plus an acceptable judgment by CIR.

Overall, the factors of contact sensitization potential and unfavorable publicity have led to a gradual decrease in the use of formaldehyde. For example, the number of shampoo and bubble bath products registered with FDA in the last 3 years has grown from 1384 to 1455, whereas the frequency of use of formaldehyde has decreased from 425 to 403 in these two types of products. This decline in usage is expected to continue.

Literature Cited

1. Orth, D. S. *J. Soc. Cosmet. Chem.* **1980**, *31*, 165–72.
2. Decker, R. L.; Wenninger, J. A. *Cosmet. Toiletries* **1982**, *92* (*11*), 57–59.
3. "Code of Federal Regulations," Title 21, Part 720, 1982, p. 190–95.
4. Cosmetic Ingredient Review In "Final Report of the Safety Assessment for Formaldehyde"; Washington, D.C., 1983, p. 4.
5. Eiermann, H. J., presented at the Conference of Consumer Federation of America, Washington, D.C., 1983.
6. Food and Drug Administration, "Cosmetic Handbook"; Government Printing Office: Washington, 1983; p. 13.
7. Marcy, R.; Quermonne, M. A. *J. Soc. Cosmet. Chem.* **1976**, *27*, 333–44.
8. Food and Drug Administration, *Fed. Regis.*, **1982**, *47*, 22753.
9. Proctor, N. H.; Hughes, J. P. In "Chemical Hazards of the Workplace"; Lippincott: Philadelphia, 1978; pp. 272–74.
10. Fielder, N. H. In "Toxicity Review 2—Formaldehyde"; Eyre & Spotteswoode: Portsmouth, U.K., 1981.
11. Carpenter, C. P.; Smyth, H. F. *Am. J. Ophthalmol.* **1946**, *29*, 1368–72.
12. Cosmetic Ingredient Review In "Team Report on Formaldehyde"; Washington, D.C., 1982, p. 11.
13. "Primary Skin Irritation and Sensitization Test on Guinea Pigs"; E. I. Du Pont de Nemours & Company, Inc.: Wilmington, Del., 1970.
14. "Standard Screening Tray Versus 1980 Summary"; North American Contact Dermatitis Group: 1980.
15. Food and Drug Administration, "Tabulation of Cosmetic Experience Reports Submitted to FDA under the Voluntary Cosmetics Regulatory Program, 1979–82"; Government Printing Office, Washington, 1983.
16. "Code of Federal Regulations," Title 21, Part 701.3, 1982, p. 177.
17. Hart, R. "Report on the Consensus Workshop on Formaldehyde," Jefferson, Ark., 1983.

RECEIVED for review September 28, 1984. ACCEPTED December 26, 1984.

Formaldehyde: The Food and Drug Administration's Perspective

ROBERT J. SCHEUPLEIN

Office of Toxicological Sciences, Bureau of Foods, Food and Drug Administration, Washington, DC 20204

Formaldehyde is used in mostly minor quantities of foods, drugs, and cosmetics. It has recently been found to be carcinogenic in rats when inhaled continuously at high doses (>2 ppm) for a lifetime. Formaldehyde has not been found to be carcinogenic in rodents when orally ingested at high doses (~5%) for a lifetime. A great deal of biological and chemical findings corroborate the view that the results from ingestion studies are more relevant to food use. Aside from the similarity in the route of administration, support for this view includes the short biological half-life of formaldehyde, the anomalous effects of locally high concentrations, and the highly curvilinear shape of the inhalation dose–response curve. On the basis of the data now available, the Food and Drug Administration does not believe that the very low levels that are used in food or cosmetics present a significant safety concern.

FORMALDEHYDE IS ONE OF THE MOST WIDELY USED of all synthetic chemicals. In 1979 the U.S. production reported in terms of formalin (37% aqueous solution) was 2900 million kg, or over 12 kg for every person in the country. The Food and Drug Administration's (FDA) interest in formaldehyde's safety is derived from formaldehyde's use in foods, drugs, and cosmetics. Compared to its major uses in plastics, resin manufacture, and production of chemical intermediates, formaldehyde's direct use in foods, drugs, and cosmetics is very small.

Following the announcement in January 1980 that the inhalation of formaldehyde apparently induced tumors in the nasal passages of rats, FDA took a good look at its inventory of approved uses of formaldehyde. On the basis of the exposure levels and the routes of administration, FDA concluded that no regulatory action against the use of formaldehyde in the products it regulates was necessary at that time to protect the public health (*1*). Since then the Chemical Industry Institute of Toxicology (CIIT) study has been completed, published (*2*), and confirmed in a similar study conducted by Albert et al. (*3*). Formaldehyde is now generally recognized to be carcinogenic in rats, or more accurately to rat nasal turbinates when inhaled at high-dose levels for 5 days per week for the lifetime of the animals. FDA's regulatory position on the safety of food, drug, and cosmetic uses of formaldehyde remains the same. In this chapter I will try to explain the toxicological basis of that position, but first I will give an overview of formaldehyde's major FDA-regulated uses.

Drugs

In human drug products, formaldehyde (HCHO) is used as a densensitizing agent in dentifrices at concentrations of 1.4% and as a preservative at concentrations of 0.1% or lower (4). It is also used in the manufacture of various vaccines to inactivate bacteria or viruses or to detoxify bacterial toxins. Depending on the intended use of the vaccines, the HCHO content may vary from less than 2 ppm to more than 100 ppm.

Cosmetics (5)

Perhaps formaldehyde's widest use, in terms of amount, is mainly as a preservative in shampoos and other hair products. HCHO is an excellent preservative against Gram-negative microorganisms, particularly *Pseudomonas aeruginosa*, a potentially pathogenic microorganism. According to voluntarily registered product formulation data, approximately 35% of the marketed cosmetic shampoos, 23% of bubble bath formulations, and 20% of hair rinses and hair conditioners contain HCHO as an antimicrobial preservative. The typical concentration of HCHO in preservative use is approximately 0.1%. Formaldehyde is a popular preservative, and attempts to replace it on a wide scale have not yet been successful.

A few years ago a manufacturer replaced HCHO in his shampoo with paraben preservatives. Almost a half million bottles of shampoos were recalled because of contamination with *Pseudomonas*. The other effective preservatives against Gram-negative microorganisms are organic mercurials. However, the use of mercury compounds in cosmetics is limited by regulation to use as preservatives in eye-area cosmetics because they are absorbed through the skin on topical application and tend to accumulate in the body and are capable of causing neurotoxic effects.

The agency's concern about the safety of HCHO and HCHO donors is threefold, namely, irritation, sensitization, and systemic effects. The carcinogenicity issue is subsumed in systemic effects. Skin irritation is of concern when HCHO is present in a product at relatively high concentrations, as, for example, in nail hardeners. FDA has not officially objected to its use in nail hardeners provided the product contains no more than 5% HCHO, provides the user with nail shields that restrict application to the nail tip, furnishes adequate direction for safe use, and warns the user about the consequences of misuse and the potential for causing allergic reactions in already sensitized users. FDA does not endorse the use of formaldehyde for this purpose, but insufficient evidence is available to indicate that any significant risk is present when used with care. FDA's concern about the possible harmfulness of HCHO as a skin sensitizer is directed more toward the elicitation of allergic reactions and less toward induction of allergenicity. When used as a preservative its concentration is usually too low and the exposure too short to induce sensitization; however, it may elicit a skin reaction in some HCHO-sensitive consumers. Sensitivity to HCHO is not uncommon in the United States; it is ranked among the 10 most prominent skin contact sensitizers. Nonetheless, millions of people can use a HCHO-preserved cosmetic without ever showing any sign of sensitization.

Foods

Formaldehyde is used in the food industry in several important ways but always under conditions that result in very small amounts of formaldehyde in food. It has never been a popular food additive in the United States probably because of its association with embalming and its strong odor. It has been cleared for use as a preservative in defoaming agents containing simethicone, in an amount not to exceed 1.0% of the simethicone content or 100 ppb in ready-to-eat food (6). It is also used as a component in adhesives intended for use in packaging food. In such use the adhesive is either separated from the food by a functional barrier except for possibly at seams and edges which can contribute only trace amounts at most (7).

Formaldehyde is used in the animal feed industry in ruminant feeds to improve the handling characteristics of animal fat in combination with certain oilseed meals. As formalin (37% solution), it is added to the mixture at a level of 4%. This mixture on drying contains less than 1% formaldehyde, and the feed is limited to contain less than 25% of the mixture. Thus, animals may ingest as much as 0.25% formaldehyde in their diet (8). The animals appear to thrive on it, and no evidence indicates that tissue levels of formaldehyde are any higher in these animals than in animals that do not ingest added formaldehyde. The metabolic capacity of ruminant animals seems quite sufficient to catabolize even rather high levels of formaldehyde.

Paraformaldehyde, which liberates formaldehyde when dissolved in water, is approved for controlling fungal growth in maple tree tapholes (9). It is used such that the maple syrup produced from the sap of treated maple trees does not contain more than 2 ppm of formaldehyde. A typical portion of syrup might then contain 50 μg of formaldehyde.

Hexamethylenetetramine (HMT), a complex of formaldehyde and ammonia that decomposes slowly to its constituents under acidic conditions, has been used for many years as a food additive in the Scandinavian countries. It is used in fish products such as herring and caviar that are generally prepared by hand. HMT has the advantage of exerting a good antimicrobial effect without influencing the taste and odor. Its effect is due to the gradual liberation of formaldehyde. It is permitted as a food additive in Norway at levels of 0.1–0.05%, corresponding to a daily ingestion of approximately 2.5 mg/day (10).

Discussion

How do these ingested use levels of formaldehyde compare to established clinical or toxicological effect levels? The major adverse effects of formaldehyde and the corresponding doses are given in Table I. The first four entries show generally that human adverse effects of increasing severity occur as the dose increases to a vapor concentration greater than 30 ppb. In 1976 the National Institute of Occupational Safety and Health (NIOSH) recommended lowering industrial exposures to 1.0 ppm on the basis of formaldehyde's irritation potential. The last five entries are results from animal studies. By dividing the human oral lethal doses by human body weight, the animal and human oral LD_{50} values are reasonably comparable. The most sensitive animal adverse effects are teratogenic effects in

Table I. Summary of Toxicity Data on Formaldehyde

Concentration	Clinical Symptoms and Toxicological End Points
30 ppb (vapor)	no observed acute effects
~ 1.2 ppm (vapor)	threshold of human response (irritation to nose, eyes)
4–5 ppm (vapor)	becoming intolerable, difficulty in breathing (humans)
50 ppm (vapor)	pulmonary edema, pneumonitis (humans)
~ 30 ppm (aqueous)	threshold of sensitization in sensitized human subject
1–100 g formalin, ingestion	some human fatalities
0.8 g/kg	oral LD_{50} in rats
15 mg/kg, ingestion	no effect level, fetotoxicity in dogs
0.15 mg/kg	estimated ADI in humans (17)
500 mg/kg	no-effect level, long term feeding in rodents
14 ppm (vapor)	carcinogenic in rat nasal mucosa by inhalation

Note: Table is compiled from data in Refs. 2, 3, 11, 13, 16, and 17.

dogs, and the fetoxic no-effect level is used as the basis for setting the Acceptable Daily Intake (ADI) for food use. The application of a 100-fold safety margin to this teratogenic no-effect level in dogs yields an ADI of 0.15 mg/kg or approximately 9 mg/day (*11*). Repeated subcutaneous injections on a weekly basis over 1½ to 2 years in rats produce local sarcomas at the site of injections (Table II) (*12*). Finally, the recent positive inhalation studies in rats indicate a highly nonlinear carcinogenic response that is detectable in 100 rats at 5.6 ppm and increases approximately cubically with increasing dose (*2*, *3*, *13*).

Table III shows several oral studies that have been conducted on formaldehyde or HMT (*14*). HMT breaks down gradually to formaldehyde and NH_3 under acidic conditions or in the presence of proteins, and is considered equivalent to formaldehyde for toxicological purposes. These long-term studies point to a dietary level of 5% HMT as causing no effect in rodents. On a weight percent basis a very similar no-effect level would be true for formaldehyde monomer (*15*).

Collectively these studies indicate that formaldehyde may be potentially carcinogenic to humans when inhaled at high levels. Generally when faced with both positive and negative data obtained from equally well-conducted bioassays, FDA usually will elect to err on the side of prudence and regulate on the basis of the positive study or studies. Our food laws require that a substance be shown to be safe for its intended use (to a reasonable certainty), and when studies disagree and are of equal quality and significance, a reasonable question of safety can be said to remain. But in the case of formaldehyde as a food ingredient, ingestion studies, inhalation studies, and injection studies are not of comparable biological significance or relevance.

Although production of local sarcomata occurs in rats at the site of repeated injections, and neoplastic lesions develop in the nasal passages and trachea of rats exposed to high vapor concentrations of formaldehyde, the probability of carcinogenic potential in food-additive use appears to be excluded on the basis of adequate ingestion studies.

A great deal of biological and chemical findings corroborate the view that the ingestion studies are probably correct and more relevant to food use. First, aside from the similarity in route of administration between the animal ingestion studies and formaldehyde's use as a food additive, formaldehyde is highly reactive. It is very rapidly converted to formic acid upon ingestion and has a biological half-life of approximately 1 min in a variety of species (*11*). Thus, one might expect that if high local concentrations are persistently applied to tissue, then conversion kinetics to formate may be overwhelmed. This situation would result in the retention of formaldehyde and the consequences of high concentrations of a highly reactive, irritating, and toxic substance. The results of subcutaneous injection studies using formaldehyde, in addition to the usual difficulties with their interpretation

Table II. Formaldehyde Carcinogenicity Studies with Laboratory Animals

Species	Compound	Route	Length of Study	Tumor Incidence	Ref.
Rat	0.4 % formalin	subcutaneous	15 months until tumor formation	2/10	12
Rat	9–4 % HMT	subcutaneous		7/20	22
Rabbit	3% formalin	direct application to palate	10 months	1/6 "carcinoma in situ"	21
Mouse	41–163 ppm formaldehyde	inhalation	35–70 weeks	negative	20
Hamster	10 ppm formaldehyde	inhalation	lifetime	negative	19
Rat	2.0–14.3 ppm formaldehyde	inhalation	2.5 years	103/200 at 14.3 ppm, 2/214 at 5.6 ppm	2, 13, 16
Mouse	2.0–14.3 ppm formaldehyde	inhalation	2.5 years	2/240 at 14.3 ppm	2, 13, 16
Rat	14.7 ppm formaldehyde, 10.6 ppm HCl	inhalation	lifetime	25/99	1
Rat	14.3 ppm formaldehyde, 10.0 ppm HCl	inhalation	588 days	12/100	1
Rat	14.1 ppm formaldehyde, 9.5 ppm HCl	inhalation	588 days	6/100	1
Rat	14.2 ppm formaldehyde	inhalation	588 days	10/100	1

Table III. Formaldehyde Carcinogenicity Oral Studies with
Laboratory Animals

Species	Compound	Length of Study	Ref.
Rat	0.4 g HMT	333 days	23
Mouse	0.5–5.0% HMT	110–30 weeks	14
Rat	1.0–5.0% HMT	156 weeks	24
Rat	1.0% HMT	3 generations	24
Rat	5.0% HMT	2 years	14

NOTE: Tumor incidence was negative in all studies.

and significance, would be particularly difficult to extrapolate to lower doses. In the nasal passages of the rat a similar effect may be enhanced by a concomitant inhibition of mucociliary function as has been reported to occur at concentrations greater than 2 ppm (*16*).

Higher concentrations also are accompanied in the rat inhalation studies by greater cell proliferation in response to increased cytotoxicity. This result may offer increased opportunity for the formation of DNA–protein cross-links which appear to occur linearly at concentrations greater than 6 ppm in some studies (*16*). Furthermore, higher local concentrations may also inhibit DNA repair and could potentiate the effects of DNA damage caused by formaldehyde or other agents. Also no convincing evidence for systemic tumors (i.e., those that occur in parts of the body remote from the site of application) has been reported in animals exposed subcutaneously or by inhalation.

Finally, formaldehyde at low concentrations is a normal biological intermediate that appears to be present in all biological tissues. Cells have developed specific enzymatic pathways for its removal. In the form of "active formaldehyde" (i.e., N^5,N^{10}-methylenetetrahydrofolate) it is used in mammals in the biosynthesis of purines, thymine, methionine, and serine. Table IV shows the amount of formaldehyde that might be absorbed systemically each day by the average American adult. As can be seen from this table, the use of food additives containing the low levels of formaldehyde that I have described make very little contribution to the formaldehyde that we breathe, drink, or find naturally in ordinary food.

Conclusions

For the reasons just discussed and on the basis of the data now available, FDA believes that the low levels of formaldehyde used in food or cosmetics do not present a significant safety issue.

Perhaps the most important issue raised is a scientific one. Namely, significant distinctions in the various uses of a substance can have a crucial bearing upon the possible carcinogenic risk. The very purpose of the toxico-

Table IV. Reported Human Exposures to Formaldehyde

Source	Amount Present	Daily Exposure	Ref.
Water			
Rain water	0.15 mg/kg	0.2 mg	25
Food			
Tomatoes	0.6 mg/100 g	varies	18
Apples	2.0 mg/100 g	varies	18
Cabbage	0.5 mg/100 g	varies	18
Spinach	0.5 mg/100 g	varies	18
Green onion	2.0 mg/100 g	varies	18
Carrots	0.8 mg/100 g	varies	18
Tobacco smoke			
1 cigarette	0.005 mg	1.0 mg	26
Air[a]			
Outside, New Jersey (1977)	5 ppb	0.03 mg	27
Outside, Los Angeles (1980)	15 ppb	0.1 mg	28
Indoor, conventional	50 ppb	0.3 mg	29
Indoor, particle board	1000 ppb	6.0 mg	29

[a]These levels are higher than anticipated lifetime average doses because they have not been corrected for the intermittency of exposure.

logical art is to identify, refine, and predict the consequences of significant biological and chemical distinctions between such uses. As our chemical detection methods improve and more carcinogens at smaller levels are found in more places in the marketplace, in the workplace, and in our environment, the need for more discriminating science will increase. It is important for everyone that regulatory agencies are supported by adequate funding for their research and by scientifically sophisticated policy direction that encourages the best possible science as a basis for regulatory decisions.

Literature Cited

1. FDA Talk Paper "Formaldehyde" (T82–27), May 21, 1980 and (T82–40), June 17, 1982.
2. Pavkov, K. L.; Mitchell, R. I.; Donofrio, D. J.; Kerns, W. D.; Connell, M. M.; Harroff, H. H.; Fisher, G. L.; Joiner, R. L.; Thake, D. C. "Final Report on Chronic Inhalation Toxicology Study in Rats and Mice Exposed to Formaldehyde," conducted by Battelle Columbus Laboratory for Chemical Industry Institute of Toxicology, December 31, 1981.
3. Albert, R.; Sellakumar, A.; Laskin, S.; Kuschner, M.; Nelson, N.; Synder, C. J. Natl. Cancer Inst. 1982, 68, 597–603.
4. Fed. Regist. 1980, 47 (101) 22753.
5. Eiermann, H. J., "The Safety of Formaldehyde as a Regulatory Issue," presented on May 23, 1983, in Washington, D.C.
6. "Code of Federal Regulations," Title 21, Part 173.340, 1982.
7. "Code of Federal Regulations," Title 21, Parts 175.105 and 178.3120, 1982.
8. "Code of Federal Regulations," Title 21, Part 573.460, 1982.
9. "Code of Federal Regulations," Title 21, Part 193.330, 1982.

10. Natuig, H.; Anderson, J.; Rasmusso, E. Wulff *Food Cosmet. Toxicol.* 1971, *9*, 491–500.
11. *WHO Food Addit. Ser.* 1972 *1*, 17–18; also Hurni, H.; Ohder, H. *Fd. Cisnet. Tixucik.* 1973, *11*, 459.
12. Watanabe, F.; Matunaga, T.; Iwata, Y. *Gann* 1954, *45*, 451.
13. Kerns, W. D.; Pavkov, K. L.; Donofrio, D. J.; Gralla, E. J.; Swenberg, J. A. *Cancer Res.* 1983, *43*, 4382–92.
14. Dellaporta, G.; Colnaghi, M. J.; Parmiani, G. *Food Cosmet. Toxicol.* 1968, *6*, 707–15.
15. Hollingsworth, R. S. "Summary on Formaldehyde," Internal Division of Toxicology memorandum FDA–BE, March 11, 1980.
16. Swenberg, J. A.; Barrow, C. S.; Boreiko, C. J.; Heck, H. d'A.; Levine, R. J.; Morgan, K. T.; Starr, T. B. *Carcinog.* 1983, *4*, 943–52.
17. *WHO Food Addit. Ser.* 1974, *5*, 63–73.
18. *IARC Monogr. Eval. Carcinog. Risk Chem. Man* 1982, *358*, 345.
19. Dalbey, W. E. "Toxicology in Review," 1981, OSHA Formaldehyde Docket H–225.
20. Horton, A. W.; Type, R.; Stemmer, K. L. *J. Natl. Cancer Inst.* 1963, *30*, 31–43.
21. Mueller, R.; Raabe, G.; Schumann, D. *Exp. Pathol.* 1978, *16*, 36–42.
22. Watanabe, F.; Sugimoto, S. *Gann* 1955, *46*, 365–67.
23. Brendel, R. *Arzneim. Forsch.* 1964, *14*, 51–53.
24. Dellaporta, G.; Cabral, J. R.; Tarmiani, G. *Tumori* 1970, *56*, 325.
25. Environmental Protection Agency "Investigation of Selected Potential Environmental Contaminants: Formaldehyde" by Kitchens, J. F.; Casner, R. E.; Edwards, G. S.; Harvard, W. E., III; Macri, B. J., Government Printing Office: Washington, 1976; pp. 5, 85–110, 126–32; EPA 560/2–76–009.
26. Mansfield, C. T.; Hodge, B. T.; Hege, R. B., Jr.; Hamlin, W. C. *J. Chromatogr. Sci.* 1977, *15*, 301–2.
27. Cleveland, W. S.; Graedel, T. E.; Kleiner, B. *Atmos. Environ.* 1977, *11*, 357–60.
28. Environmental Protection Agency "Human Exposure to Formaldehyde," Draft report (contract no. 68–01–5791) by Versar, Inc. for Office of Pesticides and Toxic Substances, 1980, pp. 70–72, 103, 107.
29. Anderson, I.; Lundgrist, G. R.; Molhave, L. *Atmos. Environ.* 1975, *9*, 1121–27.

RECEIVED for review October 16, 1984. ACCEPTED April 1, 1985.

Exposure to Formaldehyde

PETER W. PREUSS, RICHARD L. DAILEY, and EVA S. LEHMAN

U.S. Consumer Product Safety Commission, Washington, DC 20207

The potential for human exposure to formaldehyde from consumer products, occupational settings, and the ambient environment is discussed. Human subpopulations for each of these categories are identified, as are their estimated size and level and duration of exposure. Particular emphasis is placed on the potential for indoor exposure by the release of formaldehyde from consumer products such as pressed-wood products, urea–formaldehyde foam insulation, textiles, and biological specimens.

FORMALDEHYDE IS THE SMALLEST MEMBER OF THE ALDEHYDE GROUP; it has only a single carbon and oxygen atom and two hydrogen atoms (1). Its molecular weight is 30.03. It is a highly reactive gas with a characteristic odor. Exposure to formaldehyde can result from commercial processes and products as well as from noncommercial, indirect activities such as combustion of fossil fuels (2). As a result, exposure to formaldehyde can occur by inhalation, dermal contact, or ingestion.

A number of estimates have been made to assess the amount of formaldehyde released to the environment by direct or indirect processes (2). In addition, however, formaldehyde is released from many sources that are not amenable to assessment of the total amount of formaldehyde released. These sources include derivative chemicals containing residual levels of formaldehyde and products containing unreacted formaldehyde bonds. Releases from these sources may occur during production, during processing, during use by consumers, and during or after disposal (2). As a part of its efforts to understand the possible risks attendant to exposure to formaldehyde, the U.S. Consumer Product Safety Commission has funded a variety of laboratory and field studies and has collected, evaluated, and collated data from numerous other sources. This chapter summarizes these data, discusses levels and durations of exposure, and estimates the size of the population so exposed. It is hoped that in combination with other information about the health effects of formaldehyde, this chapter will serve as a basis for an assessment of the risks associated with exposure to formaldehyde. The chapter concentrates attention on consumer exposure, although ambient and occupational exposures are briefly discussed.

Ambient Exposure

The entire population of the United States has probably been exposed at some time to low levels of formaldehyde in the ambient air. The highest concentrations of atmospheric formaldehyde have been recorded in urban areas (2). The Los Angeles Basin has been extensively monitored, and formaldehyde concentrations have been reported in a number of studies (3–7). This geographic area tends to have higher levels than the rest of the country. Typical levels are approximately 10–30 ppb, although peaks as high as 48 ppb have been reported (8). Other highly industrialized areas, such as Bayonne, Camden, Elizabeth, and Newark, New Jersey, have typical levels of approximately 5 ppb (3). Levels of formaldehyde at Eniwetok Atoll in the South Pacific are approximately 0.4 ppb (9). Some typical data are presented in Table I.

Occupational Exposures

A number of surveys of occupational exposures of formaldehyde have been performed. These have shown that many workers are employed in manufacturing, processing, or sales activities that result in exposure to formaldehyde (2). Although it is not clear how comprehensive, characteristic, or representative any of these surveys have been, the results indicate that a majority of the workers are likely exposed to levels between 500 and 1000 ppb (Table II).

Exposure to formaldehyde can occur in many different occupational environments (2) including those listed in Table II.

Table I. Ambient Formaldehyde Levels

Type of Exposure	Number of Observations[a]	Exposure Level (ppb) Mean	Exposure Level (ppb) Max.
Rural areas (background)[b]			
Eniwetok Atoll, South Pacific	7	0.4	0.8
Urban areas[c]			
Los Angeles Basin (typical)	27	8.0	12.0
Los Angeles Basin (severe inversion)	65	24.0	48.0
Bayonne, N.J.	NR	6.1	20.0
Camden, N.J.	NR	3.8	14.0
Elizabeth, N.J.	NR	5.5	18.0
Newark, N.J.	NR	6.6	20.0

[a]NR indicates not reported.
[b]Estimated number exposed: 60,000,000.
[c]Estimated number exposed: 167,000,000.
Source: Adapted from Ref. 2.

Table II. Occupational Exposure to Formaldehyde

Exposure Source	Estimated Number Exposed	Number of Observations	Exposure Level (ppb) Mean	Exposure Level (ppb) Max.
Direct mfr. of formaldehyde	1400	135	410	2200
Resin mfr. (UF, PF)	6000	8	240	490
Plywood mfr. (UF, PF)	27,000	91	350	1200
Particle board mfr.	4000	6	920	1400
Wood furniture mfr.	60,000	6	100	140
Mobile home mfr.	32,000	—	—	—
UF foam mfr.	50	4	740	1280
UFFI installation	1000	17	420	1300
Metal molds–castings mfr.	60,000	11	390	690
Plastic products mfr.	17,000	8	350	500
Paper and paperboard mfr.	1000	64	470	990
Textile and apparel mfr.	800,000	30	250	310
Building paper and building board mfr.	4000	—	—	—
Paints and coatings mfr.	2300	—	—	—
Abrasive products mfr.	7000	—	—	—
Asbestos products mfr.	—	—	—	—
Nonresin CHO derivatives mfr.	250	—	—	—
Nitrogenous fertilizers mfr.	3000	1	500	500
Use of CH_2O-containing sanitation products	—	2	380	470
Use in agricultural pesticide applications	—	12	320	650
Biology–medical laboratories	—	—	—	—
Embalming and funeral service industry	2,600,000	6	740	1390
Metalworking machine operations	55,000	9	500	1200

NOTE: — indicates data not available.
Source: Adapted from Ref. 2.

Consumer Exposures

Formaldehyde and formaldehyde derivatives are present in a wide variety of consumer products (*10*). Most of these products release little, if any, formaldehyde into the air of residences (*11*). The concentration of formaldehyde in dwellings depends on the sources of formaldehyde that are present, the age of the source materials, the extent of natural and forced ventilation, the temperature and humidity, as well as the presence of materials that are not primary sources of formaldehyde but can absorb formaldehyde from the ambient air and then release it (sinks) (*12*). Work now underway indicates that the level of formaldehyde in a home is the result of the interaction of many factors, some of which may enhance and some of which may suppress the release of formaldehyde into the air (*12*). These factors include, in addition to those just discussed, seasonal and diurnal variations, weather, occupants' activities, and air exchange.

Laboratory studies and home-monitoring studies have shown that the major sources of formaldehyde in the indoor air are pressed-wood products and UFFI, although in specific instances other sources such as unvented heaters or other fossil fuel combustion may be important (11). Statistics show, compared to several years ago, a tendency to use larger amounts of pressed-wood products in new home construction, particularly as floor underlayments (13). As a result, higher levels of formaldehyde are present in new homes (11). Similarly, elevated levels are present in manufactured housing (mobile homes), again as a result of the extensive use of pressed-wood products (14). Measurements of homes that have had UFFI installed show elevated levels of formaldehyde as well (14).

Ambient Indoor Levels.   An analysis of formaldehyde levels in older homes without UFFI showed mean levels of 30 ppb (14, 15). A study of 400 randomly selected homes in Canada showed a mean level of 34 ppb (16). Similarly, a study in Great Britain showed average indoor levels in homes without UFFI to be 47 ppb (17).

A recent study by Oak Ridge National Laboratory that monitored formaldehyde levels in 40 homes for 1 year showed that average formaldehyde levels in homes more than 5 years old, without UFFI, ranged from 20 to 50 ppb (18).

UFFI.   Surveys done in the United States, Canada, and the United Kingdom all demonstrated that the installation of UFFI increases the levels of formaldehyde within a home (17–19). This increased level is inversely related to the age of the foam and shows a rapid decay with time, as shown in Figure 1 (14, 16).

Analysis of data in this country from many hundreds of homes showed that levels in homes with UFFI were approximately 200–300 ppb shortly after installation (although levels of 1000 ppb were not uncommon) and decreased to approximately 100 ppb after 1 year (15). These results compare very well with those of a study in the United Kingdom that showed an average level of 93 ppb in homes with UFFI, as well as levels greater than 1000 ppb in some cases. The United Kingdom study verified the data from the United States and Canada and concluded that the levels of formaldehyde in homes with UFFI were highest immediately after installation, and then decreased with age (17). The study done by Canadian government also demonstrated a significant difference between homes with UFFI and homes without UFFI (16). The difference between the two sets of homes was not as great, however, as that in the United States and the United Kingdom. Although the reasons for this difference are not entirely clear, it seems likely to be because the Canadian sample contained no homes in which the insulation was less than 2 years old, and the source strength had therefore decayed considerably.

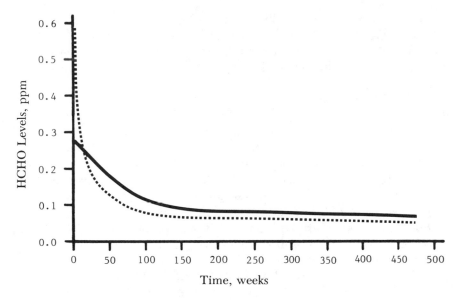

Figure 1. Average formaldehyde levels in UFFI houses in North America (----) and Britain (———).

In addition to these monitoring studies in homes, the emissions from UFFI have been studied in a variety of laboratory efforts (*11*). In one such study, simulated wall cavities were foamed by UFFI installers and manufacturers under carefully controlled, "best available technology" conditions (*20*). Formaldehyde emissions were monitored at room temperature as well as at higher temperatures, and the contributions to indoor levels were estimated. Even under these "idealized" conditions, all of the UFFI-foamed simulated wall cavities emitted formaldehyde. The relative contribution to indoor air from samples of UFFI less than 2 years old was estimated to be between 100 and 200 ppb. The emission rate was shown to increase 6- to 13-fold for a 5 to 15 °C increase in the ambient temperature (*21*).

The decay of formaldehyde emissions has also been estimated (*22*). In general, the emission of formaldehyde appears to reach a peak shortly after installation and then drops to approximately 100 ppb after 1 year and gradually approaches background levels in subsequent years.

Manufactured Housing. The levels of formaldehyde have also been measured in manufactured housing, although not as extensively as in homes with UFFI. In studies done for the Department of Housing and Urban Development, measurements were made in newly constructed dwellings that were then monitored for some time after construction. In addition, several other monitoring studies measuring the levels of formaldehyde in manufactured homes of various ages have been done. Some measurements were made of levels as high as 2000–3000 ppb; the average level was 400 ppb (22–24). As with homes with UFFI, a decay curve has been generated from the data and indicates an exponential decrease in the emission of formaldehyde into the indoor air and a half-life of approximately 4–5 years (Figure 2).

New Homes.   Data on formaldehyde levels in new homes are less extensive than are the data for manufactured housing. In the study performed by the Oak Ridge National Laboratory, the levels of formaldehyde in 40 homes in Tennessee were monitored regularly over 12 months. The levels measured varied widely and seemed to depend on the age of the home, the presence–absence of UFFI, the ambient temperature, and the season of the year. In general, this study found that the newest homes had the highest levels of formaldehyde, and that levels increased with increasing outdoor temperature both on a daily and seasonal basis. New homes and homes with UFFI installed were distinguished by their formaldehyde levels from older homes without UFFI installed (18).

Conclusions Regarding UFFI and Pressed-Wood Products.   In a recent consensus conference held to review the scientific information available, the Panel on Exposure concluded the following (11):

1. In mobile homes, new homes, and homes insulated with UFFI, mean formaldehyde levels are significantly higher than in other homes.
2. Mobile homes had the highest mean levels of formaldehyde.
3. Mean levels from existing large-scale studies can be used to

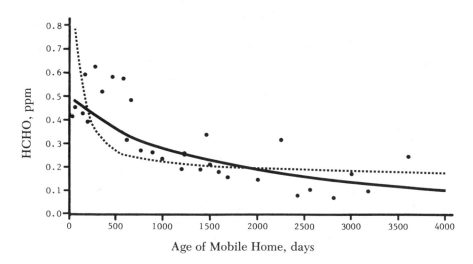

Figure 2. Mobile home age vs. average formaldehyde content: •, actual data points; – – – –, power curve fit; and ———— , exponential curve fit.

estimate long-term average exposure in various broad categories of housing types including mobile homes with UFFI.
4. Mean levels decline with the aging of the source, such as UFFI and pressed-wood products.

A summary of the data in the preceding discussion on formaldehyde in indoor air is presented in Tables III and IV.

Other Products Emitting Formaldehyde into the Indoor Air. In a study done at the Inhalation Toxicology Research Institute (ITRI), a broad spectrum of consumer products was investigated to determine the relative ranking of these products with regard to rates of formaldehyde release (25). The results of this study indicated that pressed-wood products had the highest release rates of formaldehyde, and that wearing apparel, fibrous-glass insulation, and ceiling tiles were potentially of concern as well. As a consequence of this study, each of these products was investigated in greater depth to determine the emission characteristics and the potential for human exposure (Table V).

FIBROUS-GLASS INSULATION AND CEILING TILES. Several readily available types of each product were tested to determine the emissions of

Table III. Indoor Exposure to Formaldehyde

Source of Exposure	Sampling and Analytical Methods[a]	Number of Observations	Mean Concentration Levels (ppb)
Canadian UFFI Study			
Homes with health complaints	CA	100	139
Control (non-UFFI) homes	CA	378	34
UFFI homes (no complaints)	CA	654	40
UFFI homes (no complaints)	CA	1146	54
United Kingdom UFFI Study			
Control (non-UFFI) buildings	MBTH–CA	50	47
UFFI buildings	MBTH–CA	1143	93
Buildings before–after UFFI			
Conventional (double-wall masonry) before	MBTH–CA	7	20–9.0
Conventional (double-wall masonry) after	MBTH–CA	7	130–280[b]
Nonconventional (prefab concrete) before	MBTH–CA	2	46, 34[c]
Nonconventional (prefab concrete) after	MBTH–CA	2	380, 700[d]

[a]CA is chromotropic acid and MBTH is 3-methyl-2-benzothiazolone hydrazone.
[b]Averages over first 90 days postinstallation.
[c]Averages of two houses prior to installation.
[d]Averages of the two houses measured prior to installation over first 90 days postinstallation.
Source: Adapted from Refs. 16 and 17.

Table IV. Indoor Exposure to Formadehyde

Age of House	Season	x	s	m	n
All ages	all seasons	62	77	5903	40
0–5 Years	all seasons	84	91	3210	18
5–15 Years	all seasons	42	42	1211	11
Older than 15 years	all seasons	32	42	1482	11
0–5 Years	spring	87	93	1210	—
	summer	111	102	1069	—
	fall	47	55	931	—
5–15 Years	spring	43	40	626	—
	summer	49	48	326	—
	fall	34	35	259	—
Older than 15 years	spring	36	51	757	—
	summer	29	37	341	—
	fall	26	23	384	—
All ages	spring	62	76	2593	—
	summer	83	91	1736	—
	fall	40	47	1574	—

NOTE: x is the mean concentration (ppb), s is the standard deviation, m is the number of measurements, and n is the number of homes. (Outdoor means are less than 25 ppb.)
Source: Adapted from Ref. 18.

formaldehyde and to estimate the potential of each type of product to raise the indoor levels of formaldehyde (26). Tests of these products in environmental chambers at different temperatures, humidities, and background formaldehyde concentrations showed that emissions from fibrous glass and ceiling tiles were negligible and that formaldehyde emission rates are very sensitive to background formaldehyde concentrations and seem to cease completely when background is greater than 100 ppb.

As a result, when these results were modeled, the researchers concluded that both products would have very little impact on indoor formaldehyde levels.

PRESSED-WOOD PRODUCTS. Much of the field monitoring data obtained thus far, as well as the study performed by ITRI discussed earlier, indicated that pressed-wood products emitted formaldehyde at a sufficiently high rate so that they were likely to be the predominant source of formaldehyde in new homes without UFFI and in manufactured housing. Consequently, an effort is now underway at the Oak Ridge National Laboratory to determine the emission rates of these products; effects of temperature, humidity, and background formaldehyde concentration on the emission rate; effects of sinks (such as gypsum board) and barriers (such as rugs) on formaldehyde emissions; models that will describe the data; and decay of formaldehyde emissions from these products over time (14).

Table V. Release of Formaldehyde from Consumer Products

Type of Product	Amount Released	
	$(\mu g/g/day)^a$	$(\mu g/m^2/day)^b$
Pressed-wood products		
Particle board		
A	4.1–5.3	13,000–17,000
B	6.7–8.1	23,000–26,000
C	4.9–7.1	20,000–28,000
Plywood		
A (interior)	7.5–9.2	13,000–15,000
B (exterior)	0.03–0.03	54–66
C (exterior)	ND (0.01)	ND
Paneling		
A	19–21	32,000–36,000
B	4.6–4.7	7,100–7,500
C	6.9–7.3	6,400–6,900
D	3.9–4.3	5,200–5,600
E	0.84–0.86	1,480–1,540
New clothes not previously washed		
Men's shirts (polyester–cotton)	2.5–2.9	380–550
Ladies' dresses	3.4–4.9	380–750
Girls' dresses (polyester–cotton)	0.9–1.1	120–140
Children's clothes (polyester–cotton)	0.2–0.3	15–55
Insulation products		
0.75-in. fibrous glass	1.3–1.7	390–540
Rigid round air duct	0.66–0.72	390–430
Rigid round fibrous glass duct	0.06–0.06	150–150
Fibrous glass	1.0–2.3	260–620
3.5-in. fibrous glass	0.3–0.7	52–130
Blackface insulation sheathing	0.03–0.04	340–420
Paper plates and cups		
A	0.12–0.36	400–1,000
B	0.03–0.14	75–450
C	0.10–0.15	330–335
Fabrics		
Drapery fabric		
A (100% cotton)	2.8–3.0	330–350
B (100% cotton)	0.8–0.9	90–120
C (blend: 77% rayon, 23% cotton)	0.3–0.3	50–50
D (blend: 77% rayon, 23% cotton)	ND $(0.01)^d$	ND
Upholstery fabric		
A (100% nylon)	0.03–0.05	9–11
B (100% nylon)	0.02–0.02	6–7
C (100% olefin)	0–0.02	0–5
D (100% olefin)	ND(0.014)	ND
E (100% cotton)	ND(0.014)	ND
F (100% cotton)	ND(0.015)	ND
Latex-backed fabric		
A	0.5–0.6	90–100
B	ND(0.015)	ND

Continued on next page.

Table V (*continued*)

Type of Product	Amount Released	
	$(\mu g/g/day)^a$	$(\mu g/m^2/day)^b$
Fabrics, continued		
Blend fabric		
A	0.3–0.4	20–30
B	0.2–0.3	20–30
Carpets		
A (foam-backed)	0.05–0.06	60–65
B (foam-backed)	0.006–0.01	8–13
C (foam-backed)	0–0.002	0–2
D	0.0005–0.0009	0–4
E	0.0007–0.0009	0–1
F	0–0.0009	0–1
G	ND(0.043)	ND

NOTE: ND denotes below limit of detection. Parentheses contain limit of detection.
[a]Range of two or more measured values expressed as micrograms of formaldehyde per gram of product per day.
[b]Range of two or more measured values expressed as micrograms of formaldehyde per square meter of product per day.
Source: Adapted from Ref. 25.

TEXTILES.   The testing at ITRI showed that textiles can be a significant source of formaldehyde. Consequently, several studies have been carried out to investigate this source. In one study, cloth was treated with the resin most often used by the textile industries, and formaldehyde release was determined (27). Both free and hydrolyzable formaldehyde were shown to be present, measurable, and theoretically available for exposure even after 25 washings of the cloth, although the amount available decreased with washing. A second study, using radiolabeled ^{14}C-formaldehyde in the resin, was performed to determine the bioavailability of the formaldehyde released from the cloth (28). This study showed that in the species tested (rabbits), radiolabel was transferred from the cloth to the animal. Most of the label (less than 3%) was in the skin under the cloth, and smaller amounts (less than 0.5%) were in other organs and tissues. The radioactivity in the blood and excreted urine increased with time and reached a peak value in approximately 24 h. Complementary results were obtained by another investigator in another species (cynomolgus monkeys) (29). Finally, a study was undertaken with isolated rabbit skin in a diffusion cell to determine whether, and to what degree, formaldehyde can transfer across skin (30). This study showed that as much as 10% of the applied label penetrated the skin in 8 h. However, the chemical species on the receptor side did not give a positive result in a test for formaldehyde and was not identified.

Table VI. Formaldehyde in School Biology Labs

Source of Exposure	Number of Observations	Concentration Levels (ppb)			Comments
		Min.	Mean	Max.	
Specimens fixed with formalin	6	120	730	1360	breathing zone
but without any type of after-treatment to reduce emissions	6	30	220	360	classroom background
Specimens fixed with formalin	13	—	130	270	breathing zone
but with some type of after-treatment to reduce emissions	13	20	60	130	classroom background

Note: The sampling and analytical method used in all cases was CO (pararosaniline).
Source: Adapted from Ref. 33.

Exposure in School Laboratories

Students and teachers are exposed to formaldehyde in classes in which specimens fixed in formalin are used. Some studies have previously monitored the formaldehyde levels in pathology and anatomy laboratories (*31, 32*). A study conducted by the Consumer Product Safety Commission monitored the levels of formaldehyde in a number of high school and college classrooms (*33*). The levels were measured in the breathing zone of the students, the general classroom air, and the outdoor air. In the 27 schools monitored, general laboratory levels of formaldehyde were less than 500 ppb in all schools; in 16 schools, the levels were less than 100 ppb. Peak levels in the breathing zone were between 1000 and 1400 ppb (Table VI). Significantly lower levels of formaldehyde were found in biology classes using specimens washed in water, or for which another preservative was substituted after formalin fixation.

Acknowledgment

The authors are employed by the Consumer Product Safety Commission. Because this chapter was written in their official capacity, it is in the public domain. The opinions expressed by the authors do not necessarily represent the view of the commission.

Literature Cited

1. "CRC Handbook of Physics and Chemistry," 62d ed; CRC: Boca Raton, Fla; 1981–82.
2. Environmental Protection Agency "Exposure Assessment for Formaldehyde" by Versar, Inc., 1982.
3. Cleveland, W. S.; Graedel, T. P.; Kleiner, B. *Atmos. Environ.* **1976**, 1, 357–60.

4. Grosjean, D. *Environ. Sci. Technol.* **1982**, *16*, 254–62.
5. Environmental Protection Agency "Atmospheric Measurements of Formaldehyde and Hydrogen Peroxide—Second Quarterly Report" by Kok, G. L.; Government Printing Office, Washington, 1980; EPA Grant No. R–80662910.
6. Hollowell, C. B.; Berk, J. V.; Lin, C. I.; Turiel, I., presented at the Second Int. CIB Symp. Energy Conserv. Build. Environ., Copenhagen, May 1979.
7. "Directory of Chemical Producers—United States"; Stanford Research Institute: Menlo Park, Calif., 1980.
8. Kok, G. L. "Progress Report to the U.S. Environmental Protection Agency"; Government Printing Office: Washington, 1979.
9. Zafiriou, D. C.; Alford, J.; Herrera, M.; Peltzer, E. T.; Gagosian, R. B.; Lui, S. C. *Geophys. Res. Lett.* **1980**, *7*, 341–44.
10. Ulsamer, A. G., Beall, J. R., Kang, H. K., Frazier, J. A. In "Hazard Assessments of Chemicals: Current Developments"; Academic: New York, 1984; Vol. 3, pp. 337–400.
11. Consensus Workshop on Formaldehyde; Little Rock, October 1983; *Environ. Health Perspect.* **1984**, *58*, 323–81.
12. Gammage, R. B.; Hingerty, B. E.; Matthews, T. G.; Hawthorne, A. R.; Womack, D. R. "Temporal Fluctuations of Formaldehyde Levels Inside Residences"; Oak Ridge National Laboratory Report Conf. 830338–4, 1983.
13. Predicasts, Inc. "Industry Study: Building Board Markets," May 1981.
14. Matthews, T. B.; Hawthorne, A. R.; Daffron, C. R.; Reed, T. J.; Corey, M. D., presented at the Proc. 17th Intl. Wash. St. Univ. Particle Board Composite Materials Symp., 1983.
15. Consumer Product Safety Commission "Revised Carcinogenic Risk Assessment of Urea–Formaldehyde Foam Insulation: Estimates of Cancer Risk Due to Inhalation for Formaldehyde Released by UFFI" by Cohn, M. S.; Government Printing Office, Washington, 1981.
16. "Report of the Hazardous Product Board of Review on Urea–Formaldehyde Foam Insulation"; Ottawa, 1982.
17. Everet, L. H., presented at the Consensus Workshop on Formaldehyde, Little Rock, Ark., October 1983.
18. Gupta, K. S., memorandum to Preuss, P. W., Consumer Product Safety Commission, June 12, 1983.
19. Consumer Product Safety Commission, *Fed. Reg.* **1982**, *42*, 14366–419.
20. Osborn, S. W.; Lee, L. A.; Heller, H. L.; Hillman, E. E.; Colburn, G.; Landau, P.; Thorne, E. G.; Krevitz, K. "Urea-Formaldehyde Foam Insulation Study: Final Report," Report No. F–C5316–01 for Union Carbide Corp., January 1981.
21. Consumer Product Safety Commission "Urea–Formaldehyde Foam Insulation: Static Measurements of Gaseous Formaldehyde at Ambient and Elevated Temperature Conditions" by Pressler, C. L.; Government Printing Office, Washington, October 1981.
22. Singh, J.; Walcott, R.; St. Pierre, C. "Evaluation of the Relationship Between Formaldehyde Emissions from Particle Board Mobile Home Decking and Hardwood Plywood Wall Paneling in Experimental Mobile Homes" by Clayton Environmental Consultants; Report No. HUD–0002718; Government Printing Office: Washington, 1982; NTIS No. PB83–237404.
23. Committee on Aldehydes "Formaldehyde and Other Aldehydes"; National Academy: Washington, D.C., 1981.
24. Department of Housing and Urban Development "Evaluation of Formaldehyde Problems in Residential Mobile Homes" by Technology and Economics, Inc.; Government Printing Office: Washington, 1981.
25. Pickrell, J. A.; Griffis, L. C.; Hobbs, C. H. "Release of Formaldehyde from Various Consumer Products: Final Report to the Consumer Product Safety Commission"; Inhalation Toxicology Research Institute: Albuquerque, February 1982.

26. Consumer Product Safety Commission "Briefing Package to the Commission; Testing of Consumer Products for Formaldehyde Emissions, Second Phase: Fibrous Glass and Ceiling Tiles," Government Printing Office: Washington, 1983.

27. Harper, R. J.; Andrews, B. A. K.; Harris, J. A.; Reinhardt, R. M.; Vail, S. L. "Final Report on Work Performed at Southern Regional Research Center, New Orleans, on Interagency Agreement Between the U.S. Department of Agriculture and the U.S. Consumer Product Safety Commission," 1982, ARS 12–44–07001–1288; CPSC–IAG–80–1397.

28. Robbins, J. D.; Norreal, W. P. *Tox. Appl. Pharmacol.*, in press.

29. Jeffcoat, A. R.; Chasalow, F.; Feldman, D. B.; Matt, H. In "Formaldehyde Toxicity"; Gibson, J. E., Ed.; Hemisphere: Washington, D.C., 1983, pp. 38–50.

30. Harris, R.; Hoheisel, C., memorandum to Cohn, M., Consumer Product Safety Commission, September 14, 1983.

31. Rader, J., M.D. Dissertation, Institute of Pharmacology and Toxicology of the University of Wurzburg, Federal Republic of Germany, 1974.

32. Schwartz, G. *Educ. Res. Commun. Outreach* 1980, *1*, 4–6.

33. Consumer Product Safety Commission "Exposure to Formaldehyde from Preserved Biological Specimens," Briefing Package to the Commission, January 7, 1982.

RECEIVED for review October 15, 1984. ACCEPTED March 11, 1985.

Review of Epidemiologic Evidence Regarding Cancer and Exposure to Formaldehyde

AARON BLAIR,[1] JUDY WALRATH,[1] and HANS MALKER[2]

[1]Environmental Epidemiology Branch, National Cancer Institute, Bethesda, MD 20205
[2]National Board of Occupational Safety and Health, Solna, Sweden

Completed epidemiologic studies are reviewed to evaluate the risks of cancer among persons exposed to formaldehyde. Cohort, proportionate mortality, and case-control study designs were used. The most striking findings to date are the consistent excesses for leukemia and brain cancer among professional groups (anatomists, embalmers, and pathologists). The lack of detailed exposure histories in these studies, however, precludes a clear determination of occupational factors that may be associated with these excess risks.

THE WIDE USE OF FORMALDEHYDE in industrial and commercial products coupled with its carcinogenicity in laboratory animals (1) has raised concern over its effects on human health. Exposures may occur among workers and the general public. From the National Occupational Hazards Survey (2), the National Institute of Occupational Safety and Health estimates that in the United States 1.6 million workers may be exposed to formaldehyde during the manufacture of formaldehyde resins, textiles, particle board, plywood, insulating materials, dinnerware, and paper, and in health-related professions. Outgassing from certain consumer products may result in the exposure of many persons in the general population to formaldehyde.

In response to the concern over potential health effects from formaldehyde exposure, several epidemiologic studies have been initiated over the past 4 years, although not all the studies are complete. The purpose of this chapter is to review the current epidemiologic literature regarding evidence for carcinogenicity in humans.

0065–2393/85/0210/0261$06.00/0

Study Designs

The study of populations exposed to formaldehyde could be used very effectively for instructional purposes in epidemiology because study methods employed include proportionate mortality ratio (PMR), case-control, and cohort (standardized mortality ratio [SMR] and standardized incidence ratio [SIR]) study designs. This combination of methods should provide a thorough evaluation of the issue of formaldehyde and cancer. SMR, SIR, and PMR studies allow an effective investigation of all causes of death except for rare causes such as cancer of the nasal cavity and sinuses. Case-control studies, however, effectively fill this gap. The locations of completed and ongoing studies listed in Tables I and II testify to the world-wide interest in the formaldehyde issue.

Many of the early hypothesis-generating studies suffer from the lack of information on exposure. In several of the recent or soon to be completed studies, however, considerable effort has been expended to obtain detailed estimates on level and duration of exposure. These studies should allow a clearer assessment of the cancer risk associated with formaldehyde exposure.

Results to Date

Studies discussed in this paper include those that were available to the Epidemiology Panel of "Consensus Workshop on Formaldehyde" held in October 1983 and more recently completed projects. The completed SMR, SIR, and PMR studies fall into two major groups: (1) professional occupations including embalmers, pathologists, and anatomists and (2) workers in the chemical, textile, and other industries. This dichotomy is important because the disease patterns between the two groupings differ considerably.

Although many causes of death have been evaluated including the major cancer sites, only causes that are of intrinsic interest, or that were elevated in one or more of the studies, will be discussed here.

Nasal Cancer.   The occurrence of nasal cancer in laboratory animals exposed to formaldehyde makes the nasal cavity a site of particular interest. No nasal cancers have occurred in any of the SMR and PMR studies completed (Table III). However, the expected number in any individual study is quite small.

Several case-control studies have also evaluated the relation between formaldehyde and nasal cancer (Table IV). Evidence for exposure to formaldehyde in these studies was based on occupation–industry titles or self-reports of exposure. No persuasive evidence for an association is evident. Only the Swedish study shows an association (22). In this study, Hardell (22) reported a crude relative risk of approximately 4 based on two cases

Table I. SMR, SIR, and PMR Studies of Formaldehyde

Population Location and Investigator (Ref.)	No. of Subjects	No. of Deaths	Calendar Year	Estimate of Exposure
PMR studies				
N.Y. embalmers Walrath, U.S. (3)	1132	1132	1925–80	duration of licensure
Calif. embalmers Walrath, U.S. (4)	1109	1109	1925–80	duration of licensure
Chemical workers Marsh, U.S. (5)	392	592	1950–76	duration and type of exposure
Garment workers Stayner, U.S. (6)	257	257	1959–82	latency and duration
SMR and SIR studies				
British pathologists Harrington, U.K. (7)	2079	156	1955–73	none
British pathologists Harrington, U.K. (8)	2720	126	1974–80	none
Embalmers Levine, Canada (9)	1477	337	1928–77	duration of licensure
U.S. Anatomists Stroup, U.S. (10)	2239	799	1889–1979	subspecialty and duration of professional activity
Potentially exposed workers Malker, Sweden (11)	67,378	7044[a]	1961–79	occupation at census
Pathologists Matanoski, U.S. (12)	2775	562	1915–74	none
Chemical workers Tabershaw, U.S. (13)	2084	151	1945–77	exposure level (some vs. none)
Chemical workers Acheson, U.K. (14)	7680	1626	1941–81	level and duration
Industrial workers Partanen, Finland (15)	unknown	unknown	1945–80	level and duration

[a] Number of cancer cases.

and four controls employed in particle board manufacturing, an industry in which formaldehyde exposure may occur. Contact with wood dust, a known nasal carcinogen, could be related to this excess.

In Sweden, the Cancer and Environment Registry was used to evaluate the risk of cancer among persons in selected occupations with potential for exposure to formaldehyde (physicians, biologists, seamstresses, textile workers, apparel workers, pulp and paper workers, and metal molders) (11). Among workers in these occupations, 14 nasal cancers occurred, and

Table II. Case-Control Studies of Formaldehyde

Investigator, Location (Ref.)	Source of Cases (No.)	Cancer Sites	Source of Controls (No.)	Time Frame
Tola, Finland (16)	Finnish cancer registry (51)	cancer of the nasal cavity and sinuses among persons >35 years of age	other cancers (51)	1970–73
Brinton, U.S. (17)	hospitals in Va. and N.C. (160)	nasal cavity and sinuses	hospital records and death certificates (290)	1971–82
Roush, U.S. (18)	Conn. tumor registry (216)	nasal cavity and sinuses	state mortality files (662)	1935–72
EPA, U.S. (19)	Northwest Wash. state tumor registry (345)	cancers of the pharynx, and sinus and nasal cavity	population based	1980–84
Hernberg, Finland (20)	tumor registries of Denmark, Finland, Sweden (167)	living nasal cancer cases	colorectal cancer cases (167?)[a]	1978–80
Fayerweather, U.S. (21)	deaths among Du Pont employees (481)	all cancers	employees active during last year of employment	1957–82
Hardell, Sweden (22)	Swedish tumor registry (71)	nasopharyngeal and nasal (54)	controls from herbicide studies	1970–79
Olson, Denmark (23)	Danish tumor registry (839)	nasal	other cancers	1970-82

[a]The number of cases is not clear.

7.6 were expected. The excess, however, depends mainly on one group (female seamstresses with an SIR of 5.3). The level of exposure to formaldehyde among seamstresses, however, may be low.

In summary, there is little evidence from case-control, PMR, and cohort studies that persons exposed to formaldehyde have a high risk of nasal cancer.[1]

Buccal Cavity and Pharynx. Rats are obligate nose breathers, and all cancers in exposed animals occurred at the site of first contact, that is,

---

[1]Positive findings from the case-control study in Denmark (23) have been reported since the American Chemical Society "Symposium on Formaldehyde" in St. Louis. Relative risks for nasal cancer were 2.5 among persons exposed to wood dusts, 2.8 among those exposed to formaldehyde, and 4.1 among those exposed to both.

Table III. Mortality from Nasal Cancer

Study (Ref.)	Observed Number	Expected Number
Walrath		
N.Y. embalmers—PMR (3)	0	0.5
Calif. embalmers—PMR (4)	0	0.6
Levine		
Ontario morticians—SMR (9)	0	0.2
Stroup		
U.S. anatomists—SMR (10)	0	0.4
Acheson		
British chemical workers—SMR (14)	0	1.3
Stayner		
U.S. garment workers—PMR (6)	0	< 0.2
Malker		
Potentially exposed Swedish workers—SIR (11)	14	7.6

Table IV. Case-Control Studies of Nasal Cancer

Study (Ref.)	Relative Risk	Number of Cases Exposed	Comment
Hernberg—Sweden, Denmark, Finland (20)	no association	not reported	—
Tola—Finland (16)	association with textiles	2	possible formaldehyde exposure
Hardell—Sweden (22)	≃ 4	2 cases (4.5%) 4 controls (0.8%)	all worked in particle board factories
Brinton—U.S. (17)	1.2	4 cases	RR for textiles = 1.7
Fayerweather—U.S. (21)	—	—	no nasal cancer deaths

the nasal area. Humans, however, breathe through the mouth and the nose. For this reason, cancers of the mouth and pharynx have been hypothesized as sites also likely to be affected by formaldehyde exposure. No consistent excess for cancers of the buccal cavity and pharynx has occurred (Table V). Among N.Y. embalmers and funeral directors (3), however, a twofold excess was found among persons licensed only as embalmers (the group likely to have heavier exposures to formaldehyde). In the Cancer and Environment Registry from Sweden (11) no overall excess occurred for cancer of the buccal cavity and pharynx (63 observed vs. 68 expected). However, 23 cancers of the mouth and other buccal parts (ICD 144) were reported for persons who may have had exposure to formaldehyde; 10.9 cases were expected. A much smaller excess occurred for tongue, floor of the mouth, and mesopharynx (16 observed vs. 14.1 expected), and no excess occurred for other cancers of the pharynx (10 observed vs. 13.1 expected).

Table V. Mortality from Cancers of the Buccal Cavity and Pharynx

	All Subjects	
Study (Ref.)	Observed	Expected
Walrath		
N.Y. embalmers—PMR (3)	8	7.1
Calif. embalmers—PMR (4)	8	6.1
Stroup		
U.S. anatomists—SMR (10)	1	6.8
Levine		
Ontario morticians—SMR (9)	1	2.1
Acheson		
British chemical workers—SMR (14)	5	6.1
Liebling		
Mass. chemical workers—PMR (24)	2	.2
Stayner[a]		
U.S. garment workers—PMR (6)	3	1.3
Fayerweather		
U.S. chemical workers—case-control (21)	no association	
Malker		
Potentially exposed Swedish workers—SIR (11)	63	68.0

[a]Duration and induction was >20 years.

This result suggests that in future studies special attention should be paid to subsites within the buccal and pharyngeal cavities. Among garment workers (6), three cancers of the mouth occurred, and 0.4 were expected. All three cancers, however, were cancers of the parotid gland, which would not seem to be a likely site. The location of cancers within the buccal cavity and pharynx among embalmers did not resemble the pattern seen in Sweden and did not appear to be unusual (25). Although only seven cases occurred, the Du Pont case-control study revealed no association between these cancers and formaldehyde exposure, nor did any of the other studies of industrial workers (21).

Lung Cancer. Only one of the studies in Table VI showed an excess of lung cancer among formaldehyde-exposed groups. Indeed, several studies including those of pathologists and anatomists showed striking deficits. Deficits among professional groups are undoubtedly largely because of their lower frequency of tobacco use. The risk of lung cancer, however, was not associated with cumulative exposure to formaldehyde in Du Pont workers (21) for whom information on smoking habits was available. In the study of chemical workers in Great Britain (14), the SMR for lung cancer was 95 among all subjects, and 119 among those with a 25-year latency. In one plant (the largest) in this study the risk of lung cancer rose with increasing levels of formaldehyde exposure, but not with cumulative exposure. In another study from Great Britain (26) using a job-exposure matrix

Table VI. Mortality from Lung Cancer

Study (Ref.)	All Subjects		Induction Time Restriction		Induction Time
	Obs.	Exp.	Obs.	Exp.	
Walrath					
N.Y. embalmers—PMR (3)	72	66.8	35	34.7	>35 yr
Calif. embalmers—PMR (4)	41	42.8	28	30.8	>15 yr
Levine					
Ontario morticians—SMR (9)	19	20.2	19	18.9	>20 yr
Harrington					
British pathologists no. 1—SMR (7)	10	27.4	no data		—
British pathologists no. 2—SMR (8)	9	22.0	no data		—
Stroup					
U.S. anatomists—SMR (10)	12	43.0	6	29.3	—
Matanoski					
U.S. pathologists—SMR (12)	17	38.6	no data		—
Marsh					
Chemical workers—PMR (5)	6	7.1	3	3.7	>20 yr
Tabershaw					
Chemical workers—SMR (13)	3	5.2	no data		—
Acheson					
British chemical workers—SMR (14)	205	215.0	86	72.1	>25 yr
Stayner					
U.S. garment workers—case-control (6)	no association		—		—
Malker					
Potentially exposed Swedish workers (11)	214	183.4	no data		—

to estimate formaldehyde exposure, a significant overall excess of lung cancer was noted. The risk of lung cancer was not elevated, however, among occupations judged to have the highest exposure to formaldehyde. Finally, in a follow-up of persons from the Kaiser–Permanente Medical Center in San Francisco who had formalin applied topically on warts, four lung cancers occurred, and 0.7 were expected (27). Two of the cancers were adenocarcinomas, one was of the large cell undifferentiated type, and one was undefined. Although all four persons with lung cancer were smokers, in most pathologic series less than 20% of lung cancers are of the adenocarcinoma or large-cell type (28). The occurrence of adenocarcinomas may, therefore, be an important lead.

Brain Cancer. Mortality from brain cancer was consistently elevated among professional groups (embalmers, pathologists, and anatomists), but not among industrial workers (Table VII). When time since first exposure is considered, the brain cancer excess among professionals is even more striking. Among anatomists who worked for at least 20 years after receiving their first exposure, the SMR was 350. The diagnosis of brain cancer, however, can be difficult, and access to better health care

Table VII. Mortality from Brain Cancer

Study (Ref.)	All Subjects		Induction Time Restriction		Induction Time
	Obs.	Exp.	Obs.	Exp.	
Walrath					
N.Y. embalmers—PMR (3)	9	5.8	3	1.8	>35 yr
Calif. embalmers—PMR (4)	9	4.7	7	2.7	>15 yr
Levine					
Ontario morticians—SMR (9)	3	1.2	no data		—
Harrington					
British pathologists—SMR (7)	4	2.6	no data		—
Stroup					
U.S. anatomists—SMR (10)	10	3.7	6	1.7	>20 yr
Matanoski					
U.S. pathologists—SMR (12)	6	3.8	no data		—
Acheson					
British chemical workers—SMR (14)	5	12.5	no data		—
Tabershaw					
U.S. chemical workers—SMR (13)	1	0.7	no data		—
Stayner					
Garment workers—PMR (6)	1	2.1	no data		—
Malker					
Potentially exposed Swedish workers (11)	89	101.8	no data		—

may result in a more complete ascertainment of brain cancer cases particularly among the socially privileged. Greenwald et al. (29) provided an excellent illustration of this situation in an evaluation of the diagnosis of brain cancer among employees of a large corporation with an excellent corporate health care program. Because professional groups, particularly those in health professions, are likely to receive better health care than the general population, diagnostic bias is a special concern. In the study of anatomists (10), several procedures were used to evaluate this potential bias. First, the SMR among anatomists actually increased when psychiatrists (a group with a socioeconomic status comparable to that of anatomists) were used as the reference (SMR = 597). In another approach, mortality rates from Omstead County, Minn., were used to generate expected numbers. Residents of Omstead County have access to excellent medical care because of the presence of the Mayo Clinic, and the ascertainment of brain cancer in this population should be relatively complete. The brain cancer excess among anatomists remained when expected rates were derived from this population. Finally, the SMR decreased, but was not eliminated, when expected deaths from benign tumors of the nervous system (conditions likely to be misdiagnosed) were included in the estimate of expected brain cancers. Among anatomists, therefore, diagnostic bias does not appear to account entirely for the excess of brain cancer observed.

Leukemia.   Five of the studies listed in Table VIII show a slight excess of leukemia; mortality ratios range from 1.4 to 1.8. Although the excess of leukemia tends to occur among the professional groups, a deficit occurred among one group of pathologists, and an excess occurred among garment workers. The mortality ratios are slightly larger among subcategories for which time since first exposure is considered. Among anatomists (*10*) and New York and California embalmers (*3, 4*), the leukemia excess was greatest for the myeloid leukemias. For leukemia, as for brain cancer, diagnostic practices may result in more complete ascertainment among the socially privileged. The excess among anatomists, however, remained when compared to mortality rates observed among psychiatrists. This result suggested that diagnostic bias is unlikely to account entirely for the differences observed.

Prostate Cancer.   Although few reports showed statistically significant elevations of cancer of the prostate, slight excesses were reported in four studies (Table IX): embalmers (*3*), anatomists (*10*), and U.S. chemical workers (*13, 21*). The magnitude of the excesses is not very impressive, but

Table VIII. Mortality from Leukemia

	All Subjects		Induction Time Restriction		Induction Time
Study (Ref.)	Obs.	Exp.	Obs.	Exp.	
Walrath					
N.Y. embalmers—PMR (*3*)	12	8.5	6	3.9	>35 yr
Calif. embalmers—PMR (*4*)	12	6.9	10	4.3	>15 yr
Levine					
Ontario morticians—SMR (*9*)	4	2.5	no data		—
Harrington					
British pathologists no. 1—SMR (*7*)	1	1.5	no data		—
British pathologists no. 2—SMR (*8*)	1	1.1	no data		—
Stroup					
U.S. anatomists—SMR (*10*)	10	6.7	7	4.1	>20 yr
Matanoski					
U.S. pathologists—SMR (AABP members only) (*12*)	2	3.7	no data		—
Acheson					
British chemical workers—SMR (*14*)	9	11.4	no data		—
Stayner					
U.S. garment workers—PMR (*6*)	4	2.4	2	1.0	duration and latency >10 yr
Malker					
Potentially exposed Swedish workers (*11*)	94	82.5	no data		—

Table IX. Mortality from Prostate Cancer

Study (Ref.)	All Subjects		Induction Time Restriction		Induction Time
	Obs.	Exp.	Obs.	Exp.	
Walrath					
N.Y. embalmers—PMR (3)	15	16.4	10	10.8	>35 yr
Calif. embalmers—PMR (4)	23	13.1	16	9.5	>15 yr
Levine					
Ontario morticians—SMR (9)	3	3.4	no data		>20 yr
Stroup					
U.S. anatomists—SMR (10)	20	18.7	no data		—
Tabershaw					
U.S. chemical workers—SMR (13)	2	0.6	no data		—
Fayerweather					
U.S. chemical workers—case-control (21)	rising odds ratio with cumulative exposure				—
Malker					
Potentially exposed Swedish workers (11)	228	231.4	no data		—

the occurrence of excesses among both professional and industrial populations differs from the pattern seen for most other cancers.

Other Cancer Sites.   Several other cancer sites have elevated risks in one or more of the studies, but show no overall consistency. Cancer of the bladder showed an odds ratio that rose with cumulative exposure (on the basis of six exposed cases) among U.S. chemical workers (21), and a slight excess occurred among California embalmers (4) and in the Swedish Cancer and Environment Registry study (11). Opportunities for dermal exposure to formaldehyde could occur in various occupations. Because of the potential for skin contact, mortality from skin cancer is of interest. Excesses of skin cancer occurred, however, only among N.Y. embalmers (3) and U.S. garment workers (6), and the excess among U.S. garment workers was based on only two deaths. Cancer of the kidney was slightly elevated among N.Y. embalmers (3) and U.S. pathologists (12), but neither excess was statistically significant.

Discussion

Reports from several epidemiologic studies of varying designs are now available to evaluate the risk of cancer among persons exposed to formaldehyde. Most of the studies to date, however, did not obtain exposure estimates and thus do not allow risk analyses by exposure level. The Du Pont case-control (21) and British chemical worker (14) studies are exceptions. Information on potential confounding factors such as tobacco use is lacking in all SMR, SIR, and PMR studies, although several of the case-control studies (*16, 17, 20–22*) obtained such information. In addition, results from several of the studies are available only from abstracts or in preliminary form, and further analysis may prove more illuminating. Despite such limitations, these studies provide useful data from which to obtain a crude indication of carcinogenic risks that humans may incur from formaldehyde exposure.

One illuminating feature of this review is the absence of deaths from nasal cancer from the PMR and SMR studies. Although this cancer is rare in humans, the available evidence from these studies suggests that formaldehyde exposure is unlikely to result in a large increase of nasal cancer. The case-control studies generally reinforce this conclusion, although the recent report of an association between formaldehyde and nasal cancer in Denmark (*23*) requires further scrutiny. The results from the ongoing EPA case-control study in which formaldehyde exposure will be assessed in detail will be a welcome addition.

The mortality pattern for cancers of the buccal cavity and pharynx differs from that among other sites in that the excess occurs among industrial, rather than professional, workers. Because humans are not obligate nose breathers, as are rats, the mouth may be a likely location of effect. The findings of an excess for cancer of the mouth in the Swedish Cancer and Environment Registry study provides some support for this contention. Tobacco use is a major risk factor for cancer of the buccal cavity and pharynx; therefore, a lower mortality rate of professional groups than of the general population is not surprising. Cancers of the mouth should receive special attention in future studies.

Lower smoking rates among professionals than among the general population severely restricts the usefulness of comparing lung cancer mortality when information on tobacco use is unavailable. Although neither professional nor industrial workers overall showed excesses for cancer of the lung, an association with formaldehyde levels in the largest plant in the British study of chemical workers suggests that further evaluation including information on tobacco use is needed. Special attention should also be paid to histologic type in future studies.

The excess of deaths from brain cancer among professionals and the deficit among industrial workers cannot be explained at the present time. Diagnostic bias is a concern, but mortality ratios among anatomists remain high when mortality among psychiatrists or the population from Omstead County, Minn., is used for comparison. Both are populations for which ascertainment is thought to be relatively complete. The etiology of brain cancer is largely unknown, but the epidemiologic literature does suggest associations with occupations in which exposure to chemicals is likely (*30*). The size and consistency of the excess among professionals suggest that a spurious association is unlikely. Embalmers, pathologists, and anatomists, however, are exposed to glycols, glycerols, phenol, dyes and stains, and other substances in addition to formaldehyde. Contact with biologic tissues may also provide the opportunity for exposure to viruses that may be involved in the etiology of these tumors. Identification of the factors that may be associated with the excess risk is not possible from current studies. The excess of leukemia among professionals, although less consistent than that for brain cancer, is also intriguing. Leukemia is known to be caused by exposure to benzene (31), and involvement of other chemical exposures is suspected (32).

The deficits for leukemia and brain cancer among industrial workers suggest that the excesses among professionals may be due to factors other than formaldehyde. However, information on level and duration of exposure to formaldehyde and other substances is needed to resolve the contradiction. Finally, neither cancer would seem to be a site likely to respond to formaldehyde exposure because formaldehyde appears to be metabolized rapidly in tissues (33). Our understanding, however, of the individual steps between exposure and development of cancer is elementary, even for well-known carcinogens.

In summary, current epidemiologic studies provide little evidence that the risk of nasal cancer is increased by exposure to formaldehyde. The most interesting findings are the excesses of leukemia and cancers of the brain (where consistent excesses occur among all professional groups studied) and mouth (a location where direct contact is likely). These sites deserve special attention in studies that are underway or in the planning stage.

Literature Cited

1. Swenberg, J. A.; Kerns, W. D.; Mitchell, R. I. Cancer Res. **1980**, 40, 3398–402.
2. "National Occupational Hazards Survey"; National Institute of Occupational Safety and Health: Cincinnati, 1974; DHEW Publication No. 74–127.
3. Walrath, J.; Fraumeni, J. F., Jr. Int. J. Cancer **1983**, 31, 407–11.
4. Walrath, J.; Fraumeni, J. F., Jr. Cancer Res. **1984**, 44, 4638–41.
5. Marsh, G. M. In "The Third CIIT Conference in Toxicity: Formaldehyde Toxicity"; Gibson, J. E., Ed.; Hemisphere: New York, 1983; pp. 227–36.
6. Stayner, L.; Smith, A. B.; Reeve, G.; Blade, L.; Elliot, L.; Keenlyside, R.; Frumin, E.; Halperin, W. Am. J. Epidemiol. **1984**, 120, 458–59.
7. Harrington, J. M.; Shannon, H. S. Br. Med. J. **1975**, 4, 329–34.
8. Harrington, J. M., Oakes, D. Br. J. Ind. Med. **1984**, 41, 188–91.
9. Levine, R. J.; Andjelkovich, D. A.; Shaw, L. K.; Dal Corso, D. In "Formaldehyde: Toxicology, Epidemiology, and Mechanisms"; Clary, J. J.; Gibson, J. E.; Waritz, R. S., Eds.; Marcel Dekker: New York, 1983; pp. 127–40.
10. Stroup, N. E.; Blair, A.; Erikson, G. E. Am. J. Epidemiol. **1984**, 120, 500.
11. Malker, H.; Weiner, J. Arbele Och Halsa **1984**, 9, 80–83.
12. Matanoski, G. M., personal communication.
13. Tabershaw Associates, personal communication.
14. Acheson, E. D.; Barnes, H. R.; Gardner, M. J.; Osmond, D.; Pannett, B.; Taylor, C. P. Lancet **1984**, I, 611–16.
15. Partanen, T., personal communication.

16. Tola, S.; Hernberg, S.; Collan, Y.; Kinderborg, R.; Korkala, M. *Int. Arch. Occup. Environ. Health* **1980**, *46*, 79–85.
17. Brinton, L.; Winn, D.; Browder, J.; Farmer, J.; Becker, J.; Blot, W.; Fraumeni, J. F., Jr. *Am. J. Epidemiol.* **1984**, *119*, 896–906.
18. Roush, G. G., personal communication.
19. JRB Associates, personal communication.
20. Hernberg, S.; Collan, Y.; Degerth, R.; Englund, A.; Engzell, U.; Kousma, E.; Mutanen, P.; Nordlinder, H.; Schultz-Larsen, K.; Sogaard, H.; Westerholm, P. *Scand. J. Work. Environ. Health* **1983**, *9*, 208–13.
21. Fayerweather, W. E.; Pell, S.; Bender, J. R.; Clary, J. J.; Gibson, J. E.; Waritz, R. S. In "Formaldehyde: Toxicology, Epidemiology, and Mechanisms"; Marcel Dekker: New York, 1983; pp. 47–113.
22. Hardell, L.; Johansson, B.; Axelson, O. *Ind. Med.* **1982**, *3*, 247–57.
23. Olson, J.; Jensen, O. *Int. J. Cancer* **1984**, *34*, 639–44.
24. Liebling, T.; Rosenman, K.; Pastides, H.; Leneshaw, S. *Am. J. Epidemiol.* **1982**, *118*, 570.
25. Walrath, J., unpublished data.
26. Coggon, D.; Pannett, B.; Acheson, E. D. *J. Natl. Cancer. Inst.* **1984**, *72*, 61–65.
27. Friedman, G. D.; Ury, H. K. *J. Natl. Cancer Inst.* **1983**, *71*, 1165–75.
28. Morton, W. E.; Treyve, E. L. *Am. J. Ind. Med.* **1982**, *3*, 441–57.
29. Greenwald, P.; Friedlander, B. R.; Lawrence, C. E.; Hearne, T.; Earle, K. *J. Occup. Med.* **1981**, *23*, 690–94.
30. Selikoff, I. J.; Hammond, E. C. *Ann. N.Y. Acad. Sci.* **1982**, *381*, 364.
31. *IARC Monogr. Eval. Carcinogen. Risk Chem. Man* **1982**, *29*, 416.
32. *IARC Monogr. Eval. Carcinogen. Risk Chem. Man* **1982**, *28*, 486.
33. Swenberg, J. A.; Burrow, C. S.; Boreiko, C. J.; Heck, D.; Levine, R. J.; Morgan, K. T.; Starr, T. B. *Carcinog.* **1983**, *4*, 945–52.

RECEIVED for review September 28, 1984. ACCEPTED January 14, 1985.

ADVANCES IN CHEMISTRY SERIES 210
FORMALDEHYDE: ANALYTICAL CHEMISTRY AND TOXICOLOGY

ERRATUM

On page 275, the first sentence of the abstract should read as follows:

Exposure to formaldehyde is associated with a variety of effects in dialysis patients, including sensitization, eosinophilia, and possibly chromosomal damage.

Formaldehyde in Dialysis Patients

A Review

JAMES R. BEALL

Health Effects Research Division, U.S. Department of Energy, Washington, DC 20545

Exposure to formaldehyde is associated with a variety of effects in chromosomal damage. Most notably formaldehyde stimulates antigenic changes in erythrocytes that cause the development of antibodies. With new and reused filters, residual formaldehyde left after sterilization is leached from the filter during dialysis and enters the patient. As formaldehyde contacts the erythrocytes, it apparently forms an active hapten that stimulates the production of antiformaldehyde and anti-N-like antibodies. Anti-N-like antibodies may develop in more than 30% of the patients who are exposed to formaldehyde during dialysis. Antibodies related to formaldehyde exposure have been associated with hemolysis, anemia, and changes in the hematocrit. In a few patients who had received renal transplants, erythrocyte agglutination, caused by the antigen–antibody reactions, probably blocked microcirculation in the kidney and caused its rejection by the host. Perhaps by understanding the ways that patients are exposed to and affected by formaldehyde during dialysis, systems for dialysis and patient protection may be improved. This information may also help elucidate formaldehyde's potential to elicit reactions in healthy people when the exposure occurs by other routes.

T HE PROPENSITY OF FORMALDEHYDE (HCHO) to cause irritation, sensitization, cancer, and mutations following dermal or respiratory contact has received much attention (1–5). By contrast, the potential of HCHO to cause organ changes or effects by other routes of exposure has received little attention (6). Since 1972, information has been developed about the effects of HCHO in patients who receive dialysis therapy (7). This therapy may result in exposure to HCHO by intraperitoneal and intravenous injection. Although personnel who administer dialysis therapy, as well as the patients themselves, may touch or breath HCHO (8, 9), this chapter focuses on the effects in patients following the injection of it during dialysis. Perhaps by reviewing studies of these patients, new insights may be gained into HCHO toxicity and ways to improve dialysis therapy.

W. Kolff developed the first artificial kidney for human use in 1943; it was successfully used in 1945 (10). During the 1950s and 1960s, dialysis as a therapeutic procedure was conducted on a limited scale. In 1961, development of the Teflon shunt for repeated circulatory access (hollow fiber artificial kidney) permitted therapeutic dialysis of patients with renal failure to become more common. By 1970, dialysis was generally available for commercial use (11). Since 1970, although hemodialysis therapy has been simplified and extensively applied, the hollow fiber dialyzer has remained commonly employed. In December 1982, 65,765 patients received regular dialysis therapy in the United States at an annual cost of more than $1.6 billion or approximately $25,000 per year per patient (12, 13). By assuming that the incidence of new patients in the United States is similar to that of Australia, some 7000 people each year may start on dialysis therapy for the first time (14).

In the United States, patients either self-administer therapy at home or receive it in one of the 1218 service or health centers (13, 15). In both situations, the therapy is expensive. Because it is expensive, health centers and patients search for ways to reduce the costs of dialysis (11, 12, 16). One common way to save money is by reusing dialyzers. Although reuse of dialysis filters started before 1964 (12), it is becoming more common because of economic pressure (17). For example, in 1978 and 1979, approximately 15% of the patients reused dialyzers; in the fall of 1981, 27.5% reused them; and current estimates are that 50% of patients now reuse filters (18). Because a new dialysis filter may cost up to $30 and recycling a filter costs $4–10 (18), reuse has the potential to save significant sums. The more times a filter can be reused, the more money is saved (19). Some filters have been successfully reused for 3 years (12). To further illustrate this point, one health center with 45 regular hemodialysis patients saved approximately $85,000 annually or $2000 per patient per year (14). Others report similar savings (16). Most patients receive treatments three times per week. Therefore, if the number of new filters purchased was reduced by 50% (14, 16), and if each patient saved $25 per treatment, potential savings for the United States alone might exceed $250 million per year. If the cost of new filters decreases, savings from reuse may also decrease.

Sterilization

Whether a dialysis filter is new or reused it must be sterile. Without proper maintenance of sterility, infectors (bacteria and virons) might be introduced directly into the patient. Methods for sterilizing dialyzers that have been tried include the use of cold storage; γ-radiation (20); proteolytic enzymes (11); and solutions of benzalkonium chloride (21), ethylene oxide

(*22*), hydrogen peroxide (*11*), hypochlorite, and formalin or formaldehyde (*23, 24*). Of these, HCHO was recommended as the sterilant with several advantages in 1965 (*21*). It remains widely used today (*25*). The current trend is to use sterilizing solutions having concentrations of 2.0–4.0% HCHO. However, concentrations of up to 12% formaldehyde (30% formalin) have been used (*21, 26, 27*).

New and used dialyzers contain materials that operate as "chemical sinks." These may collect formaldehyde during sterilization and storage and release it during use (*25*). The commonly used hollow fiber dialyzer illustrates this point. During its fabrication the cellulosic fibers interact chemically with the polyurethane potting material and partially inhibit the hardening or curing of the polyurethane. The situation causes a thin film or cuff of polymethane gel to form around each fiber (*25*). During storage, HCHO diffuses into the gel film (*25*); during use, it leaches out and enters the patient. Although polymethane gel appears to be the primary chemical sink, dialyzers contain others, such as gaskets, potting material, tubing, and fibers (*24, 25*). HCHO may enter a patient from nondialyzer sources in the dialysis system as well. For example, in one hospital a water filter containing cotton fibers bonded with melamine–HCHO resin was inserted between the water tap and the dialyzer. In this instance, HCHO from the resin in the water filter leached through the dialyzer into the patients (*28*).

Exposure Concentrations

Easy, accurate, and reliable methods for measuring low concentrations of HCHO in blood or in dialyzer compartments have not been generally available (*23, 29*). Consequently, exposure concentrations that have been reported in the literature were either obtained in laboratory experiments and then used to predict exposure during dialysis or represented less accurate measurements at bedside of residual HCHO in dialyzers prior to use. Primarily because of its convenience, the clinitest has been commonly used at bedside to measure residual HCHO in dialyzers to which patients would be exposed (*30*). Its use did little to protect patients from exposure to formaldehyde because the lowest concentration that it can accurately measure may exceed 50 ppm (*29, 31*).

By using methods other than the clinitest, detection of HCHO at the concentration of 5.0 $\mu g/mL$ (5 ppm) is done in some health clinics (*16*). These methods are not generally available to patients who dialyze at home. Methods for routinely and accurately measuring HCHO concentrations of 1.0 ppm at bedside have recently been developed and should gain wider use soon (*32*).

Many factors affect the amount and concentration of HCHO to which a patient is exposed during dialysis. These include the type of filter used and the number and frequency of dialysis treatments. These three factors depend in part on the patient's needs and availability of service resources.

However, additional factors include the concentration of formalin that is used to sterilize and store the equipment, the extent to which HCHO is rinsed from the equipment prior to use, and the length of time that flow through the dialyzer is stopped between rinsing and use, or during use itself (15, 23–25). Although these additional factors may be controlled, proper rinsing of equipment requires consideration of more than just removing the excess HCHO or sterilant.

If the sterilant is removed over too long a period of time or if inadequate concentrations of sterilant are used, potentially harmful infectors may grow in the filter or equipment (15, 27). If the rinse is inadequate and too little HCHO is removed, the residual amount may be sufficient to cause toxicity.

To rinse all HCHO from the sinks within a dialyzer is extremely difficult. Lewis et al. (24) flushed a dialyzer with saline for 3 h and found that even after the procedure HCHO was leached from it. Shaldon et al. (33) found that 100 L of H_2O failed to rinse all ^{14}C-formaldehyde from a dialyzer that had been sterilized with it by a standard method that they used to prepare dialyzers for patients. In addition to the difficulties in rinsing HCHO from the dialyzer, the time that patients will devote to rinsing it is limited. Lewis et al. (15) suggest that a patient should not be expected to spend more than 1 h rinsing a dialyzer before each use.

If flow through the dialyzer stops, the concentration of HCHO that is available to the patient increases. This result occurs because HCHO from the sink equilibrates in time with that in the blood and dialysate compartments. The extent of the increase depends partially on how long the flow is stopped. When flow is restarted, a bolus of HCHO enters the patient in the first few hundred milliliters. In this situation, exposure concentrations reach easily 40 ppm (24). Koch et al. (34) studied the HCHO concentration in the effluent of Kiil dialyzers at the start of 220 dialyses during home use. They found the concentration ranged from 0.3 to 108 mg/dL (mean = 6.7 mg/dL). This finding means that some patients were infused with more than 100 ppm of HCHO. Lewis et al. (24) estimated that even after a complete rinsing process, 13 mg of HCHO was leached from a hollow fiber dialyzer during a routine cycle of use.

Newer rinsing procedures and sensitive convenient methods to measure HCHO have helped reduce most exposure concentrations to the range of 2.0–5.0 ppm (25). Perhaps more sensitive detection methods and better construction of dialyzers can reduce this exposure to HCHO even more in the future.

Effects of Exposure

For many patients, exposure to low concentrations of HCHO during dialysis has not caused any observable effects. Indeed, hemodialysis was once used to maintain blood pH levels by removing excess formic acid from a 58-

year-old man who drank 8 oz of formalin in a suicide attempt (35); it probably saved his life. For other patients, exposure to HCHO has been associated with a variety of toxic effects. These include a burning sensation at the site of injection (24), possible cytogenetic damage (36, 37), inhibition of ATP production by erythrocytes (RBC) (28), development of anti-N-like (ANL) and antiformaldehyde (anti-F) antibodies (38, 39), hemolysis of RBC, decrease in the life of RBC (T $\frac{1}{2}$), and changes in the hematocrit (40). In a few patients, the exposure to HCHO was associated with eosinophilia, hypersensitivity, possible anaphylactoid reactions (41, 42) or formalin reactions (12) and, at high concentrations even death (43).

Physicians with experience in dialysis report that hepatomegaly and/ or persistently high concentrations of liver-related enzymes develop in the sera of some patients (32, 44, 45). These changes in seemingly healthy dialysis patients may be due to several factors including, in part, a direct or indirect effect of formaldehyde on the liver (6).

Chromosomal damage in dialysis patients has been related by Goh and Cestero (36, 37) to exposure to HCHO. These workers examined 1187 metaphase specimens of cells that they took directly from the bone marrow of 40 dialysis patients. Preparations obtained from relatives of the patients served as controls. They found a "marked" increase in chromosomal abnormalities including aneuploides, breaks, and structural changes in dialysis patients. Measurements made during a mock sterilization of a dialyzer in the laboratory indicated that patients had received 126.75 ± 50.84 mg of HCHO during each treatment (36, 37). Because their studies did not include groups of similar dialysis patients without exposure to HCHO, more research is needed to understand the possible relationship between HCHO and chromosomal damage in dialysis patients.

The effects of HCHO on RBC probably occur through at least two processes: (1) changes in their metabolism and (2) changes in their immunogenic potential. Orringer and Mattern (28) associated the installation of a water filter between a tap water outlet and several dialyzers with an outbreak of hemolytic anemia among hemodialysis patients. Because the water filter's construction included melamine–formaldehyde resin, they investigated the effects of HCHO on RBC metabolism. They exposed RBC to HCHO for 5 min and then incubated them in vitro for 2 h with inosine as the only substrate. Pretreatment of RBC with HCHO inhibited glycolysis by reducing nadide (NAD) to NADH and thereby caused a 90% reduction in cellular adenosine triphosphate (ATP) concentrations during the 2-h incubation. Exposure to as little as 0.1 mM HCHO was able to reduce glycolysis and ATP content in RBC. When pyruvate was also present, a HCHO-related decline in ATP did not occur. The maximum effective amount of HCHO was 1.0 mM. According to Orringer and Mattern, this amount was only one-tenth of the concentration of HCHO that was in 1 L of fluid that they obtained from a dialyzer filter. These workers also showed that, using the same systems, melamine did not affect RBC metabolism.

Belzer et al. (46) described a medical case involving a man in whom RBC "cold" agglutinins caused localized infarcts and rejection of a transplanted kidney. The patient had received dialysis therapy for a year before the renal transplant was attempted. The antibodies that caused the infarcts reacted with N-positive RBC. The next year Howell and Perkins (7) described for the first time the development of ANL antibodies in patients who received chronic hemodialysis. They contrasted the incidence of 12 in 416 patients who had ANL antibodies with an extremely rare occurrence of anti-N antibodies per se in healthy people. Several researchers have subsequently confirmed the frequent presence of ANL antibodies in dialysis patients (29, 38, 39, 47–49).

Workers also subsequently substantiated the work of Belzer et al. (46). For example, Gorst et al. (50) related formaldehyde–induced ANL antibodies to renal graft failure.

Howell and Perkins (7) listed several potential causes for ANL antibody production and included exposure to HCHO as one possibility. Although they did not specifically establish HCHO as the cause, they eliminated pregnancy and prior transfusions as possible stimuli for ANL formation. Crosson et al. (49) eliminated other chemicals, bovine implant materials, prior serum transfusions, and bacterial and viral infections as stimuli for ANL antibody production. Ultimately several workers showed that HCHO alone stimulated the production of ANL antibodies (Table I) (33, 48, 49).

Table I. Anti-N-Like Antibodies in Dialysis Patients Exposed to Formaldehyde

No. Studied	No. with Anti-N-Like Antibodies	Percent	Reference
416	12	3	7
40	6	15	48
430	38	9	49
288	37	13	33
111	18	16	51
117	42	36	34
239	14	6	75
22	6[a]	27	39
71[b]	3[c]	16	38
82	15	18	15
196	60	31	29

[a] Twenty patients (91%) showed a separate antiformaldehyde antibody.
[b] Nineteen patients were exposed during resterilization with HCHO.
[c] Seventeen patients (89%) showed a separate antiformaldehyde antibody.

Howell and Perkins (7) speculated correctly that the incidence of 12 in 416 underestimated the proportion of patients who would develop ANL antibodies. Subsequent studies report a 12–24% incidence of ANL antibodies in patients who were dialyzed at health centers (48, 51, 52). Moreover, the incidence of patients dialyzed at home is generally greater than that of patients in dialysis centers and may reach nearly 50% (15, 53). Lynen et al. (54) showed that the incidence of patients with formaldehyde-dependent antibodies increased with time on dialysis therapy, and that all patients who had been treated for 5 years or longer had the antibodies. Table II shows the incidence of anti-N antibodies in people with normal renal function. In one study, only 8 of 45,000 people had auto-anti-N antibodies (55). Other researchers project that approximately 0.3% of a normal population would possess auto-anti-N antibodies (33).

ANL antibodies are producible by patients having MM, MN, or NN antigenic RBC (7, 48). The order of the potential for agglutination with ANL is NN > MN > MM (54). Because ANL antibodies react with N antigen and because MM-type RBC also react with ANL, HCHO seems to be capable of altering M antigens to become N-like. Also, it seems as if either N or N-like antigens may stimulate ANL production.

Little is known about the characteristics of ANL antibodies. Kaehny et al. (51) suggested that inactivation of ANL antibodies by 2-mercaptoethanol suggests that they may be of the immunoglobulin M (IgM) class. More recently, Lynen et al. (54) found that the antibodies that agglutinate native NN cells are exclusively of the IgM fraction of immunoglobulins, whereas antibodies directed against formaldehyde-altered NN red cells are mainly immunoglobulin G (IgG) in addition to IgM. Depending upon the titer, the ANL antibodies will agglutinate RBC at temperatures ranging from 4 to 37 °C (49, 56, 57). Some have found that the optimal reaction temperature range for the agglutination of ANL with RBC is between 12 and 18 °C (57). However, recent studies demonstrate a considerable amount of warmer antibodies (IgG) that could react at body temperature in dialysis

Table II. Incidence of Auto-Anti-N Antibodies in People Who Did Not Receive Dialysis Therapy

Cases Examined	No. with Anti-N	Percent	Reference
45,000	8	0.0178	55
50	0	0	33
71	0	0	38
74	0	0	29
1366[a]	19	1.39	74

[a] People with abnormal antibodies.

patients (54). Although ANL antibodies are probably not species specific, they may be specific for RBC (47).

In one study, only 6 of the 22 patients who were exposed to HCHO developed anti-N-like activity, but 20 of the 22 specifically agglutinated HCHO-treated RBC. Thus, the agglutination of HCHO-treated RBC did not depend only on the formation of ANL (39). This result raised the possibility that another factor was involved in a progression of immunogenic changes in RBC. Sandler et al. (38) named this new agglutinating factor antiformaldehyde antibody (anti-F). To these workers anti-F seemed to be a high-titer IgG immunoglobulin that reacted with formaldehyde-treated RBC independently of whether they were of the MM, MN, or NN phenotype (38). In 1981, Sharon et al. found that the removal of ANL antibodies by absorption onto RBC antigens with ONN did not affect the activity of anti-F (47).

The mechanism by which HCHO causes ANL antibodies to form involves a multi-step process and the MN antigen system on the RBC membrane (7, 54, 58). Lynen et al. (54) described a three-stage time-related process for the development of formaldehyde-dependent antibodies. The stages were defined according to the agglutination of different cell types by the patient's sera. In Stage I, the patients own RBC agglutinated only after pretreatment with HCHO, and the reaction had no relation to the MN system. In Stage II, NN RBC also agglutinated if they had been pretreated with HCHO. In Stage III, agglutination of native NN RBC also occurred (54). Undoubtedly, HCHO reacts with the N antigens on the RBC surface and probably also reacts at other sites on the RBC (38). In 1981, Sharon speculated that formaldehyde might exert an effect by neutralizing a negative charge on the RBC membrane. Because HCHO induces ANL antibodies in MM-type patients, it apparently has the ability to convert the antigenicity of MM on the RBC membrane (29). RBC M and N antigens behave as simple codominant alleles at a single locus (59). An important difference between the two antigens is the existence of a terminal sialic acid on the M antigen, but not on the N one. Recent studies show that in healthy people, HCHO reacts with the terminal sialic acid moiety on the RBC M antigen and thereby converts it to an N-like antigen (60). Perhaps the sialic acid is the source of negative charge on the RBC membrane that becomes neutralized, as Sharon (47) speculated.

The fact that HCHO-N RBC are agglutinated by anti-N antibodies in dialysis patients but not in healthy people indicates that differences in the N and N-like antigens are found (61). The HCHO-modified N and/or M antigens apparently stimulate the production of or develop in association with anti-F, an IgG antibody (47). The production of anti-F apparently precedes production of ANL antibodies by approximately 6 months (47). This finding means that during the process of immunization, a shift in production from IgM- to IgG-type antibodies occurs (51); Lynen et al. (54)

suggest that this shift may occur by their Stage II. Anti-F appear to cross-react with the N-antigen on the RBC membranes (39, 47). The cross-reaction develops slowly, but leads to a type of "spreading sensitivity." Larger titers of anti-F seem to yield a greater extent of cross-reactions. Anti-N antibodies may also cross-react with the M antigen sites on the RBC (62). The extent to which this cross-reaction between M antigen and ANL antibodies occurs is not known.

The in vitro incubation of sera with HCHO not only stimulates the production of specific antibodies but also reduces the activity of other antibodies. Specifically, a 1-h incubation of sera with a 1.0% solution of HCHO at a dilution of 1 : 1 (sera to HCHO) reduced the titers of anti-A and anti-B isoagglutinins (47). In this study, 110 of 200 sera samples showed a HCHO-induced decrease in selected antibodies (47). The effect was generally more pronounced in sera with inherently low antibody titers, although the response depended in part on the specific antibody that was agglutinated. This effect of HCHO may already be important in some patients because antibodies may play a role in inhibiting infections and promoting healing. Even without exposure to HCHO, these health-promoting events may be less than desirable in patients with renal failure.

In 1975, Cestero et al. (42) reported that anaphylactoid-type reactions occurred in two otherwise stable dialysis patients. These reactions included nasal congestion, rhinorrhea, conjunctival injection, circumoral paresthesias, pallar, dyspnea, laryngeal constriction, and marked hypotension that was unresponsive to volume replacement. Both patients had marked eosinophilia, and both had been dialyzed chronically with hollow fiber filters that were originally sterilized at the factory and, later, between uses with formaldehyde. The reactions did not occur when the same patients were dialyzed on two different coil filters that eliminated exposure to formaldehyde (42). These researchers related the eosinophilia and reactions to repeated exposure to formaldehyde. In 1979, Hoy and Cestero (41) again reported that anaphylactoid-type reactions that were related to formaldehyde occurred in two patients. These were the same two patients who were in the earlier report by Cestero et al. (42, 45). Nevertheless, in one patient, the anaphylactoid reactions did not develop until the man had received dialysis therapy for 3 years. Then the reactions became progressively more marked with time, as did his eosinophilia. This patient has subsequently developed severe reactions on dialyzers that were not sterilized with formaldehyde (45).

Hakim et al. (63) reported that two patients suffered cardiovascular collapse within 2 min after the start of dialysis. They related the occurrence of chest pain, dyspnea, and hypotension in certain dialysis patients to new cuprophane-membrane dialyzers and complement activation (63). Hakim et al. (64) found that the reuse of filters decreased the capacity of the cuprophane membrane to activate complement, but did not alter the capacity of cellulose acetate membranes to activate complement. Thus, complement activation in their studies did not increase, as does the formation of ANL antibodies, with repeated reuse of dialysis filters.

Charytan et al. (65) reported that allergic-type reactions occurred in 5% of dialysis patients without eosinophilia, but in 22% of the patients with it. Hoy and Cestero (41) found that 20 of 37 patients who used hollow fiber filters and formaldehyde resterilization had eosinophilia. In contrast, none of the nine patients who used coil filters and were therefore unexposed to formaldehyde had eosinophilia. These workers later documented a 38% incidence of eosinophilia in a group of dialysis patients who were exposed to formaldehyde. This incidence was significantly greater than that in either a group of azotemic patients or in a group of control patients who were not exposed to HCHO (41). The incidence of eosinophilia in HCHO-exposed patients increased with time. Several potential causes were found for eosinophilia in chronic dialysis patients including exposure to ethylene oxide, plasticizers, poly(vinyl chloride), and various drugs (22, 41). But, for some patients, exposure to formaldehyde seems to be the cause (41).

Discussion

Some of the effects in dialysis patients that occur after their exposure to formaldehyde seem similar to those that occur after exposure to it by other routes. For example, separate reports in 1982 by Spear (66) and by Suskov and Sazonova (67) associate the exposure of humans to formaldehyde by inhalation with increased incidences of cytogenetic abnormalities. Formaldehyde is also mutagenic to human cells that are cultured in vitro (68). These data are consistent with the findings by Goh and Cestero (36, 37) of an unusually high incidence of chromosomal abnormalities in patients who were exposed during dialysis to formaldehyde. Together these data support the proposition that formaldehyde may be mutagenic in humans under certain circumstances.

Several reports discuss the development of dermal and respiratory sensitization reactions upon exposure to formaldehyde or related products (3, 69). A recent report (69) suggests that dermal sensitization reactions to formaldehyde are Type I allergic reactions. Based on such reports, one might predict that sensitization reactions would develop in people whose blood is exposed to formaldehyde. The studies of dialysis patients substantiate the development of such immunologically based changes. The agglutination of ANL antibodies with RBC evidences a Type II allergic reaction, and the anaphylactic changes suggest Type I allergic reactions (69). It would be interesting to know whether or not immune responses involving ANL antibodies, anti-F antibodies, or eosinophilia might also develop after chronic exposure to formaldehyde by inhalation.

The chronic exposure of rats and mice to 5.6 and 2.0 ppm of formaldehyde by inhalation is associated with nasal carcinoma, metaplasia, or adenomas (70, 71). Although detailed mechanisms of the development of nasal cancer have not been described, formaldehyde initiates and promotes certain carcinogenesis processes in vitro (72, 73). Many dialysis patients have been chronically exposed to formaldehyde in concentrations exceeding 5.6 ppm. These data raise the possibility that exposure to formaldehyde places dialysis patients at an increased risk of developing cancer. Additional research is needed to help define the nature and extent of that risk as well as the risks associated with its mutagenic potential.

Exposure to formaldehyde by inhalation has been associated with systemic changes in laboratory animals and humans; these include changes in the reproductive and central nervous systems and various organs (2, 3, 4, 6). Some untoward changes that seem to occur without apparent cause in dialysis patients could be related to their exposure to formaldehyde. One example of such changes is the unexplained hepatomegaly and/or elevated concentrations of liver enzymes in the sera of dialysis patients. Additional research could help elucidate formaldehyde's role, if any, in such change.

Perhaps the development and widespread use of accurate and sensitive methods of measuring residual formaldehyde in dialyzers will help answer some of these questions and provide safer therapy for dialysis patients. Other questions may be answered only by additional research.

Acknowledgments

The opinions expressed in this chapter are those of the author and do not necessarily reflect official policies or positions of the U.S. Department of Energy. I thank Sheila Palmer for her dedicated secretarial support.

Literature Cited

1. Report of the Federal Panel on Formaldehyde *Environ. Health Perspect.* **1982**, *43*, 139–68.
2. National Research Council "Formaldehyde: An Assessment of Its Health Effects"; National Academy: Washington, D.C., 1980.
3. National Research Council "Formaldehyde and Other Aldehydes"; National Academy: Washington, D.C., 1981.
4. Ulsamer, A.; Beall, J.; Kang, H.; Frazier, J. In "Hazard Assessment of Chemicals"; Academic: New York, 1984; Vol. 3, pp. 337–400.
5. *IARC Monogr. Eval. Carcinog. Risk Chem. Man* **1982**, *29*, 345–89.
6. Beall, J.; Ulsamer, A. *J. Toxicol. Environ. Health* **1984**, *14*, 1–21.
7. Howell, E.; Perkins, H. *Vox. Sang.* **1972**, *23*, 291–99.
8. Hendrick, D. J.; Lane, D. J. *Br. Med. J.* **1975**, *1*, 607–8.
9. Hendrick, D. J.; Rando, R. J.; Lane, D. J.; Morris, M. J. *J. Occup. Med.* **1982**, *24*, 893–97.
10. McBride, P. *NAPHT News*, **1979**, *Nov.* 20–21.
11. Lazarus, J.; Friedrich, R.; Merrill, J. *Dial. Transplant.* **1973**, *2*, 14–16, 32.
12. Craske, H.; Dahrowiecki, I.; Manull, M.; Knight, S.; Porrett, E.; Woods, F.; Uldall, P. *Artif. Organs* **1982**, *6*, 208–13.
13. Health Care Financing Administration, "End-Stage Renal Disease Program Highlights," DDHS, Washington, D.C., 1982, pp. 1–2.
14. Mathew, T.; Fazzalari, R.; Disney, A.; MacIntyre, D. *Nephron* **1981**, *27*, 222–25.

15. Lewis, K.; Dewar, R.; Ward, M.; Kerr, D. *Clin. Nephrol.* **1981**, *15*, 39–43.
16. Bok, D.; Levin, N. W. *Int. J. Artif. Organs* **1982**, *5*, 4–5.
17. Easterling, R. E. In "Reuse of Disposables: Implications for Quality Health Care and Cost Containment"; AAMI Technology Assessment Report No. 6-83; Association for the Advancement of Medical Instrumentation: Arlington, Va., 1983; pp. 3–10.
18. Newmann, J. *NAPHT News* **1982**, *May 6–8*.
19. Fawcett, K.; Mangles, M. *Dial. Transplant.* **1974**, *3*, 38–40.
20. Man, N.; Lebkiri, B.; Polo, P.; de Sainte-Lorette, E.; Lemaire, A.; Funck-Brentano, L. "Proc. Dial. Transplant. Forum," 1980, pp. 18–21.
21. Johnson, W.; Zahransky, R.; Mueller, G.; Wagoner, R.; Maher, F. *Mayo Clin. Proc.* **1965**, *40*, 462–72.
22. Novello, A. C.; Port, F. K. *Int. J. Artif. Organs* **1982**, *5*, 6–7.
23. Ogden, D.; Myers, L.; Eskelson, C.; Zieller, E. *Trans. Dial. Transplant. Forum* **1973**, *30*, 141–46.
24. Lewis, K.; Ward, M.; Kerr, D. *Artif. Organs* **1981**, *5*, 269–77.
25. Gotch, F.; Keen, M. *Trans. Am. Soc. Artif. Intern. Organs* **1983**, *29*, 396–401.
26. Favero, M. S. In "Reuse of Disposables: Implications for Quality Health Care and Cost Containment"; AAMI Technology Report No. 6-83; Association for the Advancement of Medical Instrumentation: Arlington, Va., 1983; pp. 19–23.
27. Murray, M.; Dathan, J.; Goodwin, F. *Br. Med. J.* **1969**, *4*, 235.
28. Orringer, E.; Mattern, W. *N. Engl. J. Med.* **1976**, *294*, 1416–20.
29. Fassbinder, W.; Koch, K. *Contrib. Nephrol.* **1983**, *36*, 51–67.
30. Pollard, T.; Barnett, M.; Eschbach, J.; Scribner, B. *Trans. Am. Soc. Artif. Intern. Organs* **1967**, *13*, 24–28.
31. Levin, N. *Dial. Renal Transplant.* **1980**, *9*, 40–46.
32. Gotch, F., personal communication.
33. Shaldon, S.; Chevallet, M.; Maraoui, M.; Mion, C. *Proc. Eur. Dial. Transplant Assoc.* **1976**, *13*, 339–47.
34. Koch, K.; Frei, U.; Fassbinder, W. *Trans. Am. Soc. Artif. Intern. Organs* **1978**, *24*, 709–13.
35. Spellman, G. *J. Iowa Med. Soc.* **1983**, *May*, 175–76.
36. Goh, K.; Cestero, R. *J. Med.* **1979**, *10*, 167–174.
37. Goh, K.; Cestero, R. *JAMA* **1982**, *247*, 2778.
38. Sandler, S.; Sharon, R.; Stessman, J.; Czaczkes, J. *Isr. J. Med. Sci.* **1978**, *14*, 1177–80.
39. Sandler, S.; Sharon, R.; Bush, M.; Stroup, M.; Sabo, B. *Transfusion* **1979**, *19*, 682–87.
40. Fassbinder, W.; Frei, U.; Koch, K. *Klin. Wochenschr.* **1979**, *57*, 673–79.
41. Hoy, W.; Cestero, R. *J. Dial.* **1979**, *3*, 73–87.
42. Cestero, R.; Hoy, W.; Freeman, R. *Abstr. Am. Soc. Artif. Intern. Organs* **1975**, *4*, 9.
43. Erkrath, K. D.; Adebaho, G.; Kloppel, A. C. *Z. Rechtsmed.* **1981**, *87*, 233–36.
44. Spinowitz, B.; Simpson, M.; Mann, P.; Charytan, C. *Trans. Am. Soc. Artif. Intern. Organs* **1981**, *27*, 161–67.
45. Cestero, R., personal communication.
46. Belzer, R.; Kountz, S.; Perkins, H. *Transplantation* **1971**, *11*, 422–24.
47. Sharon, R. *Transfusion* **1981**, *19*, 74–75.
48. McLeish, W.; Brathwarte, A.; Peterson, P. *Transfusion* **1975**, *15*, 43–45.
49. Crosson, J.; Moulds, J.; Comty, C.; Polesky, H. *Kidney Int.* **1976**, *10*, 463–70.
50. Gorst, D.; Riches, R.; Renton, P. *J. Clin. Pathol.* **1977**, *30*, 956–59.
51. Kaehny, W.; Miller, G.; White, W. *Kidney Int.* **1977**, *12*, 59–65.
52. White, W.; Miller, G.; Kaehny, W. *Transfusion* **1977**, *17*, 443–47.
53. Harrison, P.; Jansson, K.; Kronenberg, H.; Mahony, J.; Tiller, D. *N.Z. Med. J.* **1975**, *5*, 195–97.
54. Lynen, R.; Rothe, M.; Gallasch, E. *Vox. Sang.* **1983**, *44*, 81–89.
55. Perrault, R. *Vox. Sang.* **1973**, *24*, 134–49.

56. Bowman, H.; Marsh, W.; Schumacher, H.; Oyen, R.; Reihart, J. *Am. J. Clin. Pathol.* **1974**, *61*, 465–72.
57. Metaxas-Buhler, M.; Ikin, E.; Romanski, J. *Vox. Sang.* **1961**, *6*, 574–82.
58. Landsteiner, K.; Levine, P. *Proc. Soc. Exp. Biol. Med.* **1927**, *24*, 600–602.
59. Walker, M.; Rubinstein, P.; Allen, F. *Vox. Sang.* **1977**, *32*, 111–20.
60. Dahr, W.; Moulds, J. *Immunol. Commun.* **1981**, *10*, 173–83.
61. Bird, G. W.; Wingham, J. *Lancet* **1977**, *June 4*, 1218.
62. Hinz, C.; Boyer, J. *N. Engl. J. Med.* **1963**, *269*, 1329–35.
63. Hakim, R.; Fearnon, D.; Lazarus, J. *Kidney Int.* **1984**, *26*, 194–200.
64. Hakim, R.; Breillatt, J.; Lazarus, J.; Port, F. *N. Engl. J. Med.* **1984**, *311*, 878–82.
65. Charytan, C.; Mance, P.; Spinowitz, B.; Simpson, M. *Abstr. Am. Soc. Artif. Intern. Organs* **1981**, *10*, 40.
66. Spear, R. *New Physician* **1982**, *6*, 17.
67. Suskov, I.; Sazonova, I. *Mutat. Res.* **1982**, *104*, 137–40.
68. Goldmacher, V.; Thilly, W. *Mutat. Res.* **1983**, *116*, 417–22.
69. "Report of the Sensitization and Irritation Panel on Formaldehydes: Deliberations of the Consensus Workshop on Formaldehyde," Little Rock, Ark., October 1983, pp. 71–89.
70. Swenberg, J. A.; Kerns, W.; Mitchell, R. *Cancer Res.* **1980**, *40*, 3398–402.
71. Battelle Columbus Laboratory "Final report on a Chronic Inhalation Toxicology Study in Rats and Mice Exposed to Formaldehyde"; Chemical Industry Institute of Toxicology: Raleigh, N.C., 1981; Vol. 1–4.
72. Boreiko, C. J.; Regan, D. L. In "Formaldehyde Toxicity"; Gibson, J., Ed.; Hemisphere: Washington, D.C., 1983; pp. 63–71.
73. Frazelle, J. H.; Abernethy, D.; Boreiko, C. *Cancer Res.* **1983**, *43*, 3236–39.
74. McLeish, W.; Braithwaite, A.; Peterson, P. *Transfusion* **1975**, *15*, 43–45.
75. Melis, C.; Battaglini, P. *Rev. Fr. de Transfus. Immuno–Hematol.* **1978**, *4*, 953–63.

RECEIVED for review December 10, 1984. ACCEPTED March 1, 1985.

Formaldehyde and Cancer
An Epidemiologic Perspective

MAUREEN T. O'BERG

E. I. Du Pont De Nemours & Company, Wilmington, DE 19898

A review of epidemiologic findings has thus far failed to demonstrate a causal relationship between formaldehyde and cancer. Excesses of brain cancer and leukemia have occurred, mainly among anatomists and pathologists; however, diagnostic and social class bias may explain these excesses. Industrial groups have shown deficits of these cancer types. The lack of a causal relationship must be viewed cautiously, as most of the studies have low power to detect increased risks 20 or more years after first exposure.

EPIDEMIOLOGY IS THE STUDY OF THE OCCURRENCE OF DISEASE; epidemiologists look statistically at patterns of what specific diseases occur, in what groups of people, where, and when, to identify risk factors.

This review is directed at the issue of how epidemiologic research has contributed to an understanding of whether formaldehyde is a human carcinogen. The task is made simpler in that several people have looked carefully at this subject within the last few years; thus, a literature base has been established, and much careful thought and discussion can be referenced.

One major convocation on the topic was held in October 1983. At that time, a panel of epidemiologists convened in Little Rock, Arkansas, as one part of the larger "Consensus Workshop on Formaldehyde." This panel was charged with the responsibility to review the current epidemiologic research and then specifically address the question, "What evidence exists concerning the relationship between formaldehyde exposure and illness?" Furthermore, because public attention has focused on the issue of cancer, the panel spent almost all of its time considering this disease. The available literature further restricted this review primarily to mortality studies of adult working males.

The panel consisted of eight individuals from various organizations in government, academia, and private consulting, with a potpourri of perspectives. Some panel members were very familiar with the scientific literature on formaldehyde; others were not. Also, it soon became apparent

0065-2393/85/0210/0289$06.00/0

that some important studies were in progress, with late-breaking preliminary findings. These studies, previously not publicly released, were presented from the floor at the workshop.

After an update of the current research, the panel made an attempt to
pool results from the various studies and reach consensus on their overall
meaning. The highlights of their conclusions about cancer are as follows:

> The data are sparse and conflicting and do not yet provide persuasive evi-
> dence of a causal relation between exposure to formaldehyde and cancer
> in people. As far as nasal cancer is concerned, the evidence is against a
> substantial (e.g., tenfold) immediate increase in risk, but sufficient infor-
> mation is not yet available to exclude such an effect if risk starts to in-
> crease 20 or more years after first exposure. An increase in risk of brain
> cancer and leukemia is noted among each of three professional groups
> who preserve human tissues with solutions containing formaldehyde and
> other chemicals.
>
> In view of the small numbers of person-years of follow-up in subjects
> followed for 20 years or more and various methodological limitations of
> the studies, it is not possible from the available epidemiological data to
> exclude the possibility that formaldehyde is a human carcinogen (1a).

Nasal cancer was of special interest to this panel because the nose was
the target organ in the animal studies. No cases of nasal cancer were seen in
the dozen or so epidemiologic studies the panel reviewed. A total of approximately three cases would have been expected across these studies. Although the absence of nasal cancer is encouraging, the findings must be
viewed cautiously because very few of the studies have traced persons for
more than 20 years.

The panel was aware of some case reports of nasal cancer in persons
with potential exposure to formaldehyde. However, these case reports were
dismissed as not informative on the issue of cause and effect.

For brain cancer and leukemia, significant excesses were seen in several professional groups having potential formaldehyde exposures (1b–7).
Industrial groups did not experience these excesses. In fact, chemical workers had a marked deficit of brain cancer. A major study of industrial workers from Great Britain revealed 5 brain cancer deaths, with more than 12
expected (8). The Du Pont case control study also showed decreased risks
on the basis of smaller numbers (9).

The excess of brain cancer and leukemia in professionals represents a
new finding, seen mainly in three large studies, two of which were then
unpublished.

The first is a mortality study of anatomists, conducted by Stroup at the
Centers for Disease Control (CDC) (5). Stroup presented the preliminary
unpublished results of her study at the workshop. This study included 2239
male anatomists enrolled in the American Anatomists Society for at least 1
year between 1889 and 1969. These men were all physicians or had PhDs

and thus represented a high social class. They were traced during the years 1925–79 to determine whether they were dead or alive. As of the end of 1979, approximately 35% had died. These 738 deaths compare with 1130.3 expected on the basis of U.S. general population rates for the years 1925–75. The ratio of observed to expected deaths reveals approximately a 30% deficit. A deficit of this order is commonly seen when workers are compared to the general population and is referred to as the "healthy worker effect."

Among the deaths, 10 were due to brain cancer, and 3.7 were expected. This result is a statistically significant finding of nearly a threefold increase. Leukemia deaths numbered 10, and 6.7 were expected. This 50% excess is not statistically significant.

The two other major studies were proportionate mortality studies of embalmers done by Walrath at the National Cancer Institute (NCI) (6–7). Walrath's results from the New York study had been published, but the California data were, again, presented publicly for the first time at this meeting.

In each study, the names of deceased white male embalmers were obtained from the State Bureau of Licenses, Division of Embalmers, for the years 1925–80. In New York, 1132 deaths had occurred; in California, 1007 deaths had occurred.

The number of deaths from brain cancer and leukemia are shown in Table I for each group. Also shown are the numbers expected on the basis of the proportions seen in the U.S. population for the years 1925–75.

Nine deaths from brain cancer occurred in each group; 5.8 were expected in the New York study and 4.7 were expected in the California study. These numbers represent a 60–90% increase, as shown in the ratio of observed to expected deaths.

Each study had 12 deaths from leukemia; 8.5 were expected in New York, and 6.9 were expected in California. These numbers, while not statistically significant, do represent a consistent 50% excess.

In reviewing these results, one must recognize that proportionate analyses have limitations. In particular, in these studies one cannot deter-

Table I. Deaths from Brain Cancer and Leukemia of White Male Embalmers in New York and California, 1925–80

Cause of Death	New York			California		
	Observed	Expected[a]	Ratio[b]	Observed	Expected[a]	Ratio[b]
Brain cancer	9	5.8	1.6	9	4.7	1.9
Leukemia	12	8.5	1.4	12	6.9	1.7

[a]Expected deaths are based on the proportions seen in the U.S. population for the years 1925–75.
[b]Ratio of observed to expected deaths.

mine whether a specific excess is indicative of a real increase in risk, or is instead the result of deficits in other causes of death. Walrath acknowledges these limitations, but the lack of a readily available and suitable database forced her to use proportionate analyses.

Tables II and III summarize the findings from the three studies just reviewed, plus a few other smaller studies of embalmers and pathologists. The results of the studies shown in Tables II and III were pooled in a crude fashion by adding the columns of observed and expected numbers. On the basis of the total observed versus expected comparison, the panel made the statement that excesses of brain cancer and leukemia were seen in professional groups.

A few important concepts may explain why such excesses occurred in these studies of professionals.

The first issue is diagnostic bias. Evidence of diagnostic bias can be seen in a 1981 publication by Greenwald who suggested that a "diagnostic sensitivity bias" exists for brain cancer (10). He stated that an apparent initial excess of diagnosed brain tumors may have resulted from a diagnostic sensitivity bias arising from the more complete medical evaluation of Kodak employees, as compared with the general population. In a follow-

Table II. Summary of Deaths from Brain Cancer in Professional Groups

Professional Group	Researcher (Ref.)	Observed	Expected
Anatomists	Stroup (5)	10	3.7
Embalmers, N.Y.	Walrath (6)	9	5.8
Embalmers, Calif.	Walrath (7)	9	4.7
Embalmers, Ontario	Levine (3)	3	2.6
Pathologists	Harrington (2)	4	1.2
Pathologists	Matanoski (4)	5	4.7

Table III. Summary of Deaths from Leukemia in Professional Groups

Professional Group	Researcher (Ref.)	Observed	Expected
Anatomists	Stroup (5)	10	6.7
Embalmers, N.Y.	Walrath (6)	12	8.5
Embalmers, Calif.	Walrath (7)	12	6.9
Embalmers, Ontario	Levine (3)	4	2.5
Pathologists	Harrington (1, 2)	2	2.6

up article in 1982, he reinforced his original idea and stated the following (*11*):

> The possibility of a diagnostic sensitivity bias remains an important consideration, particularly where the study population has liberal access to high-quality diagnostic testing, employee insurance coverage, occupational medical referrals, and the impact of neurologic and neurosurgical specialists in a university school setting.

A social class bias may also influence brain cancer and leukemia. Support can be found in published data from the Registrar General's reports from Great Britain (*12*). Shown in Table IV are the rates for men below and above 64 years of age by social class for all cancers and for brain cancer and leukemia. Social Class I comprises professional occupations; Social Class V consists of unskilled occupations. A decreasing social class gradient exists between Social Class I and Social Class V. One can see trends in ratios for both standardized mortality ratio (SMR) and proportionate mortality ratio (PMR) analyses. For brain cancer and leukemia, higher social classes show increased risks; lower social classes show decreased risks. This result is in marked contrast with cancer overall, where higher social classes have decreased risks.

Looking back to the studies in question, one notes that they primarily included anatomists and pathologists. These people are highly educated, well-paid professionals. They work in a medically related specialty; many even work in medical facilities. Their awareness of medical diagnostics and their access to high-quality medical care is certainly well above average. As a group, they have shown excesses of brain cancer and leukemia. This result is not surprising because of their greater access to medical care due to their income, their education, and their regular interface with the medical environment.

A third source of bias may result from the increased diagnosis of brain cancer over the past decade. This increase is partially due to improved diagnostic capabilities. CAT scans and other new technologic advances now make it easier to identify brain tumors that in the past may have been misdiagnosed as strokes or neurologic disorders.

Table IV. Mortality Among Males in Great Britain and Wales, 1970–72

Cause of Death	Ages 16–64, SMRs by Social Class[a]						Ages 65–74, PMRs by Social Class[a]					
	I	II	III N	III M	IV	V	I	II	III N	III M	IV	V
All cancers	75	80	91	113	116	131	96	98	99	102	100	101
Brain cancer	108	101	111	105	100	92	225	137	109	99	85	56
Leukemia	113	100	107	101	104	95	138	124	108	98	90	77

[a]I is the highest social class; V is the lowest.

Because the computerized statistical packages used by epidemiologists had rates current only through 1975, they have underestimated the numbers expected for the late 1970s. Deaths, however, continued to accrue through 1979 or 1980 in the major studies reported. For this reason, the risk of brain cancer in the studies reviewed may not really be as high as reported. Leukemia rates have also increased during the past decade, though not as dramatically, so the same bias may apply to leukemia risks.

Without more information on the studies reported, it is not possible to determine to what extent the sources of bias just described contribute to the excesses of brain cancer and leukemia seen in professionals, but they clearly could account for much of it.

And finally, even if one ignores the potential biases and accepts the findings of increased brain cancer and leukemia in professionals, it must be noted that these persons were exposed to many substances other than formaldehyde.

The issue of lung cancer has also been a topic of much discussion. Only one of the several studies discussed by the panel showed any increase in lung cancer. In the British study conducted by Acheson, the excess was seen in one (the largest) of five plants analyzed.

Furthermore, the number of lung cancer deaths was high compared with national rates, but not compared with local rates. Local rates are likely to be more representative of the plant population with regard to cigarette smoking patterns, which clearly influence lung cancer. The professional groups discussed earlier showed a significant deficit of lung cancer. This decrease is probably due to less cigarette smoking in these highly educated, medically oriented groups.

This summary has highlighted several specific areas of focus that emerged at the consensus workshop regarding cancer. The present lack of a causal relationship between formaldehyde and cancer was stated with guarded optimism because not enough is known about formaldehyde-exposed populations that have been followed for more than 20 years.

The consensus workshop in Little Rock has provided a thorough review of epidemiologic research on formaldehyde. A few months prior to the consensus workshop, another group of experts convened in July 1983 in Oxford, England, to review the relevant available epidemiologic studies on formaldehyde, as well as several other agents. Sir Richard Doll chaired this "Symposium on Interpretation of Epidemiological Evidence." The conclusions of this symposium, although not quite as current as those of the Little Rock consensus workshop, were similar. The symposium participants stated that the evidence was inadequate to permit firm conclusions, but that it suggested that formaldehyde is unlikely to have produced a quantitatively large increased risk of cancer under the conditions of exposure that have operated in the past.

Looking ahead, new results continue to be reported as additional studies are completed. One project on the horizon is a large study of approximately 30,000 workers exposed to formaldehyde at 10 plant locations in the United States. This interindustry mortality study is being conducted cooperatively by the National Cancer Institute and the Formaldehyde Institute. Because this is a large, statistically powerful study of formaldehyde workers, we can anticipate that it will provide important input to the question of whether formaldehyde causes cancer in humans.

Literature Cited

1a. Report on the Consensus Workshop on Formaldehyde *Environ. Health Perspect.* **1984**, *58*, 339.
1b. Harrington, J. M.; Shannon, H. S. *Br. Med. J.* **1974**, *2*, 329–34.
2. Harrington, J. M.; Oakes, D. *Br. J. Ind. Med.* **1984**, *41*, 188–91.
3. Levine, R. J.; Andjelkovich, D. A.; Shaw, L. K. In "Formaldehyde Toxicology, Epidemiology and Mechanisms"; Clary, J. J.; Gibson, J. E.; Waritz, R. S., Eds.; Marcel Dekker: New York, 1983; pp. 127–40.
4. Matanoski, G. M. Preliminary studies. Letter to Martonik, John, Occupational Safety and Health Administration, 1982.
5. Stroup, N., personal communication, 1983.
6. Walrath, J.; Fraumeni, J. F. *Int. J. Cancer* **1983**, *31*, 407–11.
7. Walrath, J.; Fraumeni, J. F. *Cancer Res.* **1984**, *44*, 4638–41.
8. Acheson, E. D.; Gardner, M. J.; Pannett, B.; Barnes, H. R.; Osmond, C.; Taylor, C. P. *Lancet* **1984**, 611–16.
9. Fayerweather, William E. In "Formaldehyde: Toxicity, Epidemiology and Mechanisms"; Clary, J. J.; Gibson, J. E.; Waritz, R. S., Eds.; Marcel Dekker: New York, 1983; 47–121.
10. Greenwald, Peter; Greenwald, P.; Friedlander, B. R.; Lawrence, C. E.; Hearne, T.; Earle, K. *J. Occup. Med.* **1981**, *23*, 690–94.
11. Greenwald, Peter; Greenwald, P.; Friedlander, B. R.; Lawrence, C. E.; Hearne, T.; Earle, K. *J. Occup. Med.* **1982**, *24*, 428–32.
12. "Registrar General's Decennial Supplement for England and Wales, 1970–72: Occupational Mortality"; Her Majesty's Stationery Office: London, Series DS No. 1.

RECEIVED for review September 28, 1984. ACCEPTED February 21, 1985.

RISK ASSESSMENT

Estimating Human Cancer Risk from Formaldehyde: Critical Issues

THOMAS B. STARR, JAMES E. GIBSON, CRAIG S. BARROW,
CRAIG J. BOREIKO, HENRY d'A. HECK, RICHARD J. LEVINE,
KEVIN T. MORGAN, and JAMES A. SWENBERG

Chemical Industry Institute of Toxicology, Research Triangle Park, NC 27709

Evidence that the chronic inhalation of formaldehyde (HCHO) gas induces nasal cancer in rats has provoked widespread concern that this ubiquitous chemical may pose a significant human health hazard. However, critical issues of mechanism must be considered to accurately assess the human cancer risk from HCHO exposure. Important factors include the effects of the sensory irritation response and the mucociliary clearance mechanism on delivery of HCHO to target tissues, the disposition of delivered HCHO in target cells via metabolism and macromolecular binding, and the cellular proliferative response to cytotoxicity. These issues are directly relevant to both the low-dose and interspecies extrapolation problems. Incorporation of available mechanistic data into dose–response models can provide a rational alternative to "worst case" estimates of risk.

THE ESTIMATED 1983 U.S. PRODUCTION OF FORMALDEHYDE as a 37% solution was 5.7 billion lb (*1*). Formaldehyde's major end uses include adhesives (60%) and plastics (15%), and the main derivatives are urea–formaldehyde resins, phenol–formaldehyde resins, polyacetal, and butanediol. Formaldehyde-derived resins are used primarily in manufacturing particle board, plywood, insulation, appliances, and automobiles. Offgassing of formaldehyde from some of these products has raised concerns about possible health effects of formaldehyde in humans.

In 1980 the Chemical Industry Institute of Toxicology (CIIT) issued preliminary results from a chronic toxicity and carcinogenicity study of inhaled formaldehyde in rats and mice. These preliminary results, which pertained to 18 months of formaldehyde inhalation exposure, demonstrated that formaldehyde was carcinogenic for rats (*2*). A complete report of this study was published in 1983 (*3*) and confirmed the preliminary result that inhalation exposure to formaldehyde concentrations of 5.6 or 14.3 ppm, 6 h/day, 5 days/week for 24 months caused squamous cell carcinomas in the nasal cavities of approximately 50% of the rats in the high-exposure group and 1% of the rats exposed to 5.6 ppm. Mice also proved susceptible to formaldehyde's carcinogenicity, however, only at the highest exposure concentration and with an incidence of 1%.

This chapter reviews the design, conduct, and findings of the chronic inhalation bioassay as well as additional mechanistic research that is immediately applicable to assessing the risk of formaldehyde exposure to humans. Throughout, attention has been focused on the critical issues that must be resolved if the estimation of human cancer risk from formaldehyde exposure is to be placed on a sound, scientifically defensible footing.

Review of the CIIT Chronic Formaldehyde (HCHO) Inhalation Bioassay

CIIT commissioned the Battelle Memorial Institute to undertake a 24-month toxicity and carcinogenicity study of inhaled formaldehyde in male and female B6C3F1 mice and Fischer-344 rats. One hundred twenty animals of each sex and species were started on inhalation exposure to formaldehyde at target concentrations of either 0, 2, 6, or 15 ppm, 6 h/day, 5 days/week. The mean formaldehyde concentrations in the test chambers over the 24-month exposure period were as follows: 2.0 ± 0.6, 5.6 ± 1.2, and 14.3 ± 2.8 ppm. Interim necropsies of randomly selected animals were completed at 6, 12, 18, and 24 months after beginning exposure. Some female mice and male and female rats were followed for an additional 3 or 6 months after the completion of the planned 24-month exposures.

Throughout the study clinical signs of toxic effects, body weight, and mortality were noted at regular intervals. Before necropsy, blood samples were collected for hematology and clinical chemical analyses, and urine specimens were collected for urinalysis. At necropsy animals were examined for gross pathological changes, and tissues were collected, preselected organs were weighed, and all collected tissues were examined for histopathological change (controls and high dose, target organs at all doses).

The major toxicological finding from this study of inhaled formaldehyde was the induction of squamous cell carcinomas in the nasal cavities of 2 male mice in the 15-ppm group, 2 rats in the 6-ppm group, and 103 rats in the 15-ppm group (Table I). Two nasal carcinomas, one carcinosarcoma, one undifferentiated carcinoma, and one undifferentiated sarcoma were also observed in rats in the 15-ppm exposure group (Table I). An exposure-related induction of squamous metaplasia also occurred in the respiratory epithelium of the anterior nasal passages of rats in all formaldehyde-exposed groups. In mice, however, irritant-induced effects were essentially limited to the group exposed to 15 ppm, and no effect was observed at lower concentrations. In those animals allowed to recover after 24 months of formaldehyde exposure, an apparent regression of metaplasia

Table I. Summary of Neoplastic Lesions in the Nasal Cavity of Fischer-344 Rats and B6C3F1 Mice Exposed to Formaldehyde Gas

Diagnosis	0 ppm				2.0 ppm				5.6 ppm				14.3 ppm			
	Mouse		Rat		Mouse		Rat		Mouse		Rat		Mouse		Rat	
	M	F	M	F	M	F	M	F	M	F	M	F	M	F	M	F
Number of nasal cavities evaluated	109	114	118	114	100	114	118	118	106	112	119	116	106	109	117	115
Squamous cell carcinoma	0	0	0	0	0	0	0	0	0	0	1	1	2	0	51	52
Nasal carcinoma	0	0	0	0	0	0	0	0	0	0	0	0	0	0	1[a]	1
Undifferentiated carcinoma or sarcoma	0	0	0	0	0	0	0	0	0	0	0	0	0	0	2[a]	0
Carcinosarcoma	0	0	0	0	0	0	0	0	0	0	0	0	0	0	1	0
Polypoid adenoma	0	0	1	0	0	0	4	4	0	0	6	0	0	0	4	1
Osteochondroma	0	0	1	0	0	0	0	0	0	0	0	0	0	0	0	0

[a] One rat in this group also had a squamous cell carcinoma.
Source: Reproduced with permission from Ref. 32. Copyright 1983 IRL Press.

occurred in all affected sites of the nasal cavity in animals from the low- and intermediate-exposure groups.

Survival of rats during the course of the study was adversely affected in the 15-ppm exposure group for both sexes, whereas mouse survival was not statistically affected at any concentration. Careful examination of the eyes and simple measures of neurofunction did not reveal any formaldehyde-induced changes. Although not statistically significant, mice and rats of both sexes from the highest exposure group showed a diminution in body weights relative to controls. A similar trend was noticed in rats of both sexes exposed to 6 ppm of formaldehyde.

In rats formaldehyde caused a yellow discoloration of the coat in a concentration-dependent manner. Dyspnea was observed in rats exposed to the highest concentration. However, no exposure-related effects of formaldehyde were detected in any parameter of clinical chemistry, hematology, or urinalysis. Additionally, only sporadic changes were noted in absolute or relative organ weights, and no clear association of these changes with formaldehyde exposure was observed.

A number of polypoid adenomas were also observed in the nasal cavities of treated and control rats (Table I and Ref. 3). Although the incidence of these benign lesions in treated animals was not significantly elevated over that of controls (adjusted pairwise analysis), an adjusted trend test indicated that formaldehyde exposure increased the incidence of this lesion. The incidence of adenomas did not, however, increase as a function of formaldehyde dose. Adenomas were found in 3.4, 2.6, and 2.2% of the animals exposed to 2.0, 5.6, and 14.3 ppm of formaldehyde, respectively. This result compares to an incidence of 0.4% (one case) in control animals.

In an independent review of these findings (4), a consensus was reached among the participating pathologists that the polypoid adenomas were indeed benign. Moreover, these reviewers concluded that "there was no morphological evidence that these lesions progressed to squamous cell carcinomas." Thus, squamous cell carcinomas and polypoid adenomas were thought to be readily separable lesions that should not be combined solely for statistical purposes.

Although Takano et al. (5) have also found that papillomas of the Fischer-344 rat nasal cavity induced by 1,4-dinitrosopiperazine did not progress to adenocarcinomas, little additional information specific to tumors of the rodent nasal cavity is available in regard to relationships between benign lesions and possible malignant counterparts. Observations in other rodent tissues must therefore be considered in assessing the significance of these benign lesions.

The relationship between benign papillomas and squamous cell carcinomas has been extensively studied in mouse-skin model systems of the multistage carcinogenic process. In these systems, the application of complete carcinogens or initiating agents and tumor promoters will produce squamous cell carcinomas. The formation of carcinomas is usually preceded by the appearance of numerous benign papillomas. Some evidence suggests that papillomas are preneoplastic lesions that can progress to squamous cell carcinomas (6–8). Other experiments suggest that papillomas are terminal lesions with little, if any, capacity to progress to malignancy (9).

Whatever the actual mechanistic relationship between these two skin lesions, one-to-one correspondence does not exist between benign and malignant skin tumors (6–9). Experimental carcinogenesis protocols produce far more papillomas than carcinomas. Estimates of the quantitative relationship between papillomas and carcinomas vary widely as a function of the carcinogen studied and the treatment protocol employed. Typically, one carcinoma will arise for every 20–100 papillomas induced (6–9). Larger carcinoma-to-papilloma ratios have been suggested under certain experimental conditions (7, 8). Quantitative studies of "preneoplastic" enzyme-altered foci and hyperplastic nodules in rat liver similarly suggest that these lesions far outnumber subsequent malignancies (10).

Two interpretations of the polypoid adenomas in the CIIT study are thus possible. First, the polypoid adenomas may represent neoplastic lesions with a finite potential to progress to malignancy. In support of this view is the observation that several nonsquamous carcinomas were observed in rats exposed to 14.3 ppm of formaldehyde (3). However, conversion of polypoid adenomas to carcinomas would probably occur with low frequency. Alternatively, the polypoid adenoma lesions may be terminal in nature, that is, they may be by-products of formaldehyde's strong irritant or weak genotoxic properties and not directly involved in a multistage carcinogenic process.

Whichever interpretation is correct, polypoid adenomas should not be equated on a one-to-one basis with malignant tumors of any type for purposes of quantitative risk assessment. Nor is such an assumption needed because unequivocal evidence of carcinogenicity that is well suited for risk assessment is provided by the incidence of squamous cell carcinomas in rats exposed to 5.6 and 14.3 ppm of formaldehyde. The increase in polypoid adenomas in formaldehyde-treated rats should, however, be recognized as evidence that 2 ppm of formaldehyde interacts with rat nasal tissue.

Effects of Responses to Sensory Irritation on Inhaled HCHO Dose

The membranes of the respiratory tract contain a wide variety of sensory nerve endings that are capable of responding to chemical and/or physical stimuli. One group of nerve endings in the nasal mucosa is associated with the maxillary and ophthalmic divisions of the trigeminal nerve. Stimulation of these by airborne chemical irritants such as formaldehyde results in a painful burning sensation, a desire to withdraw from the contaminated atmosphere, and a decrease in respiratory rate (11). This response has been termed the "common chemical sense" to separate it from more specialized chemical senses such as olfaction and gustation (*12*).

The term "sensory irritation" is synonymous with the common chemical sense. It has been used to describe one effect of chemicals that, when inhaled via the nose, stimulate the trigeminal nerve endings, evoke a burning sensation of the nasal passages, and inhibit respiration (*11*). All substances that excite the common chemical sense are potentially noxious and lung damaging. The reflex responses to sensory irritation comprise an important respiratory tract defense mechanism that serves to minimize inhalation of a noxious agent and to warn of its presence through the perception of pain (*11*, *13*, *14*).

Measurement of the decrease in respiratory rate during inhalation exposure to a chemical can be used as a quantitative measure of the sensory irritation property of chemicals (*11*). Such measurement is well documented for formaldehyde in laboratory animals (*15–17*), and the results indicate that mice are far more sensitive than rats. For example, the HCHO concentration required to elicit a 50% decrease in respiratory rate (RD_{50}) is 3.13 ppm in Swiss–Webster mice (*15*), 4.9 ppm in B6C3F1 mice (*17*), and 31.7 ppm in Fischer-344 rats (*17*). It has also been shown in both rats and mice that tidal volume does not compensate entirely for the decreased respiratory rate (*17*). As a result, the minute volumes (the product of respiratory rate and tidal volume) of these species decrease during exposure to sufficiently high concentrations of formaldehyde.

Significant differences between rats and mice in the function of this defense mechanism have been observed when animals are pretreated to 0, 2, 6, or 15 ppm of HCHO (Table II). For example, although a concentra-

Table II. RD_{50} and Effect of HCHO on Minute Volume ($\dot{V}_E$) of Naive and HCHO-Pretreated B6C3F1 Mice and F-344 Rats

Pretreatment HCHO Concentration[a] (ppm)	B6C3F1 Mice		F-344 Rats	
	$RD_{50}{}^{b}$ (ppm)	% Decrease in $\dot{V}_E$ at RD_{50}	$RD_{50}{}^{b}$ (ppm)	% Decrease in $\dot{V}_E$ at RD_{50}
Naive	4.9 (3.9–6.4)	46.8	31.7 (23.1–54.0)	45.1
2	5.9 (4.4–8.5)	51.2	29.5 (20.8–55.8)	50.1
6	2.2 (1.5–2.9)	44.4	28.6 (18.8–61.2)	42.4
15	3.6 (2.4–5.5)	54.1	22.7 (15.9–38.2)	37.5

NOTE: Values within parentheses are 95% confidence intervals for RD_{50}.
[a]Pretreatment lasted for 6 h/day for 4 days.
[b]Concentration of HCHO at which respiratory rate decreased by 50%.
Source: Adapted from Ref. 17.

tion-dependent depression of respiration in both species occurred, the amplitude of the response was always greater in mice. Consequently, the concentration–response curves for mice were shifted, relative to those for rats, to lower concentration ranges; RD_{50} values ranged from 2 to 6 ppm in the mice as compared to 23 to 32 ppm in rats. Minute volumes also decreased by approximately 50%. However, rats exposed to 15 ppm did have some degree of tidal volume compensation as evidenced by only a 37.5% decrease in minute volume following pretreatment to 15 ppm of HCHO (Table II). These results suggest that the B6C3F1 mouse respiratory tract is better protected against inhaled HCHO than is the respiratory tract of the Fischer-344 rat.

This hypothesis of differential protective capability in these two species could be tested because HCHO toxicity has been shown to be limited to the nasal cavity (2). The localization of toxicity is due in large part to the high water solubility of HCHO and the fact that rodents are obligatory nose breathers. By assuming that all inhaled HCHO is deposited in the nasal cavity, a theoretical delivered dose was derived by dividing the amount of HCHO inhaled per unit time by the surface area of the nasal cavity area:

$$\text{delivered dose} = \frac{\text{HCHO concentration } (\mu g/L) \times \text{minute volume } (L/min)}{\text{nasal cavity surface area } (cm^2)}$$

Nasal cavity surface area has been reported to be 13.44 cm^2 for 288-g rats and 2.89 cm^2 for 30-g mice (18). If the delivered dose is expressed as $\mu g/min/cm^2$ (Table III) for a 10-min exposure to 15 ppm of HCHO, the dose for rats would be twofold larger than that expected for mice (19). Although the disparity between the species in this calculated dose was consistent with a similar disparity in nasal tumor incidence following chronic exposure (3), additional studies were required to establish whether or not this difference in delivered dose was maintained for longer periods of exposure.

The persistence of this effect was evaluated in rats and mice pretreated to 6 or 15 ppm of HCHO, 6 h/day for 4 days. On the fifth day, minute volume was measured during a 6-h exposure to 6 or 15 ppm (20). During

Table III. Dose of HCHO to Nasal Mucosa During 15 ppm of HCHO Exposure

Description of Dose	Rat	Mouse
HCHO conc. ($\mu g/L$)	18.4	18.4
Minute volume (L/min)	0.114	0.012
Nasal cavity surface area (cm^2)	13.44	2.89
Dose ($\mu g/min/cm^2$)	0.156	0.076

NOTE: Rats and mice were pretreated to 15 ppm of HCHO, 6 h/day, for 4 days.

Source: Reproduced with permission from Ref. 19. Copyright 1983 Hemisphere Publishing Corp.

exposure to 15 ppm of formaldehyde, the time-weighted average dose for mice continued to be approximately one-half of that for rats (Figure 1A). At 6 ppm both species were predicted to receive similar delivered doses.

The species difference predicted at 15 ppm was further assessed by comparative autoradiography, histopathology, and cell turnover studies (20). Whole body autoradiographic (WBAR) studies of formaldehyde deposition patterns in rats and mice exposed to [14C]formaldehyde revealed that radioactivity was heavily deposited in the anterior nasal cavity, and much less was deposited in olfactory regions. This anterior–posterior gradient is consistent with the high water solubility and chemical reactivity of formaldehyde. In addition, WBAR confirmed qualitatively the difference in delivered dose in rats as compared with mice exposed to 15 ppm of formaldehyde (20). Histopathologic examination of the nasal cavities of rats and mice after 1 or 5 days of exposure to 15 ppm of HCHO demonstrated that rats had more severe lesions than mice. Cell turnover studies of nasal respiratory epithelium also revealed much higher cell proliferation in rats compared with that observed in mice (20).

Therefore, potential differences in pulmonary ventilation and nasal cavity volume to surface area relationships must be considered when comparing the responses of different species to the same concentration of airborne irritants. Furthermore, potential differences in pulmonary ventilation should be considered when comparing the responses of a single species to different concentrations of an airborne irritant. For example, rats exposed to 15 ppm of formaldehyde for 6 h inhale only twice the amount of formaldehyde per unit time as do rats similarly exposed to 6 ppm (Figure 1A). This nonlinear relationship between inspired formaldehyde and its ambient air concentration is due to the larger depression of minute volume induced in rats by exposure to 15 ppm of formaldehyde relative to that induced by exposure to 6 ppm (Figure 1B). Thus, the precipitously steep rise in squamous cell carcinoma incidence from 1% among rats chronically exposed to 5.6 ppm of formaldehyde to nearly 50% among rats similarly exposed to 14.3 ppm becomes steeper yet, that is, more severely nonlinear, when the amount of formaldehyde inhaled per unit time, rather than the ambient air concentration, is used as the measure of exposure. The respiratory tract reflexes described in this section may therefore play an important role in protecting the respiratory tract from the toxic effects of inhaled HCHO.

The Mucociliary Apparatus and HCHO-Induced Nasal Toxicity

Many factors may influence the distribution of lesions induced by irritant materials in the nasal passages (21). These include species-specific anatomy and physiology, nasal aerodynamics, mucociliary flow rate and direction, as well as exposure level and tissue-specific susceptibility. The amount of formaldehyde that reaches the nasal epithelium is dependent upon the air-

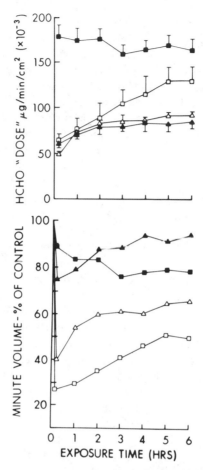

Figure 1. A: Time–response curves for minute volume from rats and mice exposed to 6 ppm (▲, rats; △, mice) or 15 ppm (■, rats; □, mice) of formaldehyde for 6 h. B: Time-weighted averages of the theoretical formaldehyde dose available for deposition on the nasal passages of rats and mice during a 6-h exposure to 6 or 15 ppm. (Reproduced with permission from Ref. 32. Copyright 1983 IRL Press.)

borne concentration and the rate at which air passes through the nose. Furthermore, airflow patterns within the nose determine, at least to some extent, the amount of gas reaching specific areas, whereas the nature and movement of the surface secretions are likely to affect absorption of formaldehyde and its subsequent fate.

Chronic exposure to high concentrations of formaldehyde induced squamous cell carcinomas in the nasal passages of rats, and exposure was also weakly associated with an increased incidence of polypoid adenomas (3). Detailed mapping of the exact locations of these neoplasms was not reported. The histologic sections from the chronic formaldehyde inhalation study were therefore reexamined to determine the locations and apparent site of origin of observed tumors. The majority of the squamous cell carcinomas occurred in two main locations. The first region was lateral to the nasoturbinate, extending from the ventral margin of this turbinate to the lateral wall just dorsal to the maxilloturbinate, at a level measured along the long axis of the nose which is just posterior to the incisor tooth. The second was composed of the ventral and middle nasal septum in a region approximately at the level of the incisive papilla. In contrast, polypoid adenomas occurred just posterior to the vestibule, on the naso- and maxilloturbinates and the adjacent lateral wall (22). All these regions of the nose are lined by respiratory epithelium, which is protected by the nasal mucociliary apparatus (23). Because of the potential importance of the mucociliary apparatus in modulating the delivery of formaldehyde to postulated target cells in the nasal passages, studies of the effects of formaldehyde on the mucociliary apparatus have been undertaken. The current state of knowledge regarding this system and the results of some initial studies are summarized in this section.

The nasal mucociliary apparatus of the rat provides a continuous layer of watery mucus covering the respiratory epithelium (24, 25) in which formaldehyde would be expected to dissolve readily, and thus would be removed from the inspired airstream. Researchers have demonstrated in dogs that almost 100 % of inspired formaldehyde is removed in the upper airways (26), but the retention efficiency of the rat nose for formaldehyde has yet to be established. The approximate thickness of the mucus (25) and its flow patterns (25, 27) and flow rate (25) have been determined in the rat nose. The layer of nasal mucus flows continuously over the surface of the nasal mucosa, and it is cleared eventually toward the nasopharynx (25, 27) to be finally swallowed with any entrapped or dissolved materials. Recent studies have demonstrated that the nasal and tracheal mucociliary clearance mechanisms in the guinea pig can respond to formaldehyde exposure with an increased rate of clearance (28). If mucus clearance does result in removal of formaldehyde from the nose, then increased clearance rates in response to exposure would be expected to increase the efficiency of this potentially protective mechanism.

However, formaldehyde is also known to have inhibitory effects upon mucociliary function. The mucociliary apparatus consists of several main components that have been reviewed in detail by Proctor (23). Each component may be influenced adversely by exposure to formaldehyde. Cilia, which are microscopic, hairlike processes of the epithelial cells, drive the mucus over the surface by their coordinated beating. Formaldehyde has been found to be ciliastatic in several species (cf 24), and it causes slowing of mucus flow in the anterior nasal passages of humans during inhalation exposure (29). Studies with a frog palate preparation indicated that formaldehyde induced slowing of mucus flow before it inhibited ciliary activity, probably as a result of reactions with components of the mucus blanket (*30*). A similar mechanism may be responsible for the slowing of mucus flow in humans.

At high concentrations formaldehyde has been found to induce both mucostasis and ciliastasis in rats following in vivo inhalation exposure (*24*). Studies with rats using a rapid postmortem assessment of nasal mucociliary function following inhalation exposures to formaldehyde revealed a clear concentration–response relationship for the inhibition of nasal mucociliary function, although 0.5 ppm was a no-effect concentration (*31*). This concentration–response relationship paralleled closely the concentration–response relationship for formaldehyde-induced lesions observed in the chronic inhalation study. It was thus postulated that localized disruption of mucociliary function could account, at least in part, for the subsequent appearance of epithelial lesions in the affected locations (*32*).

The distribution of areas of inhibition of mucociliary function and acute cytotoxicity in the nasal epithelium (*24*) involved the regions in which squamous cell carcinomas occurred. However, acute changes also consistently appeared on the medial aspect of the maxilloturbinate at 15 and 6 ppm (*24, 31*). The maxilloturbinate was rarely a site of squamous cell carcinoma development. This finding indicated possible regional differences in susceptibility of the rat nasal epithelium to carcinogenic effects of formaldehyde.

In the chronic inhalation study the polypoid adenomas occurred in the anterior nasal passages, a region that is lined by sparsely ciliated respiratory epithelium and has generally a very slow mucus flow rate (*25*). Mucus in this region is derived from mucus streams that have flowed from more dorsal or more posterior regions of the nose. Mucus flow results in translocation of materials deposited on the surface and may thus influence the final site at which the nasal epithelium is exposed to inspired materials. Formaldehyde absorbed more posteriorly might be carried forward, toward the point at which the polypoid adenomas occurred. Researchers have proposed that mucus flow patterns in the lower respiratory tract could account for the distribution of air-pollutant-induced cancer in the trachea and bronchi of humans (*33*), and similar reasoning may be applicable to the distribution of formaldehyde-induced lesions in the rat nasal cavity.

Mucus flow may play an important role in determining the distribution and frequency of neoplasia in the nasal passages of rats exposed to formaldehyde by influencing delivery of this material to the nasal epithelium in specific regions of the nose. Anatomic or physiologic characteristics of the rat nose may render this species either hypersensitive or hyposensitive with respect to formaldehyde-induced nasal cancer. Thus, comparisons between rats and humans of nasal air flow, mucus flow, and other physiologic characteristics and their effects on delivered dose may provide information that is valuable to the risk assessment process.

The Disposition of Inhaled Formaldehyde in Target and Nontarget Tissues

The disposition of inhaled HCHO in target and nontarget tissues is a very important aspect of the toxicology of HCHO. Knowledge of toxic mechanisms is an essential component of scientifically defensible assessments of risk. This section summarizes the results of investigations aimed at elucidating the mechanisms of HCHO toxicity in target tissues at the molecular level. It also addresses the question of whether or not formaldehyde can be expected to cause toxicity at distant sites.

Quantitation of the Dose of Formaldehyde Delivered to Target Macromolecules in the Rat Nasal Mucosa. Formaldehyde is known to react with DNA in cultured mammalian cells in vitro to form DNA–protein cross-links (34, 35), and this reaction may be a critical factor in the transformational (36), mutagenic (37), and carcinogenic (3) actions of HCHO. Therefore, it was of interest to determine whether inhaled HCHO could react with DNA in vivo and, if this reaction occurred, to quantify the amount of HCHO that reacted with DNA as a function of airborne concentration.

Evidence that inhaled HCHO reacts with respiratory mucosal DNA was reported recently by Casanova–Schmitz and Heck (38). In these experiments, exposure of Fischer-344 rats to HCHO at concentrations equal to or greater than 6 ppm resulted in a statistically significant decrease in the amount of DNA that could be extracted from proteins in homogenates of the respiratory mucosa. The solubilized tissue homogenates were extracted with a strongly denaturing aqueous-immiscible organic solvent mixture. When the tissue was extracted in this manner, the DNA was separated into two fractions (38). The aqueous (AQ) phase from this extraction contained DNA that by all spectrophotometric and chromatographic criteria appeared to be pure, double-stranded DNA. The interfacial (IF) layer also contained DNA, but this DNA appeared to be cross-linked to proteins because the DNA could not be released without digestion of the interface using proteinase K. Importantly, the quantity of IF DNA was dependent on the airborne HCHO concentration to which the rats had been exposed and increased as the concentrations of HCHO increased. Therefore, the conclusion was reached that HCHO does react with respiratory mucosal DNA following in vivo inhalation exposures, and that this reaction might well play a critical role in the development of nasal cancer during chronic HCHO inhalation studies.

The inability to extract DNA from proteins does not, however, constitute proof of the formation of DNA–protein cross-links. Additional evidence was required to ensure the validity of this conclusion. To obtain such evidence and to determine, if possible, the amount of HCHO that became covalently bound, experiments were undertaken to investigate the mechanisms of labeling of respiratory mucosal DNA following inhalation exposure of rats to [^{14}C]HCHO and [^{3}H]HCHO. The results of these experiments have been described in detail by Casanova–Schmitz et al. (*39*). A brief summary of the main findings will now be discussed.

The labeling of macromolecules (DNA, RNA, and protein) in the respiratory mucosa, olfactory mucosa, and bone marow was studied in rats that had been preexposed for 6 h to unlabeled HCHO on the day preceding exposure to the labeled compound. Preexposure was undertaken to stimulate cell turnover in the respiratory mucosa, a physiological response to toxic injury that appears to play an important role in the induction of nasal cancer by HCHO (*40*). The preexposure to unlabeled HCHO and the exposure to [^{14}C]HCHO and [^{3}H]HCHO were both carried out at the same airborne concentrations.

The specific activities of IF and AQ DNA obtained from the respiratory mucosa following in vivo exposure to 0.3, 2, 6, 10, or 15 ppm of [^{14}C]HCHO or [^{3}H]HCHO are shown in Figure 2. The specific activity of DNA was maximal at 6 ppm and decreased at higher concentrations. In

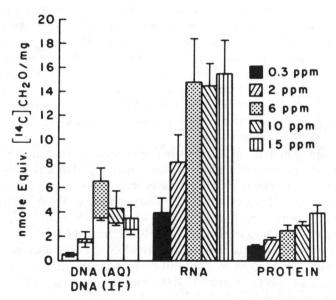

Figure 2. Concentrations of [^{14}C]HCHO equivalents in DNA, RNA, and protein of the respiratory mucosa of rats exposed for 6 h to 0.3, 2, 6, 10, or 15 ppm of [^{14}C]HCHO and [^{3}H]HCHO, 1 day after a single pre-exposure (6 h) to the same concentration of unlabeled HCHO. Bar graphs for aqueous and interfacial DNA are overlaid. Values shown are mean ± standard deviation, n = 3. (Reproduced with permission from Ref. 39. Copyright 1984 Academic Press.)

addition, the ^{14}C specific activity of the AQ DNA was found to be significantly greater than that of the IF DNA at 6 ppm, but no significant difference between the specific activities of IF and AQ DNA was found at other concentrations.

On the basis of several arguments by Casanova–Schmitz et al. (39), the major route of labeling of DNA in the respiratory mucosa was concluded to be metabolic incorporation. The maximum in the specific activity of the DNA at 6 ppm therefore implies that the incorporation of [14C]HCHO into respiratory mucosal DNA was maximal at this concentration. This result is consistent with the observation that the incorporation of [3H]thymidine into respiratory mucosal DNA after exposure to HCHO at 6 ppm was higher than after exposure to 15 ppm (41). The smaller amount of incorporation of ^{14}C into DNA that occurred at 10 and 15 ppm relative to that at 6 ppm is presumably due to the cytotoxic effects of HCHO at these high concentrations.

The finding that the ^{14}C specific activity of respiratory mucosal AQ DNA was significantly higher than that of IF DNA at 6 ppm is an extremely important one. This result implies that the AQ DNA had incorporated a significantly larger amount of ^{14}C than had the IF DNA at 6 ppm. In addition, this result implies that the two DNA fractions must have differed structurally, for otherwise they could not have been separated by solvent extraction into portions with differing specific activities. However, no difference between the specific activities of AQ and IF DNA was found at either 0.3 or 2 ppm. This finding demonstrated that the structural difference is not an inherent property of respiratory mucosal DNA but must have been induced in the DNA by exposure to HCHO at 6 ppm. A plausible explanation for this structural difference is that HCHO exposure at 6 ppm resulted in the formation of DNA–protein cross-links in the IF DNA fraction.

The specific activity of IF DNA being lower than that of AQ DNA at 6 ppm may at first seem contradictory because the IF DNA is presumed to contain covalently bound HCHO. However, this result is consistent with DNA–protein cross-linking because the major route of DNA labeling was metabolic incorporation. The formation of cross-links could possibly decrease the rate of incorporation of [14C]HCHO metabolites into DNA by preventing the dissociation of proteins from DNA that is necessary for de novo DNA synthesis to occur. An inhibition of DNA synthesis by HCHO has been shown to occur in yeast under conditions in which DNA–protein cross-links were induced (42).

The most direct evidence for the formation of covalently bound HCHO in DNA and proteins in the respiratory mucosa was provided by determining the ^{3}H-to-^{14}C ratios of respiratory mucosal macromolecules following exposure of rats to [14C]HCHO and [3H]HCHO (Figure 3). The ^{3}H-to-^{14}C ratios of the IF DNA and of the proteins increased with increas-

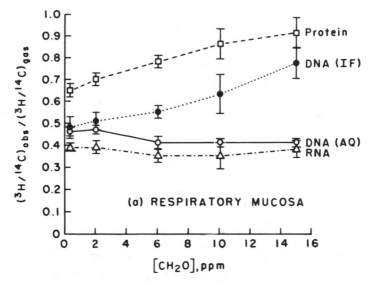

Figure 3. Normalized ^{3}H-to-^{14}C ratios (observed ratio divided by ^{3}H-to-^{14}C ratio of inhaled [^{14}C]HCHO and [^{3}H]HCHO) of AQ DNA (○), IF DNA (●), RNA (△), and protein (□) from the respiratory mucosa of rats exposed for 6 h to 0.3, 2, 6, 10, or 15 ppm of [^{14}C]HCHO and [^{3}H]HCHO, 1 day after a single pre-exposure (6 h) to the same concentration of unlabeled HCHO. Values shown are mean ± standard deviation, n = 3. (Reproduced with permission from Ref. 39. Copyright 1984 Academic Press.)

ing HCHO concentrations, but increases in the ^{3}H-to-^{14}C ratios of the AQ DNA and RNA were not observed. As discussed in detail by Casanova–Schmitz et al. (39), increased ^{3}H-to-^{14}C ratios of macromolecules with increasing HCHO concentrations are evidence of covalent binding of HCHO. The results presented in Figure 3 indicate that only the IF DNA and proteins contained measurable quantities of covalently bound HCHO.

The difference between the ^{3}H-to-^{14}C ratios of IF and AQ DNA was statistically significant at HCHO concentrations equal to or greater than 2 ppm. These differences indicate that, at these concentrations, the mechanisms of labeling of IF and AQ DNA were significantly different: AQ DNA was labeled primarily or exclusively by metabolic incorporation, whereas IF DNA was labeled by both metabolic incorporation and covalent binding. Strong support for the conclusion that the difference between the ^{3}H-to-^{14}C ratios of the two DNA fractions was due to DNA–protein cross-linking is provided by the scattergram shown in Figure 4, which relates the difference between the ^{3}H-to-^{14}C ratios of IF and AQ DNA to the percent interfacial DNA obtained in each experiment. These two variables were highly correlated. Although a significant difference between the ^{3}H-to-^{14}C ratios of IF and AQ DNA was detected at concentrations equal to or greater

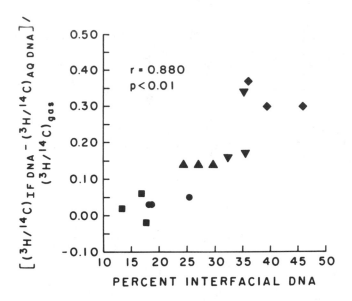

Figure 4. Plot of the difference between the normalized ^{3}H-to-^{14}C ratio of IF DNA and that of AQ DNA from the respiratory mucosa versus the percent IF DNA at selected concentrations of inhaled HCHO. Concentrations are 0.3 (■), 2 (●), 6 (▲), 10 (▼), and 15 (◇) ppm. (Reproduced with permission from Ref. 39. Copyright 1984 Academic Press.)

than 2 ppm, the percentage of interfacial DNA was significantly increased relative to controls only at concentrations equal to or greater than 6 ppm (38). This result indicates that the isotope ratio method is a more sensitive technique for the detection of DNA–protein cross-links than is the measurement of DNA extractability.

The ^{3}H-to-^{14}C ratio of a macromolecule following exposure to a toxicant labeled with both isotopes is quantitatively related to the fraction of the total ^{14}C that is due to covalent binding, f_b (39):

$$f_b = \frac{(^3H/^{14}C)_o - (^3H/^{14}C)_m}{(^3H/^{14}C)_b - (^3H/^{14}C)_m}$$

where $(^3H/^{14}C)_o$ is the observed ^{3}H-to-^{14}C ratio of macromolecule, $(^3H/^{14}C)_m$ is the ^{3}H-to-^{14}C ratio characteristic of metabolic incorporation of ^{3}H and ^{14}C (derived from [^{3}H]HCHO and [^{14}C]HCHO) into the macromolecule, and $(^3H/^{14}C)_b$ is the ^{3}H-to-^{14}C ratio characteristic of covalent binding of [^{3}H]HCHO and [^{14}C]HCHO to the macromolecule under the reaction conditions. The values of these isotope ratios can be determined by methods described by Casanova–Schmitz et al. (39). Hence, the fraction of covalently bound [^{14}C]HCHO can be calculated. Knowledge of the total ^{14}C concentration in the macromolecule permits the concentration of covalently bound HCHO to be determined. The results of calculations of the concentrations of covalently bound HCHO in respiratory mucosal DNA and proteins at 0.3, 2, 6, 10, and 15 ppm of inhaled HCHO are summarized in Figures 5A and 5B, respectively.

Figure 5A shows that the concentration–response profile for covalent binding of [^{14}C]HCHO to DNA is sigmoidal and increases gradually between 0.3 and 2 ppm, steeply between 2 and 6 ppm, and less steeply at the higher concentrations. The dashed line between 6 ppm and 0 shows the result that would be expected if covalent binding to DNA was a linear function of concentration in this concentration range. The observed concentration of covalently bound HCHO at 2 ppm was significantly lower than the value predicted by extrapolation from the concentration measured at 6 ppm. In contrast to the results obtained with respiratory mucosal DNA, covalent binding of [^{14}C]HCHO to respiratory mucosal proteins depended in an apparently linear manner on the HCHO concentration throughout the concentration range (Figure 5B).

The explanation for nonlinearity in the binding of HCHO to respiratory mucosal DNA is presently unknown. However, at least two mechanisms could explain such nonlinear behavior. First, physiological and biochemical defense mechanisms, such as mucociliary clearance, metabolism, and repair, could be inactivated or could become less efficient with increasing HCHO concentrations and result in a disproportionate increase in the concentration of DNA–protein cross-links. Second, the marked increase in cell turnover caused by HCHO exposure at 6 ppm relative to that at 2 ppm (*41*) could increase the availability of sites in the DNA for reaction with HCHO. Formaldehyde binds to single-stranded regions of DNA, but not to regions that are double-stranded (*43, 44*).

The concentration dependence of the relative disposition of HCHO in respiratory mucosal tissues is also important, as demonstrated by the data in Figure 6 which show the percentage of the total ^{14}C in respiratory mucosal DNA and proteins that was due to covalent binding. This percentage increased with concentration. If the disposition of HCHO in the respiratory mucosa was described by steady state linear kinetics, then the percentage of the total ^{14}C due to covalent binding would be constant, that is, independent of the airborne concentration.

Can Formaldehyde Cause Toxicity at Distant Sites? The possibility that inhaled HCHO might exert toxic effects in tissues remote from the site of deposition has occasionally been raised in epidemiologic studies, but no evidence for such effects has been obtained in animal studies. For HCHO to exert such effects, it would have to be carried to those sites by the circulatory system. Therefore, analyses of HCHO concentrations in the blood of rats and humans exposed to HCHO should indicate whether direct toxic effects due to HCHO inhalation are possible. The results of analyses of

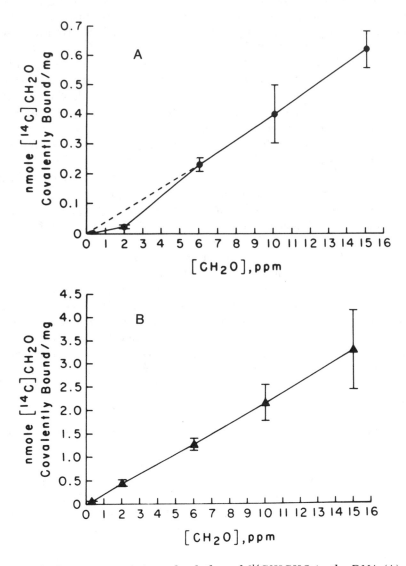

Figure 5. Concentration of covalently bound [¹⁴C]HCHO in the DNA (A) and proteins (B) from the respiratory mucosa of rats exposed for 6 h to 0.3, 2, 6, 10, or 15 ppm of [¹⁴C]HCHO and [³H]HCHO, 1 day after a single pre-exposure (6 h) to the same concentration of unlabeled HCHO. Values shown are mean ± standard error, n = 3. (Reproduced with permission from Ref. 39. Copyright 1984 Academic Press.)

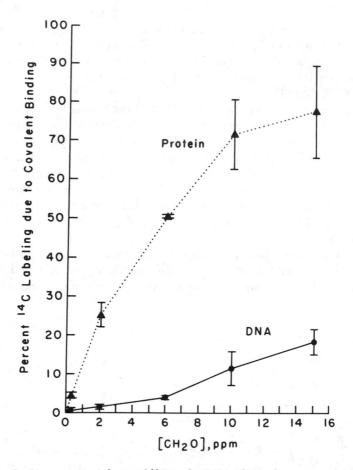

Figure 6. Percentage of the total ^{14}C in the DNA (●) and proteins (▲) from the respiratory mucosa that was due to the formation of adducts and cross-links. Rats were exposed for 6 h to 0.3, 2, 6, 10, or 15 ppm of [^{14}C]HCHO and [^{3}H]HCHO, 1 day after a single pre-exposure (6 h) to the same concentration of unlabeled HCHO. Values shown are mean ± standard error, n = 3. (Reproduced with permission from Ref. 39. Copyright 1984 Academic Press.)

HCHO in the blood of rats and humans exposed to HCHO by inhalation have recently been presented (45).

To investigate whether HCHO exposure results in an increase in the blood concentration of HCHO, the blood of eight rats exposed to 14.4 ± 2.4 ppm of HCHO for 2 h was analyzed. The rats were killed within seconds of exposure by decapitation, and the blood was collected and analyzed immediately. A similar group of eight rats unexposed to HCHO was used as controls. The analyses were performed by gas chromatography–mass spectrometry by using a stable isotope dilution technique that was developed by Heck et al. (46). The blood of the exposed group contained 2.25 ± 0.07 ppm of HCHO, and that of the control group contained 2.24 ± 0.07 ppm. Thus, no effects of HCHO exposure on the blood concentration of HCHO could be detected (one-tailed t-test, 14 degrees of freedom, $p > 0.45$) (45).

The results of blood analyses of HCHO indicating no increase in the concentration of HCHO even at a very high airborne concentration are consistent with measurements of the ^{3}H-to-^{14}C ratios of bone marrow macromolecules isolated from rats exposed to [^{14}C]HCHO and [^{3}H]HCHO. These ratios are shown in Figure 7. In contrast to the macromolecules in the respiratory mucosa (Figure 3), the ^{3}H-to-^{14}C ratios of macromolecules in the bone marrow were independent of the airborne concentration of HCHO. Therefore, no evidence of covalent binding to bone marrow macromolecules was obtained.

The blood of six human volunteers (four males, two females) exposed to 1.9 ppm of HCHO for 40 min was also analyzed. Venous blood samples were drawn shortly before and as quickly as possible after the exposure to

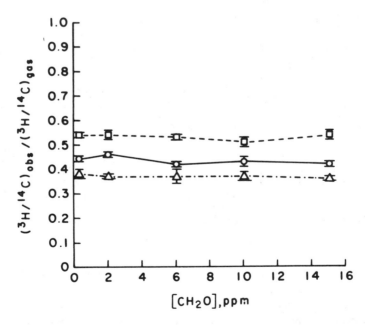

Figure 7. Normalized ^{3}H-to-^{14}C ratios (observed ratio divided by ^{3}H-to-^{14}C ratio of inhaled [^{14}C]HCHO and [^{3}H]HCHO of AQ DNA (○), RNA (△), and protein (□) from the bone marrow of rats exposed for 6 h to 0.3, 2, 6, 10, or 15 ppm of [^{14}C]HCHO and [^{3}H]HCHO, 1 day after a single pre-exposure (6 h) to the same concentration of unlabeled HCHO. Values shown are mean ± standard deviation, n = 4. (Reproduced with permission from Ref. 39. Copyright 1984 Academic Press.)

HCHO. The results of these analyses are summarized in Table IV. Analysis of variance (Table V) did not indicate a statistically significant effect of exposure on the concentration of HCHO in human blood. However, significant differences were found among the subjects with respect to their blood concentrations of HCHO, and a significant interaction of subject with exposure occurred, that is, significant differences (either an increase or a decrease) were found between the HCHO concentrations of the blood taken before and after exposure from some of the subjects. These experiments, as well as the experiments in rats, provide no evidence that HCHO exposure causes an increase in the blood concentration of HCHO. Thus, direct HCHO-induced toxicity at distant sites is very unlikely in the case of HCHO inhalation.

Table IV. Concentrations of Formaldehyde in Human Blood

Subject	Sex	Before Exposure (μg/g of Blood)	After Exposure[a] (μg/g of Blood)
1	F	3.09 ± 0.41	2.18 ± 0.09
2	F	2.56 ± 0.10	3.31 ± 0.34
3	M	2.66 ± 0.17	3.74 ± 0.13
4	M	2.61 ± 0.34	1.93 ± 0.05
5	M	2.05 ± 0.16	2.76 ± 0.21
6	M	2.73 ± 0.14	2.72 ± 0.31
Average		2.61 ± 0.14	2.77 ± 0.28

NOTE: All concentrations are mean values ± standard error; $n = 3$ for individual subjects; $n = 6$ for averages.
[a]1.90 ± 0.06 ppm, 40 min.
Source: Reproduced with permission from Ref. 45. Copyright 1985.

Table V. Analysis of Variance Table for Concentrations of Formaldehyde in Human Blood

Source of Variation	df	SS	MS	F value	Significance Level
Total	35	12.7062			
Treatment	1	0.2288	0.2288	1.3993	not sig. ($p = 0.2484$)
Subject	5	3.4835	0.6967	4.2614	sig. ($p = 0.0065$)
Subject–treatment interaction	5	5.0703	1.0141	6.2026	sig. ($p = 0.0010$)
Residual	24	3.9237	0.1635		

NOTE: df, SS, and MS denote degrees of freedom, sum of squares, and mean square, respectively.
Source: Reproduced with permission from Ref. 45. Copyright 1985.

Summary. The experiments just discussed have elucidated the kinds of reactions that HCHO can undergo in the nasal mucosa of rats during inhalation exposures. Strong evidence has been obtained that DNA–protein cross-links are formed in the respiratory mucosa at concentrations equal to or greater than 2 ppm. These experiments indicate that the formation of cross-links is a nonlinear function of concentration, and disproportionately less covalent binding is induced by exposure to low HCHO concentrations than is induced by exposure to high HCHO concentrations. In addition, these experiments establish that inhalation exposure to HCHO does not result in an increase in the HCHO concentration of the blood, and it does not result in detectable covalent binding in bone marrow macromolecules. These results strongly suggest that the toxicity induced by inhalation of HCHO gas is confined to tissues at the initial site of contact.

The Cell Proliferation Response to HCHO-Induced Tissue Injury

Morphologic changes were evident in the respiratory mucosa of Fischer-344 rat nasal passages after a single 6-h exposure to 15 ppm of formaldehyde gas. These changes consisted of acute degeneration and swelling and the formation of dense bodies and vacuoles within epithelial cells (20, 40, 41). Following three to five 6-h exposures to 15 ppm of formaldehyde, ulceration of the respiratory epithelium was evident in a high proportion of the animals. By 9 days of exposure, restorative hyperplasia and metaplasia were seen. Rats exposed to 6 ppm exhibited milder degenerative changes but prominent hyperplasia of the respiratory epithelium. No morphologic changes were evident by light microscopy in rats exposed to 0.5 or 2 ppm of formaldehyde.

In view of these cytotoxic and restorative responses to different concentrations and durations of formaldehyde exposure, a series of investigations was undertaken to identify the effects of formaldehyde exposure on cell replication in the respiratory epithelium of the nasal passages. Initial studies demonstrated that marked increases in cell proliferation were present in the second level of the nasal passages, the same region that had the acute pathology and that developed most of the squamous cell carcinomas in the 2-year bioassay. Prominent dose–response and species relationships from this study are shown in Table VI. Rats exposed to 6 or 15 ppm of formaldehyde for three 6-h exposures had a 10–20-fold increase in ^{3}H-thymidine labeling. No increase over controls was detected in rats exposed to 0.5 or 2 ppm, nor in mice exposed to 0.5, 2, or 6 ppm of formaldehyde. Mice exposed to 15 ppm of formaldehyde had a 10-fold increase in de novo DNA synthesis.

Subsequent studies showed that administration of ^{3}H-thymidine at 18 rather than 2 h after the last exposure was a more sensitive method for evaluating the effects of formaldehyde on cell proliferation (20, 41). Table VII shows that rats exposed to 15 ppm of formaldehyde for one 6-h period al-

Table VI. Effect of Formaldehyde Exposure on Cell Proliferation in Level 2 of the Nasal Passages

Exposure[a]	% of Labeled Respiratory Epithelial Cells[b]	
	Rat	Mouse
Control	0.22 ± 0.03	0.12 ± 0.02
0.5 ppm	0.38 ± 0.05	0.09 ± 0.04
2 ppm	0.33 ± 0.06	0.08 ± 0.04
6 ppm	5.40 ± 0.82	0.15 ± 0.06
15 ppm	2.83 ± 0.81	0.97 ± 0.04

[a] All animals were exposed for 6 h/day for 3 days, and ^{3}H-thymidine was administered 2 h after the third exposure.
[b] Mean ± standard error.
Source: Reproduced with permission from Ref. 41. Copyright 1983 Marcel Dekker.

Table VII. Effect of Single or Repeated Formaldehyde Exposure (15 ppm, 6 h/day) on Cell Proliferation in Level 2 Respiratory Epithelium

Exposure	% Labeled Cells[a]	
	Fischer-344 Rats	B6C3F1 Mice
Control	0.43 ± 0.05	0.27 ± 0.04
1 day	5.51 ± 0.35	2.14 ± 0.56
5 days	10.05 ± 0.27	3.42 ± 0.84

NOTE: All animals were pulsed with ^{3}H-thymidine (2 μCi/g) 18 h after the last exposure.
[a] Mean ± standard error.
Source: Reproduced with permission from Ref. 41. Copyright 1983 Marcel Dekker.

ready have more than a 10-fold increase in cell proliferation relative to controls. By 5 days of exposure, this increase exceeded 20-fold. Similar increases were demonstrable in mice.

It was of interest to determine how much of this response was due to duration of exposure and how much was due to formaldehyde concentration because markedly different results were evident in the histopathology of rats exposed for 6 months to 15 ppm of formaldehyde, 6 h/day, 5 days/week (450 ppm · h/week) (2, 3, 47) compared to that of rats exposed to 3 ppm, 22 h/day, 7 days/week (462 ppm · h/week) (48). Animals on the latter exposure regimen exhibited much less toxicity. To evaluate this discrepancy, rats and mice were exposed to 12 ppm of formaldehyde for 3 h/day, 6 ppm for 6 h/day, or 3 ppm for 12 h/day (41). Exposures were conducted for 3 or 10 days, and ^{3}H-thymidine was administered 18 h after the last exposure. Sections from the most anterior and the second level of respiratory mucosa were prepared for autoradiography, and the labeling index was determined. Table VIII shows that the most anterior level of respiratory epithelium had a similar fivefold increase in cell proliferation in all three concentration–time (C × T) groups. This result is in marked contrast to the adjacent, more posterior section (Table IX) where a distinct relationship between concentration and cell turnover was evident. Table IX also shows that the marked increase in proliferation is a somewhat transient event because its magnitude decreases with time. The C × T data are consistent with recent data on the effects of HCHO on the mucociliary clearance apparatus. The lack of a concentration effect in the anterior section reflects the fact that this region has minimal mucociliary clearance (25, 31). In contrast, the adjacent, more posterior section has a continuous flow of mucus over its surface. As the concentration of formaldehyde increases, larger areas of the mucus blanket become immobilized and thereby this protective mechanism is removed (31).

The efficacy of mucociliary clearance is likely to be greatest at low concentrations of formaldehyde. In ongoing experiments that followed a protocol similar to that shown in Table VI except that the pulse of ^{3}H-thymidine was administered 18 h after the last exposure, slight increases in cell

Table VIII. Effect of Formaldehyde Concentration Versus Cumulative Exposure on Cell Turnover in Rats (Level 1)

Exposure	% Labeled Cells After 3 days of Exposure[a]
Control	3.00 ± 1.56
3 ppm × 12 h	16.99 ± 1.50
6 ppm × 6 h	15.46 ± 10.01
12 ppm × 3 h	16.49 ± 2.02

[a]Mean ± standard error.
Source: Reproduced with permission from Ref. 41. Copyright 1983 Marcel Dekker.

Table IX. Effect of Formaldehyde Concentration Versus Cumulative Exposure on Cell Turnover in Rats (Level 2)

	% Labeled Cells[a]	
Exposure	3 days + 18 h	10 days + 18 h
Control	0.54 ± 0.03	0.26 ± 0.02
3 ppm × 12 h	1.73 ± 0.63	0.49 ± 0.19
6 ppm × 6 h	3.07 ± 1.09	0.53 ± 0.20
12 ppm × 3 h	9.00 ± 0.88	1.73 ± 0.65

[a]Mean ± standard error.
Source: Reproduced with permission from Ref. 41. Copyright 1983 Marcel Dekker.

proliferation were evident in rats exposed to 0.5 and 2 ppm of formaldehyde for one 6-h exposure, but not after three or nine such exposures. In contrast, much higher labeling indices were observed in rats exposed to 6 ppm after 1 and 3 days. The data show a distinct nonlinear dependence of the labeling indices on formaldehyde concentration. A 3-fold increase in formaldehyde concentration from 2 to 6 ppm resulted in an 8-fold increase in cell proliferation after 1 day of exposure and nearly a 25-fold increase after 3 days of exposure. These data are consistent with nonlinear data on covalent binding of HCHO to respiratory mucosal DNA (39) and carcinogenesis (32).

Cell proliferation is a critical factor in chemical carcinogenesis. Numerous studies with a broad range of chemicals have demonstrated that cell replication is required for the initiation and promotion of chemical carcinogenesis. When promutagenic DNA adducts are present during de novo DNA synthesis, the likelihood of inserting a wrong nucleotide greatly increases, and such events, if unrepaired before replication, result in permanent mutations. Cell proliferation is also responsible for expanding the clonal population of initiated cells to a cancerous mass. Furthermore, in the case of formaldehyde-induced neoplasia, cell replication is thought to be important in the initial binding of the chemical to DNA because formaldehyde is known to only bind to single-stranded DNA. The number of single-stranded sites is much greater in replicating DNA than in nonreplicating DNA. Thus, the likelihood of formaldehyde binding to DNA, formaldehyde–DNA adducts mispairing, and initiated cell populations expanding to neoplasia is all related to cell proliferation.

Mechanistic Data: Its Proper Role in Risk Estimation for HCHO

Nearly 4 years have elapsed since the first report that the chronic inhalation of gaseous formaldehyde induced nasal cancer in Fischer-344 rats (2). During that time extensive research on the mechanisms of formaldehyde toxicity has yielded a great deal of additional information that is directly relevant to concerns regarding potential adverse effects of formaldehyde exposure on human health. Some of this research, as described in the preceding sections, has focused on the biological defenses that protect organisms from toxicity at low-level formaldehyde exposures. It has also elucidated the mechanisms by which high-level formaldehyde exposures impair these defenses and thereby enhance nonlinearly the probability of irreversible toxic effects.

Nevertheless, the low-dose risk estimates that result from the typical approach to quantitative risk assessment, namely, a linearized multistage model analysis of bioassay tumor incidence versus administered dose, do not use this additional mechanistic information. Indeed, such risk estimates would be no different had none of the research just mentioned been undertaken. This apparent unresponsiveness of current risk assessment practice to pertinent scientific data is both disturbing and counterproductive. As to be discussed in this section, the risk assessment process can be readily and significantly improved in this regard by using the available mechanistic data to construct a measure of exposure that is more realistic and meaningful than administered dose.

The demonstration that formaldehyde is a rodent carcinogen and the ensuing ban of urea–formaldehyde foam insulation in the United States (49) stimulated a vigorous scientific debate regarding how to best use available formaldehyde toxicity data in assessing human cancer risk from formaldehyde exposure. One key issue concerns the form of the relationship between two distinct measures of exposure denoted by the terms "administered dose" and "delivered dose." Administered dose is the external measure of exposure that is directly controlled in laboratory studies of toxicity. For inhalation studies, it refers to the concentration of a test chemical in the inhalation chamber air. In contrast, delivered dose is an internal measure of exposure referring to the quantity or concentration of the biologically active form of a test chemical that is present in specific target tissues. Delivered dose is presumed to be the direct causative variable in mechanistic descriptions of the carcinogenic process at the cellular and molecular levels.

The relationship between administered and delivered doses reflects the entire spectrum of biological responses to exposure, ranging from physiologic responses of the whole organism to intracellular biochemical responses in target tissues. Thus, administered dose actually provides no more than an indirect, surrogate measure of delivered dose, and the relationship between these two measures of exposure need not be a simple linear one. This consideration is especially important because low-dose risk extrapolations based upon the assumption of linearity are known to yield risk estimates that are either excessively conservative (too high) or anticonservative (too low) when the true administered–delivered dose relationship is nonlinear (50).

Although the delivered-dose concept should thus play a critical role in the production of accurate assessments of human risk from chemical exposure, it has yet to be elucidated fully for any chemical agent. Such knowledge requires detailed studies of the distribution and biochemical disposition of chemical agents in whole animals, including humans. In the case of formaldehyde, the extensive mechanistically oriented studies described in preceding sections have identified four biological responses that appear to be important determinants of the formaldehyde dose delivered to target tissues in the rodent nasal cavity.

The first of these is the minute volume depression in response to sensory irritation that was described earlier. In Fischer-344 rats and B6C3F1 mice it is an important factor only at formaldehyde concentrations greater than 6 ppm. Still, the fact that it is induced at these concentrations has three important consequences. First, the amount of formaldehyde entering the rat or mouse nasal cavity is not linearly proportional to HCHO concentrations in inspired air greater than 6 ppm. Second, the precipitously steep rise in squamous cell carcinoma incidence from approximately 1% among rats chronically exposed to 5.6 ppm of formaldehyde to nearly 50% among rats similarly exposed to 14.3 ppm (Table I) is actually steeper, that is, more severely nonlinear, when the amount of HCHO inhaled per unit time, rather than the ambient air HCHO concentration, is used as the measure of exposure. Third, the marked disparity in tumor response between rats and mice identically exposed to 14.3 ppm of HCHO (Table I) can be reconciled by measuring exposure in terms of the rate, adjusted for an interspecies difference in nasal cavity surface area, at which HCHO is actually deposited in the nasal cavity.

Two other factors that need to be considered are the inhibition of mucociliary clearance and the stimulation of cell proliferation that are both induced by exposure to high but not low HCHO concentrations. Both tend to disproportionately increase the dose delivered to target tissues at high HCHO concentrations, thus counterbalancing and most likely overriding any reduction in delivered dose associated with minute volume depression. The inhibition of mucociliary clearance contributes to this effect by eliminating one pathway for removal of formaldehyde from the nasal cavity before it ever penetrates to underlying epithelial cells. Increased cell proliferation enhances the likelihood of irreversible genotoxic events once HCHO reaches target cells by increasing the number of single-stranded DNA sites at which HCHO may covalently bind, and also by decreasing the amount of time available for the repair of such lesions before they become fixed during cell replication.

Finally, studies of the disposition of HCHO in nasal cavity tissues have provided the first direct quantitative measurements of the amount of HCHO that is delivered to target cell DNA. These studies are of critical importance for several reasons. First, they provide quantitative data that demonstrate that the delivered dose–administered dose relationship is distinctly nonlinear, as would be expected from consideration of the observed spectrum of effects of inhaled HCHO on minute volume, mucociliary clearance, and cell proliferation. Second, the studies also provide evidence that metabolic incorporation, a process by which delivered formaldehyde is detoxified, is less efficient at high airborne HCHO concentrations than it is at low concentrations. Thus, another removal pathway that provides protection from HCHO toxicity at low airborne concentrations appears to be inhibited at HCHO concentrations greater than 2 ppm. Third, the data for covalent binding of HCHO to target tissue DNA are in a form that makes it possible to reanalyze the nasal tumor results from the chronic bioassay with this delivered HCHO dose, rather than airborne HCHO concentration, as the measure of exposure. Such a reanalysis was recently completed (51), and a brief summary of that study will now be given.

Tumor incidence rates nearly identical to those used by Cohn (52) were employed because his analysis of the chronic bioassay results figured prominently in the U.S. Consumer Product Safety Commission's decision to ban the sale of urea–formaldehyde foam insulation in the United States (49). Concentrations of HCHO covalently bound to respiratory mucosal DNA corresponding to the airborne HCHO concentrations employed in the chronic bioassay were derived from those reported by Casanova–Schmitz et al. (39). Four commonly used quantal response models, namely, the multistage, Weibull, logit, and probit, were used for low-dose extrapolation. Model parameters were estimated by using standard maximum likelihood (ML) techniques. Both ML estimates of risk and their upper 95% confidence bounds were calculated for three airborne HCHO concentrations, namely, 0.1, 0.5, and 1.0 ppm. For these concentrations, the delivered dose–administered dose relationship was assumed to be linear and to be given by the straight line passing from the origin through the concentration of covalently bound HCHO that was observed at 2 ppm (39). As noted by Starr and Buck (51), this assumption likely overestimates the amount of covalent binding that actually occurs at these airborne concentrations.

The maximum likelihood estimates of risk and their upper 95% confidence bounds are presented in Tables X and XI, respectively. The estimates obtained with delivered dose are unilaterally lower than the corresponding estimates obtained with administered dose. The reduction factors for ML estimates ranged from 35 (Weibull, 0.1 ppm) to more than 9 orders of magnitude (probit, 0.5 ppm). The multistage ML estimates based on delivered dose were uniformly lower by a factor of 53. Reduction factors for upper 95% confidence bounds ranged from 2.5 (multistage, 0.1 ppm) to more than 10 orders of magnitude (probit, 0.5 ppm).

Table X. Maximum Likelihood Estimates of Risk Based on Administered Dose (A) and Delivered Dose (D) at Selected Ambient Air Formaldehyde Concentrations

Concentration (ppm)	Dose Measure	Maximum Likelihood Risk Estimates			
		Probit	Logit	Weibull	Multistage
0.1	A	< 1.00 (− 26)	3.92 (− 11)	2.20 (− 10)	2.51 (− 7)
0.1	D	< 1.00 (− 26)	7.40 (− 13)	6.20 (− 12)	4.70 (− 9)
0.5	A	5.16 (− 17)	9.85 (− 8)	2.75 (− 7)	3.14 (− 5)
0.5	D	< 1.00 (− 26)	9.76 (− 10)	4.27 (− 9)	5.88 (− 7)
1.0	A	2.65 (− 11)	2.87 (− 6)	5.94 (− 6)	2.51 (− 4)
1.0	D	4.00 (− 20)	2.15 (− 8)	7.13 (− 8)	4.70 (− 6)

NOTE: Values in parentheses are powers of 10.
Source: Reproduced with permission from Ref. 51. Copyright 1984 Academic Press.

Table XI. Upper 95% Confidence Bounds on Risk Based on Administered Dose (A) and Delivered Dose (D) at Selected Ambient Air Formaldehyde Concentrations

Concentration (ppm)	Dose Measure	*Upper 95% Confidence Bounds on Risk*			
		Probit	*Logit*	*Weibull*	*Multistage*
0.1	A	< 1.00 (− 26)	2.84 (− 10)	1.57 (− 9)	1.56 (− 4)
0.1	D	< 1.00 (− 26)	6.19 (− 12)	5.12 (− 11)	6.19 (− 5)
0.5	A	7.69 (− 16)	5.13 (− 7)	1.41 (− 6)	8.09 (− 4)
0.5	D	< 1.00 (− 26)	6.31 (− 9)	2.73 (− 8)	3.10 (− 4)
1.0	A	2.58 (− 10)	1.24 (− 5)	2.54 (− 5)	1.80 (− 3)
1.0	D	7.09 (− 19)	1.22 (− 7)	3.98 (− 7)	6.24 (− 4)

NOTE: Values in parentheses are powers of 10.
Source: Reproduced with permission from Ref. 51. Copyright 1984 Academic Press.

These results demonstrate that incorporation of the delivered dose concept into low-dose extrapolation procedures leads to a unilateral reduction in estimates of cancer risk associated with exposure to low airborne HCHO concentrations. Additional research is required to further refine and elaborate the delivered-dose concept for exposure to HCHO by inhalation, especially for humans. Nevertheless, because use of the delivered-dose concept allows much of the information already obtained from mechanistic studies of formaldehyde toxicity to enter the risk assessment process in a meaningful and relevant manner, these risk estimates reflect what is known of the underlying biological reality more faithfully than previous estimates based solely on findings from the chronic bioassay. Attention is now turned to what is known in regard to the effects of human occupational exposures to formaldehyde.

Human Experience: A Review of Epidemiologic Studies of HCHO Exposure

The mortality experience of the following occupational groups with known exposure to formaldehyde has been evaluated: pathologists (53–57), anatomists (58), morticians (59–61), and chemical workers (62–67). With the exception of an elevated proportion of deaths due to skin cancer among New York undertakers, increased cancer mortality has not been observed at locations in possible contact with formaldehyde gas. Individual studies, generally of professional groups, not chemical workers, have noted two- to fourfold increases in mortality from cancer at distant sites including brain, kidney, colon, prostate, lymphatic, and hematopoietic tissues.

Tables XII and XIII summarize cohort data from studies reviewed by the Epidemiology Panel of the "Consensus Workshop on Formaldehyde" held in Little Rock, Arkansas, during October 1983 under the auspices of the National Center for Toxicological Research. These studies represent all cohorts exposed to formaldehyde for which mortality information is cur-

Table XII. Formaldehyde Exposure: Mortality of Male Workers at Eight Chemical Plants

Cause of Death				Observed/Expected (O/E) Deaths							
	A1	A2	A3	A4	A5	A6	Total A	B	C	Total A–C	O/E
All causes	77/93	98/107	49/45	845/983	104/149	446/485	1619/1862	115/—[a]	146/197	1765/2059	0.86
All cancers	19/23	32/27	18/11	251/246	21/38	114/123	455/468	20/22	37/37	512/527	0.97
Skin	—	—	—	—	—	—	2/—	—	1/1.0	1/1.0	—[a]
Buc. cav. and phar.	—	—	—	—	—	—	5/4.6	0/0.8	0/1.3	5/6.7	0.75
Respiratory	—	—	—	—	—	—	—	6/7.5	12/12.4	18/19.9	0.90
Nose	0/0.05	0/0.06	0/0.03	0/0.56	0/0.09	0/0.28	0/1.07	0/—	0/—	0/1.07	—
Larynx	—	—	—	—	—	—	4/4.5	—	—	4/4.5	0.89
Lung	6/8.8	11/15	7/6.3	128/123	7/14	46/48	205/215.1	—	11/11.7	216/227	0.95
Esophagus	—	—	—	—	—	—	—	—	—	—	—
Colon	—	—	—	—	—	—	—	—	3/3.0	3/3.0	—
Brain	—	—	—	—	—	—	5/12.5	—	3/1.6	8/14.1	0.57
Kidney	—	—	—	—	—	—	7/8.3	—	1/1.0	8/9.3	0.86
Prostate	—	—	—	—	—	—	—	—	4/1.3	4/1.3	—
Lymphopoietic	—	—	—	—	—	—	20/26.3	2/2.3	6/4.4	28/33	0.85
Leukemia	—	—	—	—	—	—	9/11.4	—	2/1.7	11/13.1	0.84

NOTE: O/E is given only when observed and/or expected deaths ≥5. A1–6: British men first employed prior to 1965. Expected lung cancer deaths computed by using local area rates (64, 69); B: white men exposed to formaldehyde >1 month (62); and C: white men (63). Skin and buccal cavity and pharynx cancer from Ref. 66.

[a]Proportional mortality study.

Source: Reproduced with permission from Ref. 61. Copyright 1984 Flournoy Publishers.

Table XIII. Formaldehyde Exposure: Cancer Mortality of Male Pathologists, Anatomists, and Morticians (Includes Combined Totals of Observed and Expected Deaths for Tables XII and XIII)

Cause of Death	Observed/Expected (O/E) Deaths									Tables XII and XIII	
	D1	D2	E	F	G1	G2	H	Total D-H	O/E	Total A-H	O/E
All causes	146/244	110/195	381/569	738/1130	1132/—[a]	1007/—[a]	319/322	1694/2460	0.69	3459/4519	0.77
All cancers	38/62	32/52	67/93	—	243/219	205/170	58/67	643/663	0.97	1155/1190	0.97
Skin	—	—	—	2/3.5	8/3.6[b]	2/3.4	0/0.9	12/11.4	1.05	13/12.4	1.05
Buc. cav. and phar.	—	—	—	1/6.8	8/7.1	8/6.1	1/2.1	18/22.1	0.81	23/28.8	0.80
Respiratory	—	—	—	13/46.3	74/70.7	43/46.0	20/21.6	150/185	0.81	168/205	0.82
Nose	—	0/0.1	—	0/0.4	0/0.5	0/0.6	0/0.2	0/1.8	—	0/2.9	—
Larynx	—	—	—	—	2/3.4	2/2.6	1/1.0	5/7.0	0.71	9/11.5	0.78
Lung	10/27.4	9/22.0	11/22.4	—	72/66.8	41/42.8	19/20.2	162/202	0.80	378/429	0.88
Esophagus	—	—	—	—	5/5.3	3/4.1	0/1.7	8/11.1	0.72	8/11.1	0.72
Colon	—	—	2/9.1	—	29/20.3	30/16.0[d]	—	61/45[b]	1.34	64/48[b]	1.32
Brain	—	4/1.2[b]	5/2.2	10/3.7[c]	9/5.8	9/4.7	3/2.6	40/20.2[e]	1.98	48/34.3[b]	1.40
Kidney	—	—	4/2.2	—	8/5.4	4/4.0	1/1.7	17/13.3	1.28	25/22.6	1.11
Prostate	—	—	—	20/18.7	15/16.4	23/13.1[c]	3/3.4	61/51.7	1.18	65/53	1.23
Lymphopoietic	8/3.8[b]	2/3.0	6/6.9	18/14.6	25/20.6	19/15.5	8/6.5	86/70.6	1.22	114/103.6	1.10
Leukemia	1/1.5	1/1.1	2/3.7	10/6.7	12/8.5	12/6.9	4/2.5	42/31	1.36	53/44	1.20

NOTE: O/E is given only when observed and/or expected deaths ≥5. D1, 2: British pathologists 1955–73 and 1974–80 (54); E: members of the American Association of Pathologists and Bacteriologists (55); F: American anatomist (58; 69); G1,2: American white morticians licensed in New York (60) or California (59); and H: Canadian morticians licensed in Ontario (61).
[a]Proportional mortality study. [b]Significant increase, $p < 0.05$. [c]Significant increase, $p < 0.01$. [d]Significant increase, $p < 0.001$. [e]Significant increase, $p < 0.0001$.

Source: Reproduced with permission from Ref. 61. Copyright 1984 Flournoy Publishers.

rently available. Data concerning the experience of the American Society for Experimental Pathology have been omitted because a considerable portion of the membership of that society belongs to the American Association of Pathologists and Bacteriologists, whose experience has been included in Table XII. Similarly, studies by Liebling et al. (65) and by Tabershaw Associates (66) provide further detail about cohorts already described by Marsh (62) and Wong (63) but do not affect overall conclusions.

The cancer mortality of men at eight chemical plants in which formaldehyde was manufactured or used is presented in Table XII. Causes of death listed include cancers at sites that may come into contact with formaldehyde gas and those at distant locations where significant excesses have been observed previously in a formaldehyde-exposed cohort. For each cause, observed and expected deaths are presented by individual plant and as a summary total. Although there are obvious limitations to inferences drawn from totals of observed and expected deaths across plants, the procedure is useful for generating hypotheses. Local area mortality rates were employed to determine the number of deaths expected from lung cancer at plants A1–6, but national rates were used for B and C.

Significant increases in mortality were not observed at individual plants, nor when the plants were combined. Although Table XII lacks complete data, apparently no important mortality excesses have been discovered. At plant C, after a latency of 20 years, deaths from prostate cancer were significantly greater than expected (4 observed, 0.9 expected, $p <$ 0.05); nevertheless, the increase in mortality did not correlate with length of employment (63).

The experience of male pathologists, anatomists, and morticians compared to the general population is summarized in Table XIII. Except for an elevated proportion of deaths due to skin cancer among New York morticians, which was observed in no other study, excess cancer mortality was not detected at sites in contact with formaldehyde gas. Several studies have recorded significant increases in deaths from cancers at distant sites, including colon (G2), brain (D2, F), prostate (G2), and lymphopoietic tissues (D1); moreover, summary totals show significant excesses of colon and brain cancers. Observed brain cancer deaths exceeded expected deaths in all seven cohorts for whom information was available (C and D2–H in Tables XII and XIII). Six of these were professional groups likely to have very low exposure to formaldehyde on the basis of a time-weighted average. Nonsignificant increases in leukemia were also found in the studies of anatomists and morticians.

Two case-control studies of formaldehyde exposure have been conducted. These detected no increased risk for lung cancer. Among physicians the relative risk of lung cancer for those with possible exposure was 1.0 (56, 57), and among chemical workers the relative risk of formaldehyde exposure for men with lung cancer was also 1.0 (67).

Because formaldehyde is highly reactive and rapidly detoxified, inhalation (or dermal) exposure is unlikely to affect an internal organ. The radioactive decay curves of intravenously administered, radiolabeled formaldehyde and formate in the blood of rats are virtually identical (68). As noted earlier, this result suggests that formaldehyde is oxidized almost immediately to formate, then the formate participates in the normal one-carbon metabolic pool. Formaldehyde concentrations in the blood of humans and Fischer-344 rats did not differ significantly following inhalation exposure of humans to 1.9 ppm for 40 min and rats to 14.4 ppm for 2 h (45). By implication, blood levels primarily reflect formaldehyde that is endogenously produced. When rats were exposed for 6 h to atmospheres containing up to 15 ppm of formaldehyde, covalent binding of formaldehyde to bone marrow macromolecules was not detectable. On the other hand, evidence was obtained for covalent binding in respiratory mucosal tissue from the nasal cavity, the principal site of formaldehyde contact and the tissue from which cancers developed in rodents (39).

Except for one instance in which the proportion of skin cancer deaths was elevated, the data to date provide no evidence for the carcinogenicity of formaldehyde in humans at sites of contact. Deaths from lung cancer were clearly not in excess among formaldehyde-exposed groups. Further information is needed, however, that takes into account latency, exposure intensity, and exposure duration. Future reports should be examined carefully for excess morbidity or mortality from skin cancer. On the basis of currently available data, it appears unlikely that formaldehyde plays a role in the etiology of cancers at distant sites. Explanations other than exposure to formaldehyde should therefore be sought for observed excesses of cancers of the brain, colon, and leukemia.

Literature Cited

1. Greek, B. F. *Chem. Eng. News* 1984, *62* (5), 10–14.
2. Swenberg, J. A.; Kerns, W. D.; Mitchell, R. E.; Gralla, E. J.; Pavkov, K. L. *Cancer Res.* 1980, *40*, 3398–3402.
3. Kerns, W. D.; Pavkov, K. L.; Donofrio, D. J.; Gralla, E. J.; Swenberg, J. A. *Cancer Res.* 1983, *43*, 4382–92.
4. Boorman, G. A., letter to Swenberg, James A., January 18, 1984.
5. Takano, T.; Shirai, T.; Ogiso, T.; Tsuda, H.; Baba, S.; Ito, N. *Cancer Res.* 1982, *42*, 4236–40.
6. Hennings, H.; Shores, R.; Wenk, M. L.; Spangler, E. F.; Tarone, R.; Yuspa, S. H. *Nature* 1983, *304*, 67–69.
7. Albert, R. E.; Burns, F. J.; Altshuler, B. *Adv. Mod. Toxicol.* 1979, *1*, 89–95.
8. Burns, F. J.; Vanderlaan, M.; Snyder, E.; Albert, R. E. In "Mechanisms of Tumor Promotion and Cocarcinogenesis"; Slaga, T. J.; Sivak, A.; Boutwell, R. K., Eds.; Raven: New York, 1978; Vol. 2, pp. 91–96.
9. Scribner, J. D.; Scribner, N. K.; McKnight, B.; Mottet, N. K. *Cancer Res.* 1983, *43*, 2034–41.
10. Farber, E. *Biochim. Biophys. Acta* 1980, *605*, 149–66.
11. Alarie, Y. *CRC Crit. Rev. Toxicol.* 1973, *2*, 299–363.
12. Keele, C. A. *Arch. Int. Pharmacodyn. Ther.* 1962, *139*, 547–57.

13. Widdicombe, J. G. In "Respiratory Physiology II"; Widdicombe, J. G., Ed.; Univ. Park: Baltimore, 1977; Vol. 14, pp. 291–315.
14. Comroe, J. H., Jr. In "Physiology of Respiration," 2d ed.; Year Book: Chicago, 1974; pp. 220–28.
15. Kane, L. E.; Alarie, Y. *Am. Ind. Hyg. Assoc. J.* 1977, *38*, 509–22.
16. Kulle, T. J.; Cooper, G. P. *Arch. Environ. Health* 1975, *30*, 237–43.
17. Chang, J. C. F.; Steinhagen, W. H.; Barrow, C. S. *Toxicol. Appl. Pharmacol.* 1981, *61*, 451–59.
18. Gross, E. A.; Swenberg, J. A.; Fields, S.; Popp, J. A. *J. Anat.* 1982, *135*, 83–88.
19. Barrow, C. S.; Steinhagen, W. H.; Chang, J. C. F. In "Formaldehyde Toxicity"; Gibson, J. E., Ed.; Hemisphere: Washington, D.C.: 1983; pp. 16–25.
20. Chang, J. C. F.; Gross, E. A.; Swenberg, J. A.; Barrow, C. S. *Toxicol. Appl. Pharmacol.* 1983, *68*, 161–76.
21. Walker, D. In "Nasal Tumors in Animals and Man"; Reznik, G.; Stinson, S. F., Eds.; CRC: New York, 1983; Vol. 3, pp. 115–35.
22. Morgan, K. T., unpublished data.
23. Proctor, D. F. In "The Nose, Upper Airway Physiology and the Atmospheric Environment"; Proctor, D. F.; Andersen, I., Eds.; Elsevier Holland: Amsterdam, 1982; pp. 245–70.
24. Morgan, K. T.; Patterson, D. L.; Gross, E. A. In "Formaldehyde: Toxicology, Epidemiology, and Mechanisms"; Clary, J. J.; Gibson, J. E.; Waritz, R. S., Eds.; Marcel Dekker: New York, 1983; pp. 193–210.
25. Morgan, K. T.; Jiang, X. Z.; Patterson, D. L.; Gross, E. A. *Am. Rev. Respir. Dis.* 1984, *130*, 275–81.
26. Egle, J. L. *Arch. Environ. Health* 1972, *25*, 119–24.
27. Lucas, A.; Douglas, L. C. *Arch. Otolaryngol.* 1934, *20*, 518–41.
28. Marshall, T. C.; Hahn, F. F.; Henderson, R. F.; Silbaugh, S. A.; Hobbs, C. H. In "Inhalation Toxicology Research Institute Annual Report, 1981–82"; Lovelace Biomedical and Environmental Research Institute: Albuquerque, 1982; LMF–102, UC–48, pp. 423–27.
29. Andersen, I.; Molhave, L. In "Formaldehyde Toxicity"; Gibson, J. E., Ed.; Hemisphere: Washington, D.C., 1983; pp. 154–65.
30. Morgan, K. T.; Patterson, D. L.; Gross, E. A. *Fundam. Appl. Toxicol.* 1984, *4*, 58–68.
31. Morgan, K. T. *Am. Rev. Respir. Dis.* 1983, *127*, 166.
32. Swenberg, J. A.; Barrow, C. S.; Boreiko, C. J.; Heck, H. d'A.; Levine, R. J.; Morgan, K. T.; Starr, T. B. *Carcinog.* 1983, *4*, 945–52.
33. Macklin, C. C. *J. Thorac. Surg.* 1956, *31*, 238–44.
34. Ross, W. E.; Shipley, N. *Mutat. Res.* 1980, *79*, 277–83.
35. Grafstrom, R. C.; Fornace, A. J., Jr.; Autrup, H.; Lechner, J. F.; Harris, C. C. *Science* 1983, *220*, 216–18.
36. Ragan, D. L.; Boreiko, C. J. *Cancer Lett.* 1981, *13*, 325–31.
37. Goldmacher, V. S.; Thilly, W. G. *Mutat. Res.* 1983, *116*, 417–22.
38. Casanova–Schmitz, M.; Heck, H. d'A. *Toxicol. Appl. Pharmacol.* 1983, *70*, 121–32.
39. Casanova–Schmitz, M.; Starr, T. B.; Heck, H. d'A. *Toxicol. Appl. Pharmacol.* 1984, *76*, 24–44.
40. Swenberg, J. A.; Gross, E. A.; Martin, J.; Popp, J. A. In "Formaldehyde Toxicity"; Gibson, J. E., Ed.; Hemisphere: Washington, D.C., 1983; pp. 132–47.
41. Swenberg, J. A.; Gross, E. A.; Randall, H. W.; Barrow, C. S. In "Formaldehyde: Toxicology, Epidemiology, and Mechanisms"; Clary, J. J.; Gibson, J. E.; Waritz, R. S., Eds.; Marcel Dekker: New York, 1983; pp. 225–36.
42. Magana-Schwenke, N.; Moustacchi, E. *Mutat. Res.* 1980, *70*, 29–35.
43. von Hippel, P. H.; Wong, K.-Y. *J. Mol. Biol.* 1971, *61*, 587–613.
44. Lukashin, A. V.; Vologodskii, A. V.; Frank-Kamenetskii, M. D.; Lyubchenko, Y. L. *J. Mol. Biol.* 1976, *108*, 665–82.

45. Heck, H. d'A.; Casanova-Schmitz, M.; Dodd, P. B.; Schachter, E. N.; Witek, T.; Tosun, T. *Am. Ind. Hyg. Assoc. J.*, **1985**, *46*, 1–3.
46. Heck, H. d'A.; White, E. L.; Casanova-Schmitz, M. *Biomed. Mass Spectrom.* **1982**, *9*, 347–53.
47. Kerns, W. D.; Donofrio, D. J.; Pavkov, K. L. In "Formaldehyde Toxicity"; Gibson, J. E., Ed.; Hemisphere: Washington, D.C., 1983; pp. 111–31.
48. Rusch, G. M.; Bolte, J. F.; Rinchart, W. E. In "Formaldehyde Toxicity"; Gibson, J. E., Ed.; Hemisphere: Washington, D.C., 1983; pp. 98–110.
49. U.S. Consumer Product Safety Commission *Fed. Regist.* **1982**, *47*, 14366–419.
50. Hoel, D. G.; Kaplan, N. L.; Anderson, M. W. *Science* **1983**, *219*, 1032–37.
51. Starr, T. B.; Buck, R. D. *Fundam. Appl. Toxicol.* **1984**, *4*, 740–53.
52. U.S. Consumer Product Safety Commission "Revised Carcinogenic Risk Assessment of Urea–Formaldehyde Foam Insulation: Estimates of Cancer Risk Due to Inhalation of Formaldehyde Released by UFFI," by Cohn, M. S.; Government Printing Office: Washington, 1981.
53. Harrington, J. M.; Shannon, H. S. *Br. Med. J.* **1975**, *4*, 329–32.
54. Harrington, J. M.; Oakes, D. *Br. J. Ind. Med.* **1984**, *41*, 188–91.
55. Matanoski, G. M., letter to Martonik, John F., March 30, 1982.
56. Andersen, S. K.; Jensen, O. M.; Oliva, D. *Ugeskr. Laeg.* **1982**, *144*, 1571–73.
57. Jensen, O. M.; Andersen, S. K. *Lancet* **1982**, *i*, 913.
58. Stroup, N. E. *Am. J. Epidemol.* **1984**, *120*, 500; also personal communication, Centers for Disease Control, Atlanta, Ga., 1983.
59. Walrath, J. *Am. J. Epidemol.* **1983**, *118*, 432; also personal communication, Environmental Epidemiology Branch, National Cancer Institute, Bethesda, Md.
60. Walrath, J.; Fraumeni, J. F., Jr. *Int. J. Cancer* **1983**, *31*, 407–11.
61. Levine, R. J.; Andjelkovich, D. A.; Shaw, L. K.; DalCorso, R. D. *J. Occup. Med.* **1984**, *26*, 740–46.
62. Marsh, G. M. *Br. J. Ind. Med.* **1982**, *39*, 313–22; also personal communication, Department of Biostatistics, Graduate School of Public Health, University of Pittsburgh, Pittsburgh, Pa.
63. Wong, O. In "Formaldehyde Toxicity"; Gibson, J. E., Ed.; Hemisphere: Washington, D.C., 1983; pp. 256–72.
64. Acheson, E. D.; Barnes, H. R., Gardner, M. J.; Osmond, C.; Pannett, B.; Taylor, C. P. *Lancet* **1984**, *i*, 611–16.
65. Liebling, T.; Rosenman, K. D.; Pastides, H.; Griffith, R. G.; Lemeshow, S. *Am. J. Ind. Med.* **1984**, *5*, 423–28.
66. Tabershaw Associates "Historical Prospective Mortality Study of Past and Present Employees of the Celanese Chemical and Plastics Plant Located in Bishop, Texas"; Tabershaw Associates: Rockville, Md., 1982; also unpublished study data courtesy of Celanese Chemical Corp., Dallas, Tx.
67. Fayerweather, W. E.; Pell, S.; Bender, J. R. In "Formaldehyde Toxicology, Epidemiology, and Mechanisms"; Clary, J. J.; Gibson, J. E.; Waritz, R. S., Eds.; Marcel Dekker: New York, 1983; pp. 47–125.
68. Heck, H. d'A.; Chin, T. Y.; Casanova-Schmitz, M. In "Formaldehyde Toxicity"; Gibson, J. E.,Ed.; Hemisphere: Washington, D.C., 1983; pp. 26–37.
69. "Deliberations of the Consensus Workshop on Formaldehyde," Little Rock, Ark., October 1983; National Center for Toxicological Research: Jefferson, Ark.

RECEIVED for review September 28, 1984. ACCEPTED December 26, 1984.

Evaluation of Potential Carcinogenic Hazard

C. JELLEFF CARR and ALBERT C. KOLBYE, JR.

The Nutrition Foundation, Inc., Washington, DC 20006

To evaluate potential hazards to public health posed by environmental chemicals, the specific configuration of biological characteristics of each chemical should be considered in the context of dose–response data and knowledge concerning mechanism of biological action. Absorption, metabolism, and excretion–storage data provide insight into toxicity and detoxification. Short-term tests in vitro and in vivo can help to clarify (within the limits of present knowledge) whether the compound in question is an initiating carcinogen (self-promoting at more substantial doses; i.e., a "complete" carcinogen) or is more likely to be a "promoter" or enhancer of carcinogenesis mediated by discernible toxicity to organs, systems, or tissues. The pattern of exposure and dosage is an important determinant of outcome as is the degree to which biological resistance can withstand or repair the biological damage that is a critical prerequisite to cancer.

EXPERIMENTS USING LIFETIME AND SUBCHRONIC EXPOSURES in laboratory animals to evaluate the toxicological characteristics of test substances have assumed an increasing scientific and societal importance. The data from these experiments are used in a variety of ways to make qualitative and quantitative judgments concerning potential hazards to human health when humans are exposed to these test substances.

The accuracy and relevance of test data derived from animals become of paramount importance when public health considerations and judgmental interpretations for safety are involved. Substantial biological differences and variations of response to carcinogenic agents exist among the species, genetic strains, sexes, and subsequent generations of laboratory animals. Their responses are governed also by environmental factors such as stress, diet, and multiple chemical exposure.

Recently, the published results of these carcinogenicity bioassays have been widely criticized on the basis of the methodology employed, including improper use of the maximum tolerated dose (MTD), excessive dosage by oil gavage of water-insoluble substances, and inattention to key nutrients in chronic animal studies lasting at least 2 years.

0065–2393/85/210/0335$06.00/0

Many of these issues have been reviewed by the Ad Hoc Panel on Chemical Carcinogenesis Testing and Evaluations to the National Toxicology Program Board of Scientific Counselors, and substantial recommendations for future changes have been made (1). However, for the immediate past and the present we are confronted with numerous difficult decisions that will, in large measure, determine current regulatory actions.

A growing concern exists in regard to the salient factors of pharmacokinetic mechanisms, target-cell concentrations, metabolism, and excretion of test substances in carcinogenicity assays. These factors are now recognized to influence significantly the outcome of these tests, including such biological processes as the formation of chemically reactive metabolites, inhibition of enzyme mechanisms, and covalent binding to cell components that may or may not account for genetic or what are said to be "nongenetic" effects.

Numerous literature references have been made to the significance of determining when the doses in the chronic toxicity tests exceed the animal's metabolic capacity. These subsequent, untoward, confounding effects have been noted in the Nutrition Foundation's review of the effects of using vegetable oils as vehicles (2). Such high doses exceed the metabolic "break point," and as a consequence high tissue concentrations of the test material are produced. This result can cause nonspecific toxic challenges that represent a series of phenomena substantially related to the process involved with tumor promotion by classical promoters. This effect may be characterized as toxic hyperplasia. Toxic hyperplasia can increase tissue susceptibility to the initiating influence of carcinogenic compounds by increasing the susceptibility of cells to electrophilic attack (3).

As has been shown, cellular injury of a nonspecific nature can impair the functioning of protective cellular enzymes, and the result is a further increase in the local concentration of the active chemical moiety. Proteins are denatured, membranes are destabilized, and normal cellular processes cease to function, such as the active and passive transport of cellular components. The net result is the potential for attack on the DNA and RNA by genotoxic agents. Unfortunately, the role of nonspecific toxicity per se in relation to carcinogenicity has been poorly appreciated in the entire field of cancer studies (3).

Scientists have little doubt that numbers of toxic substances can be shown to be carcinogenic when massive doses are administered. The significant issue is the relevance of these findings to the much smaller amounts of human exposure that can be detected by exquisitely sensitive analytical techniques.

Unfortunately, carcinogenicity studies are not terminated when the MTD dosage proves to be too high on the basis of preliminary short-term tests. Therefore, the final test data remain equivocal and are the subject of criticism from a toxicological standpoint.

The pharmacokinetics and metabolism of chemical substances may be distinctly different depending on low or high doses (4). The issues of adequate dosage schedules have been reviewed with recommendations by numerous advisory groups (1, 5). For example, the American Industrial Health Council's position was stated as follows (5):

> Bioassays using the maximum tolerated dose administered via unexpected routes of exposure have not been selective in distinguishing chemicals for regulation as carcinogens, and furthermore, such studies provide very little guidance for risk assessment.

The Potential Carcinogenic Hazard of Formaldehyde

On the basis of these general considerations, one may ask a series of questions regarding the extensive and elaborate studies conducted to estimate the human cancer risk from formaldehyde. Certainly the enormous amount of industrial, scientific, analytical chemical, and regulatory expertise that has been and continues to be devoted to this overwhelming task is worthy of our best efforts to find satisfactory answers to these penetrating questions. Our questions should concern the epidemiological, physiological, and toxicological data on formaldehyde as they pertain to the analysis and evaluation of carcinogenic risk. The answers will permit risk assessment procedures and risk management decisions to be made on the basis of all relevant biological information.

How persuasive are the data from the animal bioassays for carcinogenicity? Reference has been made to some of the criticisms of these test methods. A recent review concludes that the risk at low-level exposure would not be linearly related to the risk found at the higher levels observed to be carcinogenic in animals (6).

Animal studies demonstrated that formaldehyde is carcinogenic in the nasal cavities of rodents in cytotoxic doses inhaled and causes increased cellular proliferation. But lower levels are not carcinogenic. Major anatomical differences exist between the nasal cavities of humans and those of most animals; for example, rats and mice are obligatory nose breathers (7). Is it proper to equate inhalation studies in these rodents to humans?

Cellular Toxic Effects of Formaldehyde

Formaldehyde in high concentrations is a protoplasmic poison and is primarily an irritant as a result of its protoplasmic coagulating action. This cellular effect accounts for many of its uses, but from the standpoint of carcinogenicity, it introduces the question of a kind of nonspecific chemical burn. There would be some protection against this effect by inhalation because sensory irritation in the respiratory tract has a lower threshold than cellular alterations, and this condition would tend to avoid cytotoxicity unless animals are required to breath the vapors in high doses. From the standpoint of carcinogenicity assessment, the question remains of the significance of the cellular degeneration, necrosis, and inflammation produced by formaldehyde in adequate doses. These are levels that have been observed to cause increased cell proliferation (8) and acute degeneration, necrosis, and inflammation (9, 10), believed to be critical events in formaldehyde carcinogenesis. These characteristics of a substance have regulatory implications and must be included in the decision process because biochemical toxicity patterns vary with the specific chemical as has been pointed out.

In the body, formaldehyde is metabolized and contributes to the formate pool. It can enter into the metabolism of one-carbon compounds and give rise to methyl groups (11). In vitro preparations of liver enzymes convert the aldehyde to formic acid, and the variations of these metabolic changes largely involving aldehyde dehydrogenase have been studied in detail. Dealkylation of numerous drugs such as codeine, ephedrine, and phenacetin yield formaldehyde by the action of the microsomal enzyme systems of the liver.

In addition, formaldehyde is a normal metabolite and enters into the chain of biochemical events in humans and other animals to give rise to essential cellular substances (12). For these reasons formaldehyde is not considered a toxic cellular component in low concentrations.

The Scientific Committee of the Food Safety Council devoted 4 years to the preparation of a report entitled, "Proposed System for Food Safety Assessment" (13). This unique report has been acknowledged as a most definitive one in the field of toxicity assessment of food ingredients. The committee's system included the important decision that if a metabolite or a test substance proved to be a normal body constituent, it would be considered safe, and only the quantity consumed or formed in the body would be an issue to be resolved. It appears that formaldehyde meets this decision criterion.

Insofar as cellular toxic effects are concerned, we are confronted with the question of how shall we differentiate occasional low-level exposure to formaldehyde versus prolonged occupational exposure?

Carcinogenic Mechanisms

Several comprehensive reviews have addressed the question of the genotoxicity of formaldehyde (9, 10, 14–16). Mutations based on short-term in vitro tests have been reported, but not all tests were positive. Although conflicting results have been obtained, it is not clear whether these changes would follow noncytotoxic doses. Can formaldehyde be considered mutagenic or capable of inducing chromosomal aberrations for humans if these effects have not been demonstrated in intact mammalian systems following inhalation? In other words, is it really a truly genotoxic agent? Therefore, as pointed out by Carlborg (17), is linear risk extrapolation justified? Such a low-dose linear risk assessment can be made for truly genotoxic agents that operate to damage DNA at subtoxic doses for which no other detectable biological endpoints are observable.

Epidemiologic Evidence

Because the primary route of exposure to formaldehyde for humans is through inhalation, one might expect the upper respiratory tract to be the chief target tissue. By using ^{14}C-labeled formaldehyde in rats, the upper respiratory tract has been shown to be the major absorption site (*18*). Indeed, toxic levels induce squamous cell carcinomas in the nasal cavities of rats and mice.

For these reasons, in exposed populations one might expect to find an increased incidence of tumors of the nasal and nasopharyngeal epithelium if formaldehyde were carcinogenic to humans. This issue has been reviewed by Squire and Cameron (*6*), and they concluded that nasopharyngeal cancer is an uncommon disease in the Western world. Case-control and cohort studies have been reported by numerous investigators, and two review bodies have studied the evidence. The conclusion reached was that although formaldehyde gas can be considered carcinogenic for rats, inadequate evidence existed to evaluate its carcinogenicity to humans (*9, 10*). The Federal Panel on Formaldehyde (*19*) concluded that presumption of formaldehyde being carcinogenic in humans exists, but lack of information on exposure and confounding factors of multiple exposures hampered the interpretation of the data.

Conclusion

Certain carcinogenesis studies should be repeated to take into consideration the methodology issues now recognized to significantly influence test results. With more reasonable and verifiable scientific data, public health decisions concerning estimations of risk to humans from ingestive or inhalation exposures could be made on a much sounder scientific basis.

Humans are ingesting, and have always ingested, large amounts of many natural substances that might influence cancer risk. Everyone agrees that the public ought to be protected from new and additional significant environmental risks, but such decisions should be realistic and based on sound, agreed-upon scientific data. We cannot afford to make rash decisions or decisions by panic that erode public confidence in the regulatory process and impose tremendous economic turmoil if the actual benefit to public health is disproportionally small.

Literature Cited

1. National Toxicology Program, Ad Hoc Panel on Chemical Carcinogenesis Testing and Evaluation Report, Feb. 15, 1984.
2. The Nutrition Foundation, Ad Hoc Working Group Report on Oil–Gavage in Toxicology, July 14–15, 1983.

3. Kolbye, A. C., Jr. *Reg. Toxicol. Appl. Pharmacol.* **1982**, *2*, 232–37.
4. Watanable, P. G.; Gehring, P. J. *Environ. Health Perspect.* **1976**, *17*, 145–52.
5. Moolenaar, R. J. *Reg. Toxicol. Appl. Pharmacol.* **1983**, *3*, 381–88.
6. Squire, R. A.; Cameron, L. *Reg. Toxicol. Appl. Pharmacol.* **1984**, *4*, 107–29.
7. Proctor, D. F.; Chang, J. C. F. In "Nasal Tumors in Man and Animals"; Reznik, G.; Stinson, S. F.; Eds.; CRC: Boca Raton, Fla., 1983.
8. Swenberg, J. A.; Gross, E. A.; Martin, J.; Popp, J. A. In "Formaldehyde Toxicity"; Gibson, J. E., Ed.; Hemisphere: Washington, D.C., 1983.
9. *IARC Monogr. Eval. Carcinogen. Risk Chem. Man* **1982**, *29*, 345–89.
10. *IARC Monogr. Eval. Carcinogen. Risk Chem. Man* **1982**, *29*, 391–98.
11. Williams, R. T. "Detoxication Mechanisms: The Metabolism and Detoxication of Drugs, Toxic Substances, and Other Organic Compounds," 2d ed.; Wiley: New York, 1959; pp. 88–90.
12. Committee on Aldehydes, National Research Council "Formaldehyde and Other Aldehydes," NAS: Washington, D.C., 1981.
13. Food Safety Council "Proposed System for Food Safety Assessment," Washington, D.C., 1980.
14. Auerbach, C.; Moutschen-Dahmen, M.; Moutschen, J. *Mutat. Res.* **1977**, *39*, 317–62.
15. Formaldehyde Institute, Report on the NCTR Consensus Workshop on Formaldehyde, Scarsdale, N.Y., November 1983.
16. Clary, J. J.; Gibson, J. E.; Waritz, R. S. "Formaldehyde: Toxicology, Epidemiology, and Mechanisms"; Marcel Dekker: New York, 1983.
17. Carlborg, F. W. In "Formaldehyde: Toxicology, Epidemiology, and Mechanisms"; Marcel Dekker: New York, 1983; pp. 31–45.
18. Heck, H. D.; Chin, T. Y.; Schmitz, M. C. In "Formaldehyde Toxicity"; Gibson, J. E., Ed.; Hemisphere: Washington, D.C., 1983.
19. Federal Panel on Formaldehyde *EHP, Environ. Health Perspect.* **1982**, *43*, 139–68.

RECEIVED for review September 28, 1984. ACCEPTED December 19, 1984.

Formaldehyde Risk Analysis

JOHN J. CLARY
Celanese Corporation, New York, NY 10036

Formaldehyde is a ubiquitous chemical with widespread use. Potential exposure to formaldehyde is possible under many circumstances. Short-term health effects, like respiratory irritation and dermal sensitization, are well documented in humans. Potential long-term effect information comes from animal studies. Formaldehyde is weakly genotoxic in short-term in vitro tests. The major concern relates to the finding of nasal cancer in rodents exposed for a lifetime to formaldehyde. No evidence suggests any systemic health effect or effects at sites remote from the site of contact. Metabolism is rapid in both animals and humans as formaldehyde is converted to formic acid and then to CO_2. The use of mathematical models is difficult with the short-term data, although it can be used to extrapolate the data from high dose to low dose in the Chemical Industry Institute of Toxicology long-term animal studies. The most conservative approach in this area is given by a linear model. Nonlinear models, however, give a better fit to the data as they incorporate all the modifying factors. Human data is a good source of information on potential short- and long-term health effects. Although formaldehyde is an irritant and a sensitizer, epidemiological studies to date have shown formaldehyde is not a potent carcinogen in humans. However, further work is needed to determine if any carcinogenic potential exists.

FORMALDEHYDE, one of the most commonly misunderstood gases, has been produced commercially for approximately 100 years. Originally, it was used as an embalming fluid. But today it is used in a variety of different applications, such as manufacturing resins for plywood, particle board, and insulation; in plastics, cosmetics, and vaccines; and in permanent press textiles and paper products (1).

Formaldehyde is a colorless gas with a very pungent odor. This smell is important because the irritant has served as an automatic mechanism to keep human exposures low. Formaldehyde is a normal metabolite in humans and, in small quantities, is metabolized rapidly. It is broken down in the body by oxidation to formic acid, followed by further oxidation to carbon dioxide and water.

0065–2393/85/0210/0341$06.00/0
© 1985 American Chemical Society

Although formaldehyde is manufactured in modern, technological production facilities, it is also a natural by-product of smoke and exhaust from common, everyday sources such as cigarettes, power plants, and automobiles. To date, a number of estimates have been made concerning formaldehyde produced from some of these various sources. For example, direct emission of 6 million lb per year of total aldehydes from manufacturing plants has been reported. However, automobile exhaust contributes even more. An estimated 260 million lb per year (in the U.S.) results from the direct combustion of automobile fuels, and another 1 billion lb is produced from atmospheric oxidation of exhaust hydrocarbons. Power plants are another source. Industrial plants that are fired by coal, oil, and gas produce 50 million lb annually, and burning waste material in town and village dumps accounts for roughly another 13 million lb per year.

Although the amounts contributed by these various sources of formaldehyde (including industrial emissions) seem high to the average person, the ambient level of formaldehyde is actually relatively low and in most cases averages well below 0.10 ppm. In one study in Los Angeles, ambient formaldehyde levels ranged from 0.05 to 0.12 ppm over a course of 26 days. The average daily concentration was 0.06 ppm (2).

Numerous air quality standards have been established and are in the process of being revised in the United States and in Europe (3). A summary of these outdoor ambient and indoor standards is given in Table I and shows that the air standards are quite low, particularly in some European countries. Enforcing these standards presents several problems: one is ana-

Table I. Recommended and Promulgated Limits for Exposure to Formaldehyde

Exposure	Country or State	Limit (ppm)	Status
Outdoor, ambient air	United States	0.1 (ceiling)	recommended[a]
	New York State	0.0013 (annual avg.)	recommended[b]
	Federal Republic of Germany	0.025 (ceiling)	promulgated
	USSR	0.008 (ceiling)	promulgated
Indoor air	Denmark	0.12 (ceiling)	recommended
	The Netherlands	0.1 (ceiling)	promulgated
	Sweden	0.1–0.4 (ceiling)	recommended
	Federal Republic of Germany	0.1	recommended
	Wisconsin	0.4	promulgated
	Minnesota	0.5	promulgated
	California	0.2	proposed

[a]American Industrial Hygiene Association.
[b]New York Air Guide-1 (12/4/81).
Source: Reproduced from Ref. 3.

lytical; another is the broader problem of energy conservation and the number of air exchanges in the home.

The purpose of this chapter is to review all available key health effects data on formaldehyde and to put this information in some sort of perspective as it relates to human risk. Defining human risk is the bottom line of any worthwhile testing program. In the case of formaldehyde, the Chemical Industry Institute of Toxicology (CIIT) has done work that ties key observations together and makes them more useful in defining risk.

Exposure

Blade, in 1982, reviewed five sampling and analytical methods for formaldehyde evaluated by the National Institute of Occupational Safety and Health (NIOSH) (4). These included the Draeger detector tube, a CEA instrument model 555 direct-reading ambient air monitor, a midget impinger system containing 1% aqueous sodium bisulfite (followed by colorimetric analysis by the chromotropic acid), a charcoal-filled tube, and finally, a coated chromosorb 102R tube. In the last two cases, determination was made by desorbed formaldehyde, followed by gas chromatography (GG). The first three methods were primarily for area sampling, whereas the last two could be used both for areas and personal sampling. Advantages and disadvantages were discussed. The coated chromosorb 102R tubes were favored by NIOSH because the method was specific for formaldehyde and the sample stability was less of a problem with this method. The exposure panel of the National Center for Toxicological Research (NCTR) "Consensus Workshop on Formaldehyde" also reviewed analytical methods.

Acute Human Effects

Humans first sense the presence of formaldehyde by its odor or detection threshold. This response is followed by sensory irritation of the eyes, nose, and throat as the airborne concentration increases. Identification of the level of odor that is first detected is extremely difficult because it depends largely on the test environment. Sensory irritation in humans has been reported at levels as low as 0.25 ppm of formaldehyde in controlled-environment chamber studies, whereas under more normal (standard) conditions, levels of 1 ppm are linked to sensory irritation in humans. Table II lists some of the sensory irritation responses reported in the literature.

The first study to report eye, nose, and throat irritation in conjunction with airborne measurements of formaldehyde was conducted in 1955 by Ettinger and Jeremias (5). Formaldehyde levels ranged from 1 to 11 ppm. The population studied worked with fabric treated with formaldehyde resin.

Table II shows additional similar data. This information was taken from the NIOSH criteria document (1) and is arranged chronologically.

Table II.   Airborne Formaldehyde and Sensory Irritation in Humans

Concentration (ppm of HCHO)	Duration of Exposure	Responses
1–11	8 h/day	eye, nose, and throat irritation (5)
13.8	30 min	nose and eye irritation subsiding after 10 min in chamber (6)
0.13–0.45	?	temporary eye and upper respiratory tract irritation (7)
16–30	8 h/day	eye and throat irritation and skin reaction (8)
0.9–1.6	8 h/day	itching eyes, dry and sore throats, disturbed sleep, and unusual thirst upon awakening in the morning (9)
0.3–2.7	8 h/day	annoying odor, constant prickling irritation of the mucous membranes, disturbed sleep, thirst, and heavy tearing (10)
0.09–5.26 (with paraformaldehyde)	1 h	eye and upper respiratory irritation (lessened during the day) (11)
0.9–3.3	1 h	mild eye irritation and objectionable odor (12)
0.9–2.7	1 h	tearing of eyes and irritation of nasal passages and throat (irritant effects were greatest at very beginning of workday and after lunch) (13)
2.1–8.9, 0.5–3.3	daily	increased occurrence of upper respiratory irritation (14, 15)
3	?	irritation of the conjuctiva, nasopharynx, and skin (17)

The most current information tends to be more realistic as analytic methodology as well as study designs have improved since the early studies (5–17).

The fact that formaldehyde causes contact dermatitis following high-level skin exposure, primarily from occupational exposures, is well documented. However, low-level exposure, from formaldehyde's use in cosmetics and other consumer products, usually does not present a problem as only a few very sensitive individuals respond to this type of exposure (18).

Dermal sensitization from formaldehyde skin contact has been reported in many industries. The rates of dermal or contact sensitization have been reported to be between 4–6% of the work population (19). Respiratory sensitization has only been documented in one case (20).

Some individuals may be hypersensitive to formaldehyde, that is, they will respond to formaldehyde by showing dermal irritation or irritation of the eyes, nose, and throat at lower levels than a normal individual. The percentage of hypersensitive individuals in the population is unknown.

Genotoxicity

Mutagen tests can be divided into two categories: in vitro and in vivo. Of the in vitro tests, formaldehyde has shown genetic activity in *E. coli*, *Saccharomyces*, mouse lymphoma, Chinese hamster ovaries, sister chromatid exchange, unscheduled DNA repair, *Neurospora*, *Aspergillus*, *Salmonella*, and unscheduled DNA synthesis. In addition, formaldehyde has shown genetic activity in the Ames test, which is reported to be the best evaluated of any of these tests as it relates to carcinogenicity (*21*). However, formaldehyde has not shown genetic activity in chromosomal aberration.

Of the in vivo tests, the Drosophila test was positive. In two dominant lethal tests the results were mixed: one was negative and the other was positive. A positive finding for sister chromatid exchange at airborne levels greater than 25 ppm was reported, but two mouse spot tests were negative (*21*). In general, the in vivo mutagen results are unclear. From a metabolic point of view, high levels of formaldehyde are unlikely to reach the target site, so the negative results are expected.

Teratogenicity–Reproduction

Formaldehyde has been evaluated in several test systems for potential teratogenic response. The results of two inhalation studies in rats (*22*, *23*) indicate no teratogenic response at a low-level exposure, but fetal or maternal toxicity is suggested. Both studies are somewhat lacking in detail and thus make complete assessment difficult. Several studies using the oral route have also been done. Hurdi and Ohder (*24*) studied the effect of formaldehyde on reproduction in beagle dogs. They used either hexamethylenetetramine (HMT), a material that breaks down to formaldehyde in the body, or formaldehyde itself, at levels of 125 or 375 ppm between day 4 and day 56 of gestation. They concluded that no increase in malformations occurred. However, the animals were allowed to deliver, and no specific attempt was made to look for malformation in the offspring prior to delivery.

The most recent study, by Marks et al. (*25*), used albino mice incubated with 1% aqueous formaldehyde levels of 0, 74, 148, and 185 mg/kg/day on days 6 through 15 of gestation. Formaldehyde treatment did not result in any malformations, as seen in Table III, although fewer total litters and pups were produced at the high dose because of mortality to the dams due to formaldehyde. No animal tests to date show any teratogenic response following exposure to formaldehyde.

Cancer

The carcinogenic potential of formaldehyde has been investigated and reported in several recent studies. The most complete is one by CIIT (*26*). In this study, 240 rats and mice, divided evenly between male and female, were exposed by inhaling 15, 6, or 2 ppm of formaldehyde for 6 h/day, 5 days/week, for 2 years. When some animals were sacrificed at 6-, 12-, and

Table III.   Effect of Formaldehyde on the Incidence of Malformed Mouse Fetuses

Description	Dose (mg/kg/day)			
	0 (Control)	74	148	185
No. of fetuses examined externally	832	311	336	83
No. of fetuses with external malformations	3	3	0	1
No. of litters containing fetuses with external malformations	3	3	0	1
No. of fetuses examined viscerally	315	113	119	33
No. of fetuses with visceral malformations	0	0	1	0
No. of litters containing fetuses with visceral malformations	0	0	1	0
No. of fetuses examined skeletally	832	311	336	83
No. of fetuses with skeletal malformations	0	1	1	0
No. of litters containing fetuses with skeletal malformations	0	1	1	0
Total no. of malformed fetuses	3	4	2	1
Total no. of litters with malformed fetuses	3	4	2	1
Average percent of malformed fetuses	0.4	1.3	1.2	1.0

Source: Reproduced with permission from Ref. 41. Copyright 1983 Hemisphere Publishing Corp.

18-month periods, examination showed histological changes in the nasal epithelium at high formaldehyde exposure levels, primarily described as metaplasia (replacement of one cell type by another). Hyperplasia (building up of cells) was noted by the 12th month in rats exposed to 15 ppm. By the 13th or 14th month, the first squamous cell carcinoma in a rat's nasal cavity was seen. When the study was terminated after 2 years, two squamous cell carcinomas were found in the mice exposed to 15 ppm of formaldehyde, and two were found in the rat exposed to 6 ppm. Only the incidence of nasal squamous cell carcinoma in rats exposed to 15 ppm was statistically significant. The observations at 6 ppm in the rat and 15 ppm in the mouse have to be considered of biological significance. The finding of nasal cancers in rats at the high exposure level, at which a 45% incident rate was noted, was of great interest. First, 15 ppm of formaldehyde leads to significant tissue destruction. Additional CIIT research indicates that exposure to 15 ppm of formaldehyde yields rapid cell turnover in the nasal cavity of rats and mice, mainly because of the destructive nature of such a high level of formaldehyde. The rapid cell turnover results in overcoming of the DNA repair mechanisms. This type of activity (promotion) could be considered a primary mechanism of the formaldehyde action, and is not seen at low concentrations.

A second inhalation study was done by New York University (NYU) and used rats at exposures of 14 ppm. The results were similar to CIIT results with respect to cancer production (27). Other studies in mice and hamsters have shown no increases in the incidence of cancer (28).

Mechanism

These studies indicate that the carcinogenic response to formaldehyde varies from species to species. Where cancer is observed it is seen only at the site of first contact, and can be related to delivered dose. Interestingly, no evidence suggests any systemic effect from formaldehyde exposure. This finding is not unexpected because, as discussed earlier, formaldehyde is a normal metabolite both in animals and humans (29). In small quantities it is rapidly metabolized via oxidation to formic acid and then to CO_2 and water. Several enzymes are capable of breaking down formaldehyde to CO_2 and water. Formaldehyde is also incorporated in thymine, purines, and amino acids. With respect to the rodents that developed cancer in the laboratory tests, cancer was produced only at the site of contact in these rodents, and this result may be related to both the promotion activity as well as the reactivity of formaldehyde as an electrophile. This reactivity leads to the formation of adducts with nucleic acids and proteins, especially at high concentrations.

The data from the CIIT study have been used in different mathematical models to predict carcinogenic response at low exposure levels in rats. These predictions apply only to the rat; the mouse data cannot be used because only two animals showed a response at 15 ppm. If humans were similar to rats, extrapolation would perhaps have some value; if humans were similar to mice, extrapolation would be practically meaningless. Moreover, humans are not obligatory nose breathers like rodents.

Table IV shows that when most of the standard mathematical extrapolation models are used (probit, logit, multihit, Weibull, and multistage), a concentration of 1 ppm of formaldehyde is necessary to produce one tumor in 100,000 rats. The linear model gave a response quite different from the other models and is not supported by mechanistic information. Taking this extrapolation a step further and looking at the formaldehyde level necessary to produce one tumor in 100,000,000 rats (equivalent in humans to 2.3 tumors in the whole U.S. population if humans are similar to rats), one finds that the predicted formaldehyde level necessary to produce a response is well above the normal ambient air level found in the United States (Table IV).

Many factors must be considered in discussing the mechanism for for-

Table IV.   Formaldehyde Risk Estimation for Tumors in
Rats

| Model | Formaldehyde (ppm) | |
	1 : 100,000 Risk	1 : 100,000,000 Risk
Probit	2.3	1.3
Logit	1.1	0.22
Multihit	1.7	0.58
Weibull	0.98	0.17
Multistage	0.98	0.17
Linear	5.5×10^{-3}	5.5×10^{-6}

Source: Reproduced with permission from Ref. 41. Copyright
1983 Hemisphere Publishing Corp.

maldehyde action. Is concentration more important than total dose? Is the
rat a good model for humans? Is formaldehyde acting as a tumor promoter
or as a complete carcinogen? Because of the genotoxic nature of formaldehyde, one must assume that it has some potential to be an initiator. However, it behaves more like a nongenotoxic agent because serious tissue damage is needed before any cancer is seen. Recent evidence generated by CIIT
in the area of mucociliary function, DNA adduct formation, and cell proliferation supports the concept of nonlinear response at low doses. This response can be related to the delivered dose.

In the CIIT study a great difference was found in the response between the rat and the mouse. This finding can be partially explained by the
difference in each species' respiratory rates, the effect of irritants, and the
comparable dose in micrograms per square centimeter of nasal area (30).
For example, rats have a 10–20% reduction in minute volume, whereas the
mouse reduction is 70% at 15 ppm. Because of this disparity and the difference in surface areas, etc., the average exposure at 15 ppm to a mouse is
approximately one-half that of the rat. Or, in other words, the mouse's
response would be equivalent to approximately one-half of a dose, or 6
ppm in the rat. One look at the test responses between the two species indicates this situation is indeed what occurred. However, the important thing
to focus on here is actual dose at the target organ. This concept is becoming
increasingly important as more is understood about this area. At very low
levels of formaldehyde, the nasal mucociliary apparatus performs a vital
protective function that changes the response, or at least gives a practical
threshold, for formaldehyde exposure in the rat. The mucous layer is primarily water, but it does contain glycoproteins, proteins, and salts, as well
as other materials. Formaldehyde can readily react with the proteins in this
mixture.

CIIT has demonstrated the effect of low levels of formaldehyde exposure on mucociliary function (31). The first thing seen as the exposure in-

creases is the reduction of the mucus flow rate; this reduction is eventually followed by a thickening of the mucus and a cessation of the ciliary motion that moves the mucus along. This response is correlated, to some degree, with the earlier reported decrease in mucus flow rates occurring in humans exposed to concentrations as high as 1.6 ppm. At concentrations greater than 5 ppm in rats an initial stimulation of the mucus flow occurs, but eventually a thickening and then a cessation takes place, even in the ciliary function. In animals exposed to concentrations less than 0.5 ppm, no effect on mucociliary function was seen. At levels around 2 ppm, only small areas of cilia are affected, and at 6 ppm, focal effects on the whole mucociliary apparatus can be seen. At 15 ppm, complete cessation occurs. This protective mechanism is very effective at low concentrations and actually results in no formaldehyde reaching the target cell because of the reaction with the protein of the material, the mucus, and the sweeping action of the cilia to clear the potential target cells. However, once this mechanism is saturated, formaldehyde can reach the target cells. This result, then, is a demonstration of the delivered dose.

This protective mechanism is extremely important. The effectiveness of this mechanism is further demonstrated by the work of Heck from CIIT (*30–31*) in which he measured DNA covalent binding. The amount of DNA and protein reacting with formaldehyde was significantly increased at 6, 15, and 30 ppm of formaldehyde. However, the amount of formaldehyde covalently bound to DNA at low levels, 0.3 and 2 ppm, was 10 times lower than the cross-linking at 6 ppm. This dose-response data shows a sharp departure from linearity at 6 ppm and below. This break or decrease in response correlates extremely well in the mucociliary clearance mechanisms. Essentially, what appears to be happening is that the mucus and ciliary action is protecting to some degree the animal up to approximately 6 ppm. At higher levels, this protective mechanism tends to be completely swamped, but at lower levels it prevents formaldehyde from getting to the target tissue. This result gets into the very important concept of the delivered dose versus the administered dose. The delivered dose is the amount that actually reaches the target cell. The administered dose is the amount of material that is in the air.

Swenberg has shown some surprising results with cell proliferation following formaldehyde exposure. Effects have been seen as low as 0.5 ppm. This increase in cell proliferation is very transient and returns to normal even though exposures continue. However, at concentrations greater than 6 ppm a continued evolution of cell proliferation occurs. One might speculate that the first exposure results in cell proliferation and, therefore, a possibility for DNA binding (initiation?). Continued cell proliferation such as that seen at 6 ppm and greater is needed to get promotion and the carcinogenic response. These changes between 2 and 6 ppm in regard to cell proliferation are quite dramatic. Although this increase is only threefold in terms of the administered dose in 1 day, the first day when a transient response at both 2 and 6 ppm occurs, an eightfold difference in the proliferative rate is seen. By 3 days the difference in proliferative rate in animals exposed at 2 or 6 ppm of formaldehyde is 35-fold.

Chronic Inhalation

A study of rats, hamsters, and monkeys done by the Formaldehyde Institute (32) had the following objectives: to determine if any adverse effect from low-level, near-continuous exposure to formaldehyde occurs; to determine the existence and degree of any dose response from exposure to formaldehyde; to obtain input data suitable for extrapolation models; and to determine if any demonstrable species' response differences to formaldehyde exposure occur between the tested species. The animals were exposed by inhalation at four levels (0, 0.2, 1, and 3 ppm) for 22 h/day, 7 days/week, for 6 months. At each exposure level, 20 hamsters (10 male, 10 female), 40 rats (20 male, 20 female), and 6 male monkeys were used.

The only treatment-related effect on body weight was in the rat at 3 ppm. In this rat some visible indication of irritation was also present, whereas at the lower levels no compound-related effects were seen. In the monkey, nasal discharge and a hoarseness were observed at 3 ppm, but not at 1 ppm. In the hamster, no visible signs of a treatment-related response were seen at any level.

Histopathological examination indicated the nasal cavity was the only tissue showing a treatment-related effect. The effect seen was squamous cell metaplasia in rats and monkeys at 3 ppm. No effects were seen at 1 ppm or lower exposures in these two species; this result suggested a threshold.

The rat appears to be the most sensitive species of the three studied at 3 ppm, as indicated by its body weight changes as well as by histological changes in the nasal cavity classified as metaplasia. The monkey showed an intermediate response at 3 ppm of metaplasia in the nasal cavity and nasal discharge. The hamster showed no adverse effect at any level investigated. Because these studies were designed as continuous low-level exposures for 6 months, no judgment can be made about carcinogenic potential, although some interesting observations can be made about low-level continuous exposure.

Nearly continuous exposure to all levels of formaldehyde does not appear to offer any severe adverse health effects such as irreversible organ damage. Formaldehyde-induced metaplasia, following high-level exposure, has been shown to be reversible by Schreiber et al. (33) and CIIT (35). The systemic effects indicated in the test rats are, at this stage, ill-defined and could also be reversible on cessation of exposure. One of the key response differences between the species appears to be metaplasia, and this phenomenon merits some additional discussion. Table V shows differences in species' response to formaldehyde and compares concentration

Table V.   Effect of Airborne Concentration of Formaldehyde on Metaplasia in the Nasal
Cavity

Species	Concentration (ppm)	Time	Concentration × Time (ppm·h)	Effect[a]
Rats	1	22 h/day, 7 days/week for 6 months	4001	NM
	2	6 h/day, 5 days/week for 10 months	4600	M
	3	22 h/day, 7 days/week for 6 months	12,012	M
	6	6 h/day, 5 days/week for 10 months	14,010	M
	15	6 h/day, 5 days/week for 1 week	450	M
Mice	2	6 h/day, 5 days/week for 18 months	4680	NM
	6	6 h/day, 5 days/week for 18 months	14,040	M
	15	6 h/day, 5 days/week for 18 months	35,100	M

[a]NM indicates no metaplasia; M indicates metaplasia.
Source: Reproduced with permission from Ref. 41. Copyright 1983 Hemisphere Publishing Corp.

versus total dose in the Formaldehyde Institute study and the CIIT study. The total dose per week is calculated by multiplying the concentration with the total time of exposure per week (ppm·h). The rat data in Table V include exposure concentrations from 1 to 15 ppm. If one looks at the concentration effect, metaplasia is seen at 2 ppm and above. But if one looks at the total dose (concentration × time), no correlation with metaplasia is seen. This result suggests that concentration is much more important than total dose in producing metaplasia and any other chronic response. As shown in Table V, concentrations in the mouse study ranged from 2 to 15 ppm. Metaplasia was not seen in the mouse except at levels greater than 6 ppm.

In addition to the data in Table V, CIIT recently conducted a study (36) in which total exposure was compared to concentration or length of exposure. Animals were exposed at 3 ppm for 12 h, 6 ppm for 6 h, or 12 ppm for 3 h. Table VI shows the cell turnover rate in rats exposed under these various regimens. Level A responses appear to be very similar, but level B responses show a very significant difference, varying with concentrations—not the total exposure. Although this response may mean more to a pathologist, I think the 3-ppm animals look considerably different from the 12-ppm animals and, if anything, look more similar to the controls.

This type of data strongly indicates that concentration is much more important than accumulated dose. An accurate measurement of the concentration in the air becomes even more important for material of this nature.

Other important points to consider are the following: First, unlike rats, humans are not obligatory nose breathers, that is, they tend to breathe through their noses at high concentrations. Second, and equally critical, is the irritating nature of formaldehyde exposure. Tearing of the eyes, increased mucus flow, etc. are warnings of overexposure. Humans tend to avoid situations that are irritating.

Table VI.   Effect of Formaldehyde Concentration Versus Cumulative Exposure
on Cell Turnover in Rats

| | % Labeled Cells | | |
| | Level A | Level B | |
Exposure	3 days	3 days	10 days
Control	3.0 ± 1.6	0.54 ± 0.03	0.26 ± 0.02
3 ppm × 12 h	17.0 ± 1.5	1.73 ± 0.63	0.49 ± 0.19
6 ppm × 6 h	15.5 ± 10.0	3.07 ± 1.09	0.53 ± 0.20
12 ppm × 3 h	16.5 ± 2.1	9.00 ± 0.88	1.73 ± 0.65

Note: All percents are mean values ± standard error. Level A refers to sections from the
anterior portion of the nasal passages that have minimal mucociliary function. Level B sections were taken from the central portion of the respiratory epithelium.
Source: Reproduced with permission from Ref. 41. Copyright 1983 Hemisphere Publishing Corp.

Extra caution should be taken when using mathematical models to
predict a response in humans because these models are, in a sense, an extrapolation of animal data from an area of known exposure to a lower area
for which no actual data is available (34). Then this extrapolation is taken
one step further and is applied to humans.

Human data play a key part in risk analysis of both the short- and
long-term effects. Irritation has been the base of the American Conference
of Governmental Industrial Hygienists (ACGIH) for establishing formaldehyde threshold limit values (TLV) in the past (35). By establishing levels
that minimize the acute response, chronic effects have been prevented in
humans. This lack of chronic effects has been established in studies of formaldehyde-exposed workers (35).

Epidemiology

A number of studies on formaldehyde in the workplace have examined cancer rates (36–39). As might be suspected in many of these smaller studies,
decreases in certain types of cancer and increases in others were found. Basically, most of these studies demonstrate no clear-cut formaldehyde-related
effects and no pattern of response, that is, no increase in any particular
type of cancer in any of the studies. Two studies perhaps stand out above
the rest. One was done by Du Pont (40) and examined eight plants having
493 cancer cases of individuals exposed to formaldehyde. This group was
matched with an equal number or employees not exposed to formaldehyde.
The report concluded that formaldehyde exposure did not result in any significantly greater risk of cancer in any organs. Although the data are limited, the study was well conducted and many comparisons were made.

A more powerful study was done in the United Kingdom by Acheson
and focuses on 17,000 workers exposed to formaldehyde since the early

1940s. The preliminary results of this study were recently reported at a conference at Oxford in July 1983. They show one plant at which a slight excess of respiratory cancer was found when compared to the national rate. However, when compared to the local rate, no excess was found, and more importantly, as of this date, smoking histories have not been examined in the study. In both the Du Pont and Acheson studies, as well as the other studies conducted, no nasal cancers were seen, and no cancerous effects that could be related to formaldehyde exposure were demonstrated.

Currently, an additional study is underway that merits some comment. This study is being conducted by the National Cancer Institute in cooperation with the Formaldehyde Institute and covers more than 30,000 workers. When complete, this study should be the most comprehensive evaluation of formaldehyde exposure possible. No information is available currently on the status of this study and none is expected until later in 1985. But some encouragement can be taken from the smaller studies that show formaldehyde is not a potent carcinogen in humans.

When one looks at all the epidemiologic evidence and the concerns generated by some excess of brain cancer and leukemia in professional people, that is, pathologists, etc., one should go back and look at mechanisms. The work by Heck suggests that formaldehyde does not cause a measurable increase in formaldehyde in the blood following exposure, both in humans and in rats. His work has also demonstrated that formaldehyde, when given in radioactive form, is found in the bone marrow strictly from incorporation of the formaldehyde metabolite. No covalent binding indicates ^{14}C-formaldehyde itself does not reach the bone marrow. Both these facts suggest that formaldehyde does not get to distant sites, but reacts at the site of contact. This information strongly supports the idea that the brain cancer and leukemia are probably due to something that might be associated with the work done by certain professional classes, but it cannot be associated directly with formaldehyde. One would also have to strongly consider the diagnostic bias in the professional group as compared to industrial workers. Workers in an industrial setting probably had higher exposure than the professional group to formaldehyde. Dose response is still a key part of evaluation of any potential response. For these reasons it is difficult to believe that the excess brain cancer and leukemia response seen in this professional group is due to formaldehyde exposure.

Literature Cited

1. U.S. DHEW, PHS, CDC, National Institute for Occupational Safety and Health "Criteria for a Recommended Standard Occupational Exposure to Formaldehyde"; Government Printing Office: Washington, 1976. DHEW (NIOSH) Publication No. 77–126.
2. Committee on Aldehydes "Formaldehyde and Other Aldehydes"; National Academy: Washington, D.C., 1981; Chapter 5.
3. National Research Council, 1980.

4. Blade, L. M. In "Formaldehyde: Toxicology, Epidemiology, and Mechanisms"; Marcel Dekker: New York, 1983; pp. 1–23.
5. Ettinger, I.; Jeremias, M. *Ind. Hyg. Mon. Rev.* **1955**, *34*, 25.
6. Sim, V. M.; Pattle, R. E. *JAMA* **1957**, *165*, 1908.
7. Bourne, H. G.; Seferian, S. *Ind. Med. Surg.* **1959**, *28*, 232.
8. Glass, W. I. *N.Z. Med. J.* **1961**, *60*, 423.
9. Morrill, E. E. *Air. Cond. Heat. Vent.* **1961**, *58*, 94.
10. Shipkovitz, H. D. "Formaldehyde Vapor Emissions in the Permanent-Press Fabrics Industry"; U.S. DHEW: Cincinnati, 1968.
11. Kerfoot, E. J. *Am. Ind. Hyg. Assoc. J.* **1975**, *36*, 533.
12. Miller, B. H.; Bleger, H. P. "Report of an Occupational Health Study of Formaldehyde Concentrations at Maximes"; California Health and Welfare Agency: Los Angeles, 1966; Study No. S–1838.
13. Blejer, H. P.; Miller, B. H. "Occupational Health Report of Formaldehyde Concentrations and Effects on Workers at the Bayly Manufacturing Company, Visalia, Calif."; California Health and Welfare Agency: Los Angeles, 1966; Study No. S–1806.
14. Yefremov, G. G. *Zh. Ushn. Nos. Gorl. Bolezn.* **1970**, *30*, 11.
15. Zaeva, G. N.; Ulanova, I. P.; Dueva, L. A. *Gio. Tr. Prof. Zabol.* **1968**, *12*, 16.
16. National Institute for Occupational Safety and Health "Criteria for a Recommended Standard: Occupational Exposure to Formaldehyde"; Government Printing Office: Washington, 1976; pp. 40, 98; NIOSH Publication No. 77–176.
17. Kratochvil, I. *Pr. Lek.* **1971**, *23*, 374.
18. Cronin, E. "Contact Dermatitis"; Churchill Livingston: London, 1980.
19. Maibach, H. In "Formaldehyde Toxicity"; Hemisphere: Washington, D.C., 1983; pp. 166–73.
20. Hendrick, D. J.; Lane, D. J. *Br. J. Ind. Med.* **1977**, *24*, 11.
21. National Toxicology Program *NTP, December*, 1980.
22. Brusick, D. J. In "Formaldehyde Toxicity"; Hemisphere: Washington, D.C., 1983; p. 72.
23. Gofmekler, V. A. *Hyg. Sanit.* **1969**, *34*, 266.
24. Sheveleva, G. A. *Toksikol. Nov. Prom. Khim. Veshchestv.* **1971**, *12*, 78.
25. Hurni, H.; Ohder, H. *Food Cosmet. Toxicol.* **1973**, *11*, 459.
26. Marks, T. A.; Worthy, W. C.; Staples, R. E. *Teratology* **1980**, *22 (1)*, 51.
27. Swenberg, J. A.; Barrow, C. S.; Boreiko, C. J.; Heck, H. d'A.; Levine, R. J.; Morgan, K. T.; Starr, T. B. *Carcinog.* **1983**, *4 (8)*, 945.
28. Albert, R. E.; Sellakumar, A. R.; Laskin, S.; Jischer, M.; Nelson, N.; Snyder, E. A. *J. Nat. Cancer Inst.* **1982**, *68*, 597.
29. Dalbey, W. E. *Toxicology* **1982**, *24 (1)*, 9.
30. Heck, H. d'A.; Casanova-Schmitz, M. "Biochemistry Toxicology of Formaldehyde," Preprint submitted to EPA, 1983.
31. Casanova-Schmitz, M.; Starr, T. B.; Heck, H. d'A. Preprint submitted to EPA, 1983.
32. Rusch, G. M., Clary, J. J., Rinehart, W. E., Bolte, A. F. *Toxicol. Appl. Pharmacol.* **1983**, *68 (3)*, 329.
33. Schreiber, H. M.; Bibbo, G. L.; Weed, G. L.; Saccommow, G.; Nettesheim, P. *Acta. Cytol.* **1979**, *23*, 496.
34. Grafstom, G. C.; Formace, A. J.; Anlup, H.; Lecliner, J. F.; Harris, C. C. *Science*, **1983**, *220*, 216.
35. "Documentation of Threshold Limit Values: Formaldehyde"; American Conference of Governmental Industrial Hygienists: Cincinnati, 1982; p. 197.
36. Levine, R. J.; Gelkovich, D. A.; Shaw, L. K.; DaCorso, R. D. In "Formaldehyde: Toxicology, Epidemiology, and Mechanisms"; Marcel Dekker: New York, 1983; p. 127.
37. Walrath, J.; Fraumeni, J. F. In "Formaldehyde Toxicity"; Hemisphere: Washington, D.C., 1983; p. 227.

38. Marsh, G. M. In "Formaldehyde Toxicity"; Hemisphere: Washington, D.C., 1983; p. 237.
39. Wong, O. H. In "Formaldehyde Toxicity"; Hemisphere: Washington, D.C., 1983; p. 256.
40. Fayerweather, W. F.; Pell, S.; Bender, J. R. In "Formaldehyde: Toxicology, Epidemiology, and Mechanisms"; Marcel Dekker: New York, 1983; p. 47.
41. Clary, John J. In "Formaldehyde Toxicity"; Hemisphere: Washington, D.C., 1983; pp. 284–94.

RECEIVED for review September 28, 1984. ACCEPTED January 10, 1985.

Formaldehyde: Refining the Risk Assessment

JOHN A. TODHUNTER

Todhunter Associates, 918 16th Street, N.W., Washington, DC 20006

Formaldehyde presents better than normal case material for risk assessment. The design of the bioassay conducted at the Chemical Industry Institute of Toxicology and the number of supporting studies available allow the risk assessor to dissect this case well. Analysis of the various factors involved in this risk assessment suggests the following: (1) Although a scientifically sound basis for routinely making interspecies conversions of unit dosages on the basis of the two-thirds power of body weight does not exist, in the specific case of formaldehyde this basis appears to be the most biologically sound. Specific differences between rats and humans suggest, however, that on such a basis the effective exposure to rats and humans may be quite similar at a given ambient level of formaldehyde gas. (2) The dose response for formaldehyde-related risk is markedly nonlinear and therefore does not appear to be a biologically sound basis for the use of linear or linear-constrained low-dose extrapolation models for formaldehyde-related risk. (3) Formaldehyde is most likely to be a local-acting carcinogen. (4) Biological considerations suggest that a practical threshold exists for increased risk due to formaldehyde exposures. (5) Apparently a linear correction for exposure duration, in units of fractional lifetime exposure, can be used to compare exposures of different lengths, but it is likely to significantly overstate the level of risk associated with exposures of circa 100-fold less than lifetime or shorter. (6) The mean estimates of low-dose risk from the multistage, Weibull, and logit models indicate that the nonthreshold estimated risks associated with human exposures are in the 10^{-6} range or lower as lifetime risks.

IN LATE 1980, preliminary results of a Chemical Industry Institute of Toxicology (CIIT) inhalation bioassay revealed that formaldehyde was carcinogenic in the rat by this route (1). This observation was subsequently repeated at New York University (NYU) in a different strain of rat (2). A number of previous studies had not provided any clear evidence of carcinogenic activity by other routes (3–7), although Watanabe et al. (3) did ob-

0065–2393/85/0210/0357$06.00/0

serve local sarcomas at the site of multiple injections of formalin, and Mueller et al. (6) reported lesions with the suggestion of carcinoma in situ on the oral mucosa of rabbits exposed to formalin via an oral "soak tank." Methodological uncertainties precluded the use of either of these two studies as evidence for a carcinogenic potential of formaldehyde. Epidemiological studies (8–11), although limited in size, have not provided any evidence of a carcinogenic effect of observable magnitude in humans exposed to levels of formaldehyde in excess of 1 ppm as a result of occupational exposure.

Formaldehyde provides excellent material for a case study in risk assessment. The CIIT bioassay employed a three-dose design that allows for more meaningful analysis of the shape of the dose–response curve than is possible with the usual bioassay design that employs the maximum tolerated dose (MTD) and only one other, intermediate, level of exposure (12). The CIIT design also included a number of interim sacrifices that allow for the use of low-dose extrapolation models that are time as well as dose dependent (time-dependent models, however, are not used in the present assessment). In addition to these improvements in the bioassay design, the amount of biochemical and physiological data on formaldehyde is large. This abundance of data allows the assessor to dissect the toxic effects of formaldehyde and to interpret the dose–response curve in more detail than is usually possible (12).

Because risk assessment is usually seen as a "black box" process, the National Research Council (NRC) has recommended that risk assessments be explicitly laid out for peer review and that the "inference options" implicit in a risk assessment be clearly identified and the choices made be explained (13). The present case study is an attempt to put those recommendations into practice.

In any risk assessment, a variety of factors must be considered if the resulting estimates of risk are to be meaningful. For the present assessment the following factors were seen as relevant: confidence in the bioassay, site of action of formaldehyde, cellular formaldehyde metabolism, pharmacodynamic considerations, DNA repair and detoxification processes, effects of cytoxicity on tumor response, consideration of threshold processes, rat-to-human scaling, rat and human sensitivity, choice of extrapolation model, duration of exposure, and time-to-tumor effects. Several of these factors are interrelated, and the discussion that follows is not laid out strictly by these factors.

Synopsis of the CIIT Bioassay

The choice of study for the extrapolation of low-dose risk is the first critical point in a risk assessment. The CIIT bioassay used ambient air exposures to 2.1, 5.6, and 14.3 ppm of formaldehyde in addition to a control group. Both rats (Fisher-344) and mice (B6C3F1) were exposed. Interim sacrifices at 6, 12, and 18 months were scheduled with a final sacrifice at 24 months.

The design of the bioassay was such that the rats were exposed intermittently (6 h/day for 5 days/week). This design results in an aggregate of 3120 exposure hours out of the normal (ca. 25,000 h) rat lifetime and requires an adjustment if the risks derived are to be expressed as for complete lifetime exposure. This adjustment is done here on the basis of cumulative dose even though CIIT data indicate that cumulative dose may overestimate risk from formaldehyde when compared to dosing rate as the basis for adjustment (*14*).

The NYU study is not used here as a basis for risk assessment because it used too few exposure levels to allow for meaningful analysis of the shape of the dose–response curve. Combination of the NYU and CIIT incidence data is precluded because the NYU study used a different strain of rat (Sprague–Dawley) that may show a different level of response than the Fisher-344 rat at equal exposures.

The CIIT study has been thoroughly reviewed (*14, 15*) and is established as a valid bioassay for the inhalation carcinogenicity of formaldehyde gas.

Incidences of tumors in the CIIT rats are given in Table I. The table includes some animals sacrificed at 18 months and some that died or were terminated between 18 and 24 months and therefore represents a slightly reduced population at risk at 24 months. In Table I, also, carcinomas of different types and sarcomas are pooled to maximize treatment-related effects. The incidence data in Table I form the basis for the low-dose extrapolations that are presented later. Polyploid adenomas were observed in all groups, including controls, and were neither clearly treatment related nor clearly precancerous. Even when adenomas or other benign lesions do progress to malignancy, the efficiency of the process is very low and variable (*15*); therefore, benign neoplasms cannot properly be pooled with malignancies for purposes of risk estimation.

Mice in the CIIT study did not develop nasal tumors at any exposure level in significant excess of controls. Nasal tumors in rodents appear to be quite rare as a spontaneous lesion, and so the low incidence seen in the mice given the highest dose is biologically significant, even though it is not statistically significant.

Table I. Incidences of Tumors in the Fischer-344 Rat in the CIIT Bioassay (Both Sexes, All Malignancies)

Exposure (ppm)	Response	Incidence
0	0/232	0
2.1	0/236	0
5.6	2/235	0.0085
14.3	106/232	0.457

NOTE: Data are given as animals with tumors per number of animals with nasal cavities examined.

Quantitative Risk Assessment

Risk–Exposure Curves. The relationship between exposure and a toxic effect such as carcinogenic response is determined by a risk–exposure curve (or a dose–response curve). The difference between a dose–response curve and a risk–exposure curve is that the dose–response curve represents a fit to the observed-response data as a function of dose (or exposure), whereas the risk–exposure curve is the extrapolation of the dose–response curve to domains of exposure lower than those within which observable response occurs. For most toxic end points, the dose–response curve does not appear to show a response greater than zero until some "threshold" dose, greater than a zero dose, is reached. For a carcinogenic end point a population threshold dose is generally difficult to define, although for many types of carcinogens a threshold is theoretically expected (*16, 17*). Accordingly, the usual practice is to construct a risk–exposure curve for a carcinogenic effect by use of any of several available extrapolation models (*16*). No sound basis is apparent in experimental data for generally preferring one model over another (*12, 16*). The construction and interpretation of a set of risk–exposure curves require a variety of explicit and implicit assumptions. Those that are explicit in the present case are dealt with in this section. Those that are implicit are dealt with in the following section.

Figure 1 shows the risk–exposure curves generated by application of the multistage, Weibull, and logit models to the CIIT tumor incidence data of Table I. Observed tumor responses are shown on the figure. The CIIT data show a very nonlinear dose–response curve, and a number of linear model forms did not provide a good fit to the data. For reasons discussed in the following section, this nonlinear response appears to be a reflection of the way in which formaldehyde gas is delivered to and interacts with target tissues in an exposed organism. This observation indicates that appropriate models for extrapolation must be drawn from among those that are not constrained to linearity.

Of the models that fitted the data well, the multistage (Curve A) gave the most "conservative" risk–exposure function (i.e., the highest extrapolated risk at any given exposure). The multistage model's 95% confidence limits (the "linearized" multistage model) gave a particularly inappropriate function (Curve AU). The other models fitted the actual CIIT data better and generated functions that were consistent with each other (Curves B and C). The 95% upper confidence limits also fitted well and were consistent with each other (Curves BU and CU). Curves B, C, BU and CU gave lower estimates of risk at any given dose than Curves A and AU.

No one model of those shown is "best" (*12*), and, therefore, the estimates presented in this analysis are based on a mean of the estimates provided by all three models in their respective maximum likelihood forms (Curves A, B and C). By definition, the probability that the "true" risk lies at the 95% upper confidence limit is low relative to the probability that it

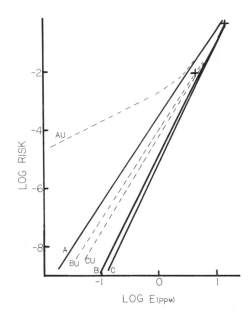

Figure 1. Risk–exposure curves derived from the CIIT rat bioassay results by application of four extrapolation models. The curves shown were generated by use of the multistage (A), Weibull (B) and logit (C) models. Curves AU, BU and CU are the respective 95% upper confidence limits on the models used. The models were applied by use of the GLOBAL 79 and RISK 81 programs, and the results shown are those obtained by EPA's Office of Toxic Substances Epidemiology Branch for the CIIT data.

lies at or near the maximum likelihood value. Similarly, the "true" risk lying at the 95% lower confidence limit is equally probable to it lying at the upper confidence limit. Accordingly, the use of risk estimates derived from the upper confidence bounds creates a distorted impression of what is most probable with respect to risk and is avoided here. A composite curve, fitted to the mean of estimates from each of the four models at various exposures, is shown in Figure 2. This composite curve is used for subsequent derivation of risk estimates. The curve shown has been adjusted to reflect risk from a full lifetime exposure in a rat (as discussed in the following paragraph). The curve has not been drawn on an exposure scaled for the differences between rats and humans with regard to effective formaldehyde dose and is thus directly applicable to only rats as shown.

Adjustment for Lifetime Exposure.   As discussed earlier, the CIIT assay design involved an intermittent exposure for an aggregate less than full lifetime exposure. This design requires that the risk estimates derived from the CIIT data be adjusted to reflect the risk that would be posed by a full lifetime of exposure (24 h/day for life). This adjustment can be done on the basis of cumulative exposure (*14*) expressed as total exposure hours as a

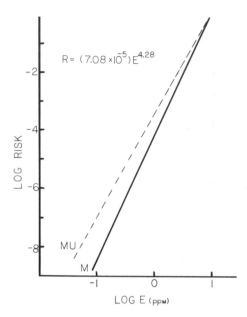

Figure 2. Risk–exposure relationship for formaldehyde in rats based on the CIIT bioassay data derived from the mean of estimates from four models and adjusted for a lifetime cumulative dose. Curve M is the mean of the estimates from the multistage, Weibull, logit, and gamma-multihit models. Curve MU is the mean upper confidence limit at the 95% level.

fraction of possible lifetime exposure hours. This quantity is the fractional lifetime exposure, f. Swenberg et al. (18) have pointed out that the dosing rate may be more dominant for exposures to formaldehyde than the cumulative dose. The use of the fractional lifetime exposure to adjust for exposures of differing durations assumes a linear relationship between length of exposure and risk and also assumes a simple additivity of numerous short exposures. This problem of dose fractionation is a vexing one in carcinogen assessment (17), and for many human cancers and carcinogens the relationship between exposure duration and incidence or risk is nonlinear (19). This result suggests that the use of a fractional lifetime adjustment will tend to overestimate the risk from exposures of significantly less than lifetime duration.

Rat-to-Human Scaling. The incremental risk posed to humans by ambient air levels of formaldehyde gas can in theory be estimated from the risk–exposure function. If the assumption of no threshold dose for this risk is made, then a putative risk at low dose is extrapolated on the basis of some low-dose extrapolation model. In this case, the composite curve of Figure 2 is suggested to best characterize the risk–exposure function. As pointed out earlier, the curve of Figure 2 is based on response data in rats. Use of this curve to estimate human risks requires that the exposure scale used reflect equivalent human exposure at given ambient levels of formaldehyde gas. (Other assumptions are required as well, and these are discussed elsewhere in this chapter.) The major assumption involved in constructing a human-equivalent exposure scale is that human and rat tumorogenesis will respond equally at the same absorbed unit doses of formaldehyde. This assumption can not, of course, be tested. Data reviewed by Meselson (14) suggest that this assumption frequently overstates the degree of human response. As any degree of overstatement in this case cannot be quantified, I will assume equivalency of response and note that the resultant risk estimates are likely to be high.

Once this operating assumption is made, an equivalent-dose scale may be approached on the basis of the mechanics and rates of formaldehyde delivery and detoxification in the target tissues. In the rat (and likely in the mouse) the target tissues for tumor response to inhalation exposure are the surfaces of the nasal turbinates. For reasons discussed later in this chapter, the nasal cavity is likely to be a principal target in humans as well, and the oral cavity is likely to be a secondary and lesser affected target (because of occasional mouth breathing in humans). A number of studies (20–22) indicate no detectable delivery of unmetabolized formaldehyde to the lungs or upper airway in rats. Thus, the load of formaldehyde in each inspired breath is efficiently scrubbed from the air in the nasal cavity and possibly the nasopharynx. The same studies suggest that absorbed formaldehyde is rapidly metabolized at the dose rates tested at the site of absorbance. Thus, an equivalent-dose scale would have to be based on the mole amount per unit time absorbed per unit of absorbing surface. An "effective"-dose scale could be derived then on the basis of adjustment for rates of metabolic conversion of formaldehyde at the absorbing site. Because minute volume and several other parameters of respiration vary as the two-thirds power of body weight (23), I can estimate that the amount of absorbing surface for formaldehyde will also vary as the two-thirds power of body weight. This estimation allows for the direct calculation of the equivalent dose of formaldehyde in humans relative to rats (using an ambient level of 1 μg of formaldehyde/L of air):

$$\text{Rats:} \quad (1 \ \mu\text{g of CH}_2\text{O/L}) \, (0.073 \ \text{L/min}) \ = \ 0.073 \ \mu\text{g of CH}_2\text{O/min}$$

$$(0.073 \ \mu\text{g/min})/(0.2 \ \text{kg})^{2/3} \ = \ 0.21 \ \mu\text{g/min} \cdot \text{kg}^{2/3}$$

$$\text{Humans:} \quad (1 \ \mu\text{g of CH}_2\text{O/L}) \, (7.5 \ \text{L/min}) \ = \ 7.5 \ \mu\text{g of CH}_2\text{O/min}$$

$$(7.5 \ \mu\text{g/min})/(70 \ \text{kg})^{2/3} \ = \ 0.44 \ \mu\text{g/min} \cdot \text{kg}^{2/3}$$

This analysis suggests that at any given ambient level of formaldehyde, the equivalent dose to target tissue in humans is twice that in rats. This conclusion, however, does not take into account differences in the thickness of the mucous blanket over the target tissues in humans and rats and the rate of mucosal flow. The studies cited earlier (20–22) suggest that a major portion of the delivered formaldehyde is trapped in the mucous blanket and swept out of the nasal cavity unless the dose is high enough to produce stasis of the mucociliary flow and/or thinning of the blanket. Both the rate of mucociliary flow and the rate of mucous production are higher in humans than in rats (23), and both of these effects combined tend to cancel out the twofold difference suggested solely on surface deposition rate. Thus, in this assessment rat and human exposures are considered roughly equivalent at given ambient levels of formaldehyde gas in air.

Exposure Estimates for Human Subpopulations. The data in the Environmental Protection Agency (EPA) Integrated Exposure Assessment for formaldehyde (available from EPA) are used for exposure in this chapter. These data have been used by both EPA and the Consumer Product Safety Commission (CPSC) in their assessments of possible formaldehyde-related risk, although CPSC used the high values of the range of exposure for any given subpopulation as opposed to EPA's use of the range mean (24). The cumulative exposure hours used to calculate f for each subpopulation are those given in the EPA exposure assessment. The value of f is derived by dividing the cumulative exposure hours by the 613,000 h in a 70-year lifetime. The exposure numbers in the EPA document are themselves at times difficult to compare because of variations in analytical methodology.

Quantitative Estimates of Putative Human Risk. The exposures of various subpopulations are presented in Table II and in Figure 3. Table III gives values of f for various combinations of hours-per-day exposure and years over which this exposure occurs. A sample calculation is shown as follows for the first entry in Table III:

$$(7.08 \times 10^{-5}) \times (1.35)^{4.28} \times 0.0339 = 8.67 \times 10^{-6}$$

Figure 3 shows the distribution of lifetime risks by the size of the subpopulations identified in Table II. For comparative purposes the lifetime risks from some other quantifiable factors are included in Figure 3 (25, 26).

The average lifetime risk of cancer in the United States is 0.25 for all sites combined except nonmelanoma skin cancer (27). The risk estimates in Table II and Figure 3 represent, therefore, an incremental risk of 0.0098% for the subpopulation at the highest putative risk and a mean increment of 2.2×10^{-6}% for all subpopulations. The lifetime risk of nasal cavity cancer in the United States is circa 4.5×10^{-4} (28). Thus, the subpopulation at the highest putative risk in Table II represents a risk increment of 5% at this site, and the mean increment for all subpopulations is circa 1.2×10^{-3}%. Nearly all estimated risks are less than 1×10^{-6}. This value represents a risk increment of 4×10^{-4}% in risk at all sites and 0.2% in nasal cavity risk.

Table II. Estimated Risks to Various Subpopulations

Group (Size)	Mean Ambient Exposure (ppm)	f	Mean Estimated Risk[a]
Formaldehyde producers (42)	1.35	0.0339	8.67×10^{-6}
Resin producers (4100)	0.9	0.0339	1.42×10^{-6}
Plywood–particle board producers (25,500)	1.75	0.0339	2.5×10^{-5}
Urea–formaldehyde foam producers (55)	0.85	0.0339	1.1×10^{-6}
Foam installers (8500)	0.02	0.0339	9.9×10^{-14}
Textile producers (3180)	0.48	0.0339	9.7×10^{-8}
Paper–paper products producers (26,100)	0.06	0.0339	1.3×10^{-11}
Fertilizer producers, applicators (700)	0.9	0.0254	1.1×10^{-6}
Embalmers, funeral workers (70,000)	0.52	0.017	6.8×10^{-8}
Pathologists (12,000)	ca. 1	0.0254	1.7×10^{-6}
Biology instructors			
College (13,000)	ca. 1	0.017	1.1×10^{-6}
High school (22,000)	ca. 1	0.0042	2.8×10^{-7}
Rubber–plastic workers (17,000)	0.03	0.0339	7.1×10^{-13}
Biology students (1,200,000)	ca. 1	0.0034	2.2×10^{-7}
Foundry workers (7700)	0.43	0.0254	4.7×10^{-8}
Mobile home residents (2,200,000)	0.4	0.053	7.3×10^{-8}
Urea–formaldehyde home residents (1,500,000)	0.72^b	0.0053	9.0×10^{-8}
	0.03	0.106	2.2×10^{-12}
	0.03	1.00	2.1×10^{-11}
U.S. population by ambient air (220,000,000)	0.005^c	1.00	9.8×10^{-15}

[a]Values are estimated from the mean function of Figure 2.
[b]Three scenarios are given. The first is 6 months of exposure in an average "complaint" home, a small subset of all insulated homes. The second and third scenarios reflect average levels of formaldehyde in homes for a 10-year and lifetime exposure, respectively.
[c]The range is from less than 0.0004 ppm to 0.03 ppm of which 97% is derived from various combustion processes.

Confidence in the Modeling

An obvious question in extrapolating outside of the limits of experimental observation is the degree to which such an extrapolation is expected to be predictively valid. In high- to low-dose extrapolation in the same species, the question is whether responses seen within the dose range tested will predict response at lower, untestable, doses. In extrapolating from animal data to possible risk in humans, the question of whether the animal data will be predictive of tumor response in humans is considered. The CIIT data appear to be at least qualitatively predictive of possible responses in humans exposed to high ambient levels of formaldehyde gas. This view is put forward because no data appear to support the view that at a sufficiently high dose humans would be insensitive to a tumorogenic effect of formaldehyde in the nasal cavity. The next part of the question becomes, then, how quantitative and how biologically valid for both rats and humans are predictions about low-dose responses to formaldehyde based on

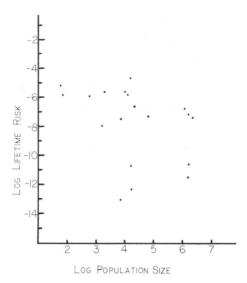

Figure 3. Population distribution of lifetime risk from various formaldehyde exposure scenarios over various identified populations from Table II. The lifetime risks given in Table II are plotted against the size of the various groups identified in Table II on a log–log plot. For purposes of reference some lifetime risks from other sources that have been quantified are shown on the figure. These are an airline pilot from extra exposure to cosmic radiation over a 30-year career, 1.2×10^{-3}; consumption of the saccharin in a 12-oz diet soda daily for 60 years, 3.4×10^{-4}; and consumption of the aflatoxin in 1 pt. of milk daily for 70 years, 1.4×10^{-4} (27, 28).

Table III. Values of *f* for Various Exposure Durations

Hours per Week	Duration (years)	f
40	10	0.0339
30	10	0.0254
20	10	0.017
5	10	0.0042
125	5	0.053
168	70	1.00

mathematical extrapolation from responses observed at a high dose (high dose here is used in the sense of a dose at or near to an MTD). In this portion of this chapter some factors that bear on this question are addressed.

Linearity of Response at Low Doses. A common policy assumption in risk assessment is that the shape of the dose–response function for a carcinogenic response is linear at low doses and does not exhibit a threshold dose. I will deal here with the question of linearity at low doses because the observed data in the CIIT bioassay are markedly nonlinear. The theoretical basis for low-dose linearity has been discussed by Crump et al. (29) and depends on the additivity of response to a pre-existing background carcinogenic process. As discussed in Ref. 16, this additivity is really an expression of the limit value of the nonlinear terms in any model as the dose approaches zero (or as exposure-related risk becomes small relative to background risk). Because chemical carcinogens are generally site specific [or act on well-defined sites characteristic of each carcinogen (17)], consideration of additivity to background at the target site as opposed to additivity to background at all sites combined is appropriate (as in the original Crump et al. treatment). Furthermore even if a carcinogen produces tumors in a tissue above a background of spontaneous tumors, the carcinogen-induced tumors may be of a different histological type than the spontaneous tumors. In such a case, one must question whether the process leading to the carcinogen-induced tumors is indeed additive to the process that produces the background tumors.

In the rat, spontaneous nasal carcinoma is very rare, having an upper bound on incidence of perhaps 5×10^{-4} (30). This value suggests that, in rats, the carcinogenic dose response to formaldehyde does not readily linearize. This situation is apparent from the CIIT data (1). As such, models that are always linear at small dose levels do not appear to be appropriate to the modeling of formaldehyde-related risk as a function of exposure.

Human nasal cancers are also rare. Annual rates for nasal carcinoma in the United States are circa 6×10^{-6}, as are those for nasopharyngeal cancers (28). Nasopharyngeal cancers may have a strong genetic component (31). The relative rarity of human nasal cancers suggests that no large background is present to which any formaldehyde-related effects would be additive. The annual incidence rates cited suggest a lifetime risk of circa 4×10^{-4} for human nasal cavity cancer. This value is comparable to the boundary estimate in rats given earlier. This human lifetime risk can also be compared to the estimates of possible exposure-related risk given in Table II.

This line of argument presumes that the nasal cavity surfaces are key target tissues for formaldehyde-induced tumorogenesis after inhalation exposure in humans. This presumption raises the question of site comparison between species in carcinogenic response. Although some hold that no site concordance occurs, any detailed review to support this view is as yet unpublished. Tomatis, on the other hand, has reviewed a number of agents adequately studied in at least two species and finds, at least qualitatively, an 80% site concordance in response (32). In addition, for formaldehyde the toxicological data base provides ample justification for treating the nasal surfaces, the nasopharynx, and possibly the oral cavity as primary target sites for any tumorogenic effect of inhaled formaldehyde gas in humans. Review of the animal studies on formaldehyde tumorogenicity reveals a pattern of response in which any neoplastic lesion seen forms only at the site of contact with formaldehyde or formalin and does not occur at distal sites (1, 2, 3, 6).

A rationale for this view is found in uptake and metabolism studies of formaldehyde in humans and animals. These studies demonstrate that formaldehyde is rapidly metabolized at the site of uptake, and unmetabolized formaldehyde is not distributed in the body or circulation (20, 21). Studies by Heck (22) on the distribution of ^{14}C-labeled formaldehyde show that the concentration of ^{14}C is 1 or 2 orders of magnitude greater in the nasal mucosa after inhalation than in other tissues, and that the ^{14}C in other tissues appears to represent metabolites rather than formaldehyde.

Effects of Cytotoxicity on the Linearity of the Dose–Response Curve. Dose-dependent cytotoxicity can affect the expression of tumor response by a variety of mechanisms. This point is articulately discussed in relation to formaldehyde and the cytotoxic effects it has on nasal tissues by Griesemer et al. (33). Cytotoxicity can both stimulate response to initiation and can allow for the increased expression of previously initiated cells. In the case of increased expression of previously initiated cells, a marked cytotoxic threshold is also a threshold for increased tumor risk. Because nasal cancer is rare, a large population of preinitiated cells in nasal tissues is not apparent. Thus, in the CIIT study, formaldehyde may be acting as both initiator and promoter. Initiating effects of formaldehyde would be enhanced because of more frequent cell turnover as a consequence of restorative hyperplasia in response to marked cytotoxicity (due to more frequent passage of the genome through the carcinogen-sensitive replicative phase). A stasis of mucociliary flow as part of a cytotoxic response would also decrease the natural barrier to formaldehyde penetration into the cells of the nasal mucosa. The restorative hyperplasia resulting from cytotoxicity would also act to enhance the expression of any cells that were initiated and would thus lead to a nonlinear increase in tumor yield as dose is increased. Such an effect has recently been confirmed in studies on urethane carcinogenesis (34).

Indeed, if a linear dose response is assumed in the absence of any cytotoxic enhancement of response, and a linear relationship between the degree of cytotoxic enhancement of response and dose is further assumed, the dose–response curve can be shown to contain both linear and nonlinear terms:

$$\text{basal response function: } R = bD$$

$$\text{cytotoxic enhancement function: } E = aD + 1$$

$$\text{enhanced response function: } R(E) = bD + abD^2$$

where R is the response, D is the dose, b is the slope of the dose–response curve in the absence of any stimuli, E is the enhancement factor, and a is the slope of the dose–enhancement function. Thus, in the presence of cytotoxic stimuli, an otherwise linear tumor response will become nonlinear. The contribution of the nonlinear term will depend on the magnitude of the slopes of both functions. This relationship suggests that a linear downward extrapolation of risk from a dose domain in which cytotoxicity is observed will generally overestimate risk at lower doses. This situation argues against the use of the "linearized" multistage model for extrapolation of formaldehyde-related risk on the basis of the CIIT data. This analysis also provides a rationale of why the dose–response curve seen in the CIIT study is the third or fourth power of dose.

The Effect of Endogenous Levels of Formaldehyde.   Formaldehyde is a normal cellular metabolite, and levels in mammalian cells range from 1.5 to 15 ppm (22). The bulk of this amount is conjugated to glutathione or tetrahydrofolate, but such conjugated material is in dynamic equilibrium with free formaldehyde and formaldehyde bound to macromolecular species. As such, the amount of free formaldehyde can influence the rate of equilibration among the various bound forms, but the total intracellular pool is available for reaction. In support of this view is the fact that the glutathione and tetrahydrofolate adducts are quite reactive. Recent work by Heck (35) indicates that exposure of up to 15 ppm of formaldehyde in air has little effect on intracellular levels and thus suggests that the rate of endogenous production and metabolism must be much greater than the rate of intracellular penetration of formaldehyde at these exposure levels [a great deal of the formaldehyde is trapped in the mucosal blanket of the nasal epithelia (35)]. These considerations suggest that the relationship between the ambient level of formaldehyde and the delivery of exogenous formaldehyde to intracellular targets is not likely to be a simple linear relationship. Furthermore, at some low level of ambient exposure, the rate of delivery of formaldehyde into the exposed cells must, in relationship to the mean rate of endogenous production, become insignificant and smaller than any diurnal variation in this endogenous production rate. This situation, of course, implies that a practical threshold must exist for any increased risk associated with such low-level exposures. The currently available data do not provide enough information to indicate at how low of a dose this practical threshold lies, but further experimentation could fill in this gap in understanding of the cellular kinetics of formaldehyde.

The Effect of a Fixation Requirement in Initiation.   The initiating event in neoplastic transformation must become "fixed" into the affected cells (15, 33, 36–38). The process known as initiation is believed to involve, at least in part, an alteration in some critical part of the genome (15, 17, 19). McCormick and coworkers have demonstrated that cellular repair processes can eliminate mutagenic insults in cells and thus prevent expression of mutation if the cells are restrained from entry into the S phase (the DNA synthesis phase) of the cell cycle until repair is completed (39, 40). Kakunaga has shown that fixation is required in the transformation of human fibroblasts and that several rounds of cell division are required for this fixation to occur (38). Thus, a cytotoxic response will increase the efficiency of initiation by increasing the rate of cell division during restorative cell proliferation and, thereby, the rate at which initiating events are fixed. These changes will alter the transforming efficiency of an administered dose in a nonlinear way. This effect, then, is a specialized mechanism whereby cytotoxicity enhances carcinogenicity (33).

With particular respect to formaldehyde, Fornace (41) has demonstrated that the repair of formaldehyde-induced DNA damage in cell culture is very efficient. Inhibition of repair polymerases is required to detect damage. Ross et al. (42, 43) have also reported efficient repair of formaldehyde-induced DNA-protein cross-links. If, as these studies suggest, formaldehyde-induced genetic damage is repaired efficiently, then the effect of increasing the cell turnover rate in response to cytotoxic injury may be significant with respect to the shape of the dose response. Also, the efficiency of mutagenic response as a function of dose becomes nonlinear if the rate of damage is allowed to exceed the repair capacity of the cell (44). This result suggests that a similar effect occurs for carcinogenicity. Swenberg has reported that formaldehyde exposures of 6 and 15 ppm in vivo produce increases in cell turnover in rats (18). Exposure to 0.5 or 2 ppm does not produce this effect. In mice, 15 ppm produces an increase in cell turnover, but 6, 2, and 0.5 ppm do not. These results suggest that the doses that produce increased cell turnover exactly parallel the doses at which tumor response is observed in each species. These considerations further support the use of a nonlinear risk extrapolation model for formaldehyde.

Effect of Dose on Latency.   The relationship between mean tumor latency and dose is well known (23, 45–47). Generally, an inverse cube relationship is found between dose and mean latent period (i.e., the time for 50% of the observed tumor response to occur). The risk estimates presented here were not adjusted for such effects because an accurate estimate of the 50% response time for formaldehyde tumors in the CIIT bioassay was not readily available, and the value of the power coefficient in a Druckery relationship for formaldehyde is not known. Also, analysis of Guess and Hoel's treatment of this effect (47) suggests that at least a five-fold increase in mean latent period is needed to make a significant impact on the portion of the tumor distribution falling within the life span of the target species. Use of a typical value of 3 for the Druckery coefficient and rough estimates of the mean latent period in the CIIT study suggests that such dose effects on latency would only become significant for formaldehyde exposures less than approximately 0.2 ppm. Sielkin has developed risk estimates for formaldehyde that attempt to account for these time-to-tumor effects (48). These estimates are in general agreement with those given here.

Discussion

Formaldehyde and attempts to assess the degree of risk that it may pose to exposed humans have engendered a great deal of study and discussion. This research is all quite beneficial to the advancement of a scientific understanding of the basis of risk assessment. The present analysis has attempted to lay out in a straightforward way some of the factors that must be considered in constructing a meaningful assessment of risk. These are the sorts of considerations that the National Research Council has termed "inference options" and has recommended for explicit discussion in the presentation of a risk assessment (13). Formaldehyde does present the opportunity to dissect the risk assessment in a way that is seldom possible. This situation is due to the extensive nature of the toxicological and biochemical data base and the design of the CIIT bioassay and supporting studies.

   With respect to formaldehyde, the present analysis suggests the following:

1. The relationship of risk to exposure is the fourth power of dose, and linear extrapolation models are inappropriate in this case.
2. Because of some of the factors discussed earlier, the risk estimates presented here will tend to overestimate the "true" risk.
3. Identifiable risks to humans are quite low and represent some $2.2 \times 10^{-6}\%$ of lifetime risk of all sites in nonsmokers and some $1.2 \times 10^{-3}\%$ of lifetime risk of nasal cancer.
4. The possibility of a practical or real threshold for formaldehyde-related risk cannot be dismissed. Because of formaldehyde's significant production rate endogenously, a practical threshold for such risk is expected, but the present data are insufficient to assess at what exposure level such a practical threshold lies.

   In the preceding presentation a number of factors that bear on the risk assessment were considered. The importance of careful analysis of such factors cannot be overstressed. The report of Starr (49) on the effects of decreased mucus flow in modulating the tissue-delivered dose of formaldehyde is a case in point. On this basis alone, the estimates produced by the linearized multistage model are predicted to be from 3 to 4 orders of magnitude too high in the low-exposure domain. The risk estimates given here are in rough agreement with what would be expected if the linearized multistage estimates were corrected for delivered dose as per Gibson.

Literature Cited

1. Chemical Industry Institute of Toxicology "Final Report on a Chronic Inhalation Toxicology Study in Rats and Mice Exposed to Formaldehyde"; CIIT: Research Triangle Park, N.C., 1981.
2. Albert, R. E.; Sellakumar, A. R.; Laskin, S.; Kuschner, M.; Nelson, N.; Snyder, C. A. *J. Natl. Cancer Inst.* 1982, 68, 597.
3. Watanabe, F.; Matsunaga, T.; Soejima, T.; Iwata, Y. *Gann* 1954, 45, 451.
4. Watanabe, F.; Sugimoto, S. *Gann* 1955, 46, 365.
5. Della Porta, G.; Cabral, G. *Tumori* 1970, 56, 325.
6. Mueller, R.; Raabe, G.; Schumann, D. *Exp. Pathol.* 1978, 16, 36.
7. Dalbey, W. E. *Toxicology* 1982, 24, 9.
8. Moss, E.; Lee, W. R. *Br. J. Ind. Med.* 1974, 31, 224.
9. Bross, I. D. J.; Viadama, E.; Houten, L. *Arch. Environ. Health* 1978, 33, 300.
10. Decoufle, P. *Arch. Environ. Health* 1979, 34, 33.
11. Matanoski, G. "Cancer Mortality Analysis of Pathologists Exposed to Formaldehyde"; Government Printing Office: Washington, 1980; JRB Contract No. 68.01.6280 (EPA) Task #3.
12. Todhunter, J. A., presented at the 2d Int. Conf. Saf. Eval. Regul. Chem., Bioresearch Institute and Boston University School of Medicine, Cambridge, Mass., 1983.
13. National Research Council "Risk Assessment in the Federal Government: Managing the Process"; NAS: Washington, D.C., 1983.
14. Meselson, M. S. "Contemporary Pest Control Practices and Prospects"; NAS: Washington, D.C., 1975; Vol. 1.
15. Pitot, H. "Fundamentals of Oncology"; Marcel Dekker: New York, 1981.
16. "Final Report of the Scientific Committee"; Food Safety Council, Washington, D.C., 1980; Chapter 11.
17. Weisburger, J.; Williams, G. M. In "Toxicology," 2d ed.; Doull, J.; Klaassen, C. D.; Amdur, M. O., Eds.; MacMillan: New York, 1980.
18. Swenberg, J. A.; Gross, E. A.; Martin, J.; Popp, J. A. "Proceedings of the Third CIIT Conference on Toxicology"; Hemisphere: New York, 1982.
19. Knudson, A. G. In "Neoplastic Transformation"; Koprowski, H., Ed.; Dahlem Konferenzen: Berlin, 1977.
20. Malorney, G.; Rietbrock, N.; Schneider, M. *Arch. Exp. Pathol. Pharmakol.* 1965, 250, 419.
21. McMartin, K. E.; Martin-Amat, G.; Noker, P. E.; Tephly, T. R. *Biochem. Pharmacol.* 1979, 28, 645.
22. Heck, H.; Chin, T. Y.; Schmitz, M. C. "*Proceedings of the Third CIIT Conference on Toxicology*"; Hemisphere: New York, 1982.
23. Calabrese, E. "Principles of Animal Extrapolation"; John Wiley: New York, 1983.
24. Gulf South Insulation vs. Consumer Product Safety Commission (1983), U.S. Court of Appeals, 5th Circuit, Docket No. 82–4218.
25. Crouch, E. A. C.; Wilson, R. "Risk Benefit Analysis"; Ballinger: Cambridge, Mass., 1982.
26. Cohen, B. L. *Nature* 1978, 271, 492.
27. Doll, R.; Peto, R. "The Causes of Cancer"; Oxford Medical: Oxford, 1981.
28. "SEER: Incidence and Mortality Data 1973–77"; National Cancer Institute, Monograph 57, 1981.
29. Crump, K. S.; Hoel, D. G.; Langley, C. H.; Peto, R. *Cancer Res.* 1976, 36, 2973.
30. Upton, A., personal communication, 1981.
31. Henderson, B. E.; Louie, E.; Jing, J. S.; Buell, P.; Gardiner, M. B. *N. Engl. J. Med.* 1976, 295, 1101.
32. Tomatis, L. *Cancer Res.* 1978, 38, 877.
33. Griesemer, R.; "Report of the Federal Formaldehyde Panel"; National Toxicology Program, NTIS: Springfield, Va., 1980.
34. Dourson, M. L.; O'Flaherty, E. J. *J. Natl. Cancer Inst.* 1982, 69, 851.

35. Heck, H. *CIIT Activities* **1982**, *3*, 3.
36. Berwald, Y.; Sachs, L. *Nature* **1963**, *200*, 1182.
37. Kakunaga, T. *Cancer Res.* **1975**, *35*, 1637.
38. Kakunaga, T. *Proc. Natl Acad. Sci. USA* **1978**, *75*, 1334.
39. Maher, V. M.; Dorney, D. J.; Mendala, A. L.; Konzi-Thomas, B.; McCormick, J. J. *Mutat. Res.* **1979**, *62*, 311.
40. Heflich, R. H.; Hazard, R. M.; Lommel, L.; Scribner, J. D.; Maher, V. M.; McCormick, J. J. *Chem. Biol. Interact.* **1980**, *29*, 43.
41. Fornace, A. J. *Cancer Res.* **1982**, *42*, 145.
42. Ross, W. E.; Shipley, N. *Mutat. Res.* **1980**, *79*, 277.
43. Ross, W. E.; McMillan, D. R.; Ross, C. F. *J. Natl. Cancer Inst.* **1981**, *67*, 217.
44. Russell, W. L.; Hunsicker, P. R.; Raymer, G. D.; Steele, M. H.; Stelzner, K. F.; Thompson, H. M. *Proc. Natl. Acad. Sci. USA* **1982**, *79*, 3589.
45. Druckery, H. In "Potential Carcinogenic Hazards from Drugs"; Truhaut, R., Ed.; UICC Monograph Series Vol. 7; Springer Verlag: New York, 1967.
46. Jones, H. B.; Grendon, A. *Food Cosmet. Toxicol.* **1975**, *13*, 251.
47. Guess, H. A.; Hoel, D. G. *J. Environ. Pathol. Toxicol.* **1977**, *1*, 279.
48. Sielken, R. L. "Incorporating Time into the Estimation of the Potential Human Cancer Risk from Formaldehyde Inhalation"; Institute of Statistics, Texas A&M University: College Station, Tex., 1983.
49. Starr, T. B.; Gibson, J. *Annu. Rev. Pharmacol. Toxicol.* **1985**, *25*, 745–67.

RECEIVED for review October 25, 1984. ACCEPTED March 13, 1985.

RISK ASSESSMENT PANEL

THE PANEL WAS COMPOSED OF Philippe Shubik, Leon Golberg, and John Higginson; and Jelleff Carr served as moderator. Carr noted that the rather morbid theme that occurred in some presentations such as cemeteries recorded as toxic waste sites, as well as the mortality rates of embalmers, morticians, and gross anatomists, did not allow for much levity.

Shubik stated that objectivity and science are the primary approaches to risk assessment. To follow up on what has been done, to benefit from lessons learned, and to have a continuum of good scientific experiments are the proper goals. One of the things Shubik stressed is that unfortunately we sometimes find ourselves forced by regulations and regulators into approaches that really are not entirely scientific. Such approaches are used by people to circumvent rules and regulations that have not been carefully formulated. He hoped that these approaches would be replaced by objective science.

Golberg cited an example of studies made 30 years ago that used iron dextran to treat iron deficiency anemia. He recalled that in 1954 researchers had conducted the usual toxicity tests, and the product was found to be nontoxic in the dosage used. But an enterprising Scottsman had given iron dextran to rats by intramuscular injection every day for their entire lives. Iron dextran was so low in toxicity that very large doses were possible. For example, ferrous sulfate has an LD_{50} of approximately 20 mg/kg. This material could be given to a mouse in a dose of 2000 mg/kg without toxic effects. The inevitable thing happened: sarcomas developed at the site of injection. What is interesting in retrospect is that manufacturers were forced to take this product off the market. But the Food and Drug Administration reinstated the product after reviewing the details of the pathogenesis that were presented later. Ferrous sulfate was also reinstated in Britain and in many other countries. Now the product has been on the market for 30 years; at least 10 million people, and possibly 20 million people, have been treated with it during this time. The entire populations of islands such as Mauritius and African countries such as Kenya have used it on an enormous scale because of its value as a single injection. These are places where you could not rely on seeing patients more than once in their lifetimes. You treat their anemia once, and they go away and remained treated. A single

0065–2393/85/0210/0375$06.00/0

authentic case of cancer or subcutaneous sarcoma has not been reported in all this time.

Nevertheless, the International Agency for Research on Cancer (IARC) lists iron dextran as a human carcinogen. But here, I think, is an example where initially we were told to just wait, and a terrible epidemic of sarcomas will occur. Thirty years has passed, which I think is a fairly reasonable latent period, and nothing of the sort has yet happened. This situation really illustrates that the study of mechanism has not, in the past, been used by government as a means for rational decision making, and I am hoping that this will not happen again. I think that the efforts being made at the Chemical Industry Institute of Toxicology (CIIT) not only have value in that direction, but they are breaking new ground in developing approaches that are applicable to a wide variety of animal carcinogens in the search for an understanding of their risk to humans. This route is the one we have to take. We have to keep applying the best science that we know. Unfortunately, some people will always look at the end result, count up the number of tumors, and then try to reach a conclusion on that basis, ignoring everything that preceded the tumor formation. Golberg hoped that we will see less and less of this type of analysis in the future.

Carr asked the group to comment on the relative hazard of low-dose exposure to formaldehyde over long periods of time. Higginson pointed out that the basic question was that if someone has been exposed for a long time to formaldehyde, does it constitute a risk? He believed for practical purposes that it does not constitute a risk at the levels that have been used for nose or lung exposures for reasons that have been presented in depth. The discussions suggested strongly that formaldehyde should not, under normal inhaled conditions, produce remote tumors. The question is that if you were exposed under unusual circumstances, could formaldehyde be a carcinogen? His belief was that it probably could, but it would require extreme circumstances. The close range or window between the carcinogenic and the noncarcinogenic dose appears somewhat different for formaldehyde than for many other compounds. This situation may be an unusual generic phenomenon, and a "usual" dose may be reasonably safe whereas very high doses may present a special situation.

Carr pointed out that one of the papers suggested that if formaldehyde itself does not prove to be an initiator, could it set the stage so that some other agent might then prove to be carcinogenic? Higginson stated that one could not prove such a situation. In rats you have to regard formaldehyde as a good carcinogen for the nasal mucosa. Mechanisms have been suggested, but reference to formaldehyde as an initiator alone is premature. Golberg observed that rats will tolerate an incredible amount of tissue injury and still appear healthy. They will withstand a state of bronchitis that no human being could tolerate for any length of time. This finding explains the rat's ability to withstand ulceration of the nasal epithelium at exposures of 15 ppm. In fact, some experiments have been done at 35 ppm. Further-more, studies using exposures of 100 or more parts per million for months at a time have been reported. Golberg did not think a human being could stand 15 ppm for any length of time. Once a pathological process or infla-mation, let alone ulceration, starts to occur in people, they will seek relief one way or another and simply remove themselves from the area causing irritation. He emphasized that trying to equate humans to rats has a built-in fallacy. Shubik quickly pointed out that some human beings enjoy taking snuff in large quantities over long time periods. He told of a personal expe-rience when he was handed some snuff. He recalled that it nearly blew the top of his head off. He added that cocaine is another substance that pro-duces all kinds of problems with the nasal mucosa; people spend a lot of money to destroy their nasal mucosa with this substance. Perhaps for rats formaldehyde is their cocaine.

At this point James Gibson of CIIT expressed his thanks for the kind remarks on CIIT's work and also for the encouragement to keep going. However, he pointed out that CIIT is an institute with limited resources and not the chemical industry institute of formaldehyde. He suggested that the critical need for definitive experiments to put the formaldehyde issue at rest is still present. A repeat is being considered of the experiments that Henry Heck did in elucidating the amount of DNA protein binding that occurred in the nasal mucosa at various concentrations of formaldehyde in the air. However, this time subhuman primates would be used instead of rats. This experiment would be expensive and tedious and would possibly give precisely the same information that we already have. But Gibson asked what the panelists thought of such a study in terms of providing more persuasive evidence that formaldehyde risk at low concentrations is really the same as we would predict now from the rat data.

Shubik recalled that some primates had developed squamous metapla-sia when exposed to formaldehyde via inhalation, and he asked Gibson for more information. Gibson explained that the Formaldehyde Institute did a 6-month exposure study of rhesus monkeys using concentrations up to 3 ppm, 22 h/day, 7 days/week. The monkeys developed just minimal squa-mous metaplasia during that period. He added that he was contemplating an experiment in which the top dose would be 3 ppm, and an exposure for some number of days to nonradioactive formaldehyde would be used, fol-lowed by an exposure to tritiated ^{14}C-labeled formaldehyde. He would de-termine the amount of binding, protein, RNA, and DNA in the tissues from a substantial region of the upper respiratory tract. He would probably also try to determine the dosimetry to the lungs or to distant sites and the possi-bility of changes in bone marrow. He would work his way down from 3 ppm, attempting to determine the shape of the dose-binding curve. The outcome presumably would be the establishment of a concentration of for-maldehyde in air that leads to no detectable binding to genetic material.

He noted that such an experiment would be based on the assumption that cancer is the issue. However, irritation is a different issue.

Shubik agreed that such data would be desirable, but he did not think it should have top priority. He wanted to see more studies on comparative models that gave positive and negative data for carcinogenicity; relating binding to squamous metaplasia was not enough. He did not think the proposed experiments would answer the questions of the many people trying to relate dose to carcinogenesis. He commented that data on the mouse, the hamster, and the rat were available at different dosage levels. However, he preferred a situation in which one could say that a certain dose does or does not relate to the occurrence of carcinogenesis in one way or another.

Higginson also expressed reservations about DNA adducts. Basically, one gets a large number of cells to form adducts. But probably only one or two of those cells undergoes neoplasia, so most of what one sees is binding to DNA that is not hitting the key site. Therefore, one must assume a rough correlation between binding and the fundamental cellular error, which never has been proven. Gibson pointed out that the key here is that this binding is a measured dosimetry approach. Higginson agreed, but stated that we really do not know enough about binding and dosimetry at the bottom of the dose curve, and is such study worth doing? He suggested that researchers may want to use humans rather than primates. A little bit of mucosa is not that difficult to take, and it can be taken after relatively modest exposures.

Gibson concluded by stating that everyone recognizes that current studies are being driven by the regulatory concern. However, rightly or wrongly, will these studies dealing with mechanisms assist policymakers in understanding the risk, or lack of it, of exposure to formaldehyde?

Suggestions were made to go to very low doses, to use the hamster as an example of a species that is resistant to formaldehyde, and to use subhuman primates because they are believed to be more similar to human beings.

At this point John Todhunter asked a question about the value of extrapolations of dose–effect curves to arrive at a reasonable conclusion regarding a safe dose. He stated that during the last year or so he had noted a tendency among certain moguls of policy to move away from using the approach of the no observable effect level and margin of safety. In other words, some policymakers are beginning to feel that if a large enough population is studied, a responder may be found. This thinking leads to setting all sorts of standards by using a linear extrapolation or some other mathematical formula. He asked for comments from the panel.

Golberg pointed out that after the 6 h the animal is exposed, some degree of recovery occurs for the next 18 h. Over the weekend a considerable recovery takes place. This situation is the kind with which we are dealing. However, when the risk assessment is made, it is worked out on a lifetime constant exposure basis, and this dosage level is different. Statisticians have not been able to evolve a model to take this kind of situation into account. The panel members agreed that the use of mathematical models had not contributed a great deal to understanding levels of toxicity or safe dose levels in general. Someone suggested that in some cases it becomes a matter of symbolism and does not add greater understanding to the results of a toxicity study.

Someone asked what would be the acceptable level of formaldehyde that would be in agreement with all the studies that have been made. Golberg replied that for the indoor environment he would place the level somewhere in the neighborhood of 0.5 ppm. Minnesota has set a high level of 0.39 ppm, which is not all that different. Many people are suggesting 0.01 ppm, and Golberg thought that was just ridiculous. Higginson agreed and observed that the political situation strongly influences regulatory decisions. The major contribution of the symposium was the compilation of data to avoid unnecessarily rigid control levels.

When Hans Graven of Georgia Tech asked the panel if anyone was concerned about low levels of formaldehyde in the water supply at concentrations of 5–10 ppm, the members replied that they did not think this level posed a health problem. How such levels might ever be reached was questioned because of ordinary chemical reactivity in the soil and water. EPA had originally decided that formaldehyde contamination in water was not a hazard because of the activity of bacteria and organic substances in the environment.

George Tiers of 3M observed that the symposium participants had agreed fairly well that safe levels of formaldehyde are present in the air. He felt this agreement had public policy implications. He pointed out the divisiveness of political positions that were favorable or unfavorable to accepting any level of formaldehyde and indicated that such a situation will not lead to any kind of rapprochement. The major contribution of the symposium, he concluded, must be the opinions of the distinguished experts in this field. A way must be found to present these opinions and information to everyone concerned and especially to those responsible for current political opinions. Sufficient data and studies will never be available to prove safety, so some point must be reached at which public officials can be convinced they can responsibly regulate for the welfare of everyone.

Carr pointed out that this goal was exactly the major thrust of the entire symposium, and he asked the panel for their comments.

Higginson observed that regulatory agencies under attack had a tendency to avoid calling upon outside consultants in recent years. He suggested that such agencies must be completely divorced from political pressures and must be objective if they are to operate meaningfully.

Shubik pointed out that the subject was not just a political one and that industrial representatives need to resort to the media programs they support. In fact, he said, these programs seem to be the cause of so many problems in society today, and he called upon the scientific and industrial communities to try to present a coordinated program with accurate and fair reporting.

Carr concluded the discussion at this point and thanked the participants and the panel members for their contributions, time, interest, and enthusiasm in developing a factual symposium.

Condensation of the ACS transcript from the Risk Assessment Panel of the Formaldehyde Symposium by C. Jelleff Carr.

RECEIVED February 1, 1985

AUTHOR INDEX

SUBJECT INDEX

Copyediting and indexing by Karen McCeney
Production by Meg Marshall
Jacket design by Pamela Lewis
Managing Editor: Janet S. Dodd

Typeset by Action Comp Co., Baltimore, Md.
and Hot Type Ltd., Washington, D.C.
Printed and bound by Maple Press Co., York, Pa.